TURBULENCE, DYNAMOS, ACCRETION DISKS, PULSARS AND
COLLECTIVE PLASMA PROCESSES

ASTROPHYSICS AND
SPACE SCIENCE PROCEEDINGS

For other titles published in this series, go to
www.springer.com/series/7395

TURBULENCE, DYNAMOS, ACCRETION DISKS, PULSARS AND COLLECTIVE PLASMA PROCESSES

FIRST KODAI-TRIESTE WORKSHOP
ON PLASMA ASTROPHYSICS HELD
AT THE KODAIKANAL OBSERVATORY
KODAIKANAL, INDIA
AUGUST 27 – SEPTEMBER 7, 2007

Edited by

S.S. HASAN
*Indian Institute of Astrophysics,
Bangalore, India*

R.T. GANGADHARA
*Indian Institute of Astrophysics,
Bangalore, India*

and

V. KRISHAN
*Indian Institute of Astrophysics,
Bangalore, India*

Editors
S.S. Hasan
Indian Institute of Astrophysics
Sarjapur Road
Bangalore-560034
IInd Block, Koramangala
India
hasan@iiap.res.in

R.T. Gangadhara
Indian Institute of Astrophysics
Sarjapur Road
Bangalore-560034
IInd Block, Koramangala
India
ganga@iiap.res.in

V. Krishan
Indian Institute of Astrophysics
Sarjapur Road
Bangalore-560034
IInd Block, Koramangala
India
vinod@iiap.res.in

ISBN 978-1-4020-8867-4   e-ISBN 978-1-4020-8868-1

Library of Congress Control Number: 2008934393

© 2009 Springer Science + Business Media B.V.
No part of this work may be reproduced, stored in a retrieval system, or transmitted in any form or by any means, electronic, mechanical, photocopying, microfilming, recording or otherwise, without the written permission from the Publisher, with the exception of any material supplied specifically for the purpose of being entered and executed on a computer system, for the exclusive use by the purchaser of the work.

Printed on acid-free paper

9 8 7 6 5 4 3 2 1

springer.com

Preface

The Kodaikanal Observatory of the Indian Institute of Astrophysics traces its origins to the end of the nineteenth century when it was decided to relocate the Madras Observatory to a high altitude site with a view to initiate observations of the Sun. Many valuable observations were made here including the discovery of outflowing material in sunspots discovered by John Evershed in 1909. The Observatory continues to provide useful solar data as well as serve as a centre for research and training programmes. Moreover, with its serene and beautiful environment, and good infrastructure it is an ideal location for the pursuit of intellectual and pedagogical activity.

In 2006 an initiative was taken to begin a series of schools and workshops in different areas of astronomy and astrophysics with a view to attract students to this field as well as to enhance excellence and greater interaction among researchers working in these areas. The first Kodai-Trieste Workshop on Plasma Astrophysics, which was held at the Kodaikanal Observatory, Kodaikanal during August 27 - September 7, 2007, was a continuation of this effort. Organized jointly by the Indian Institute of Astrophysics, Bangalore and the Abdus Salam International Centre for Theoretical Physics (ASICTP), Trieste, its aim was to provide a strong conceptual foundation in plasma astrophysics. The Workshop was conceived when Prof. K. R. Sreenivasan, Director, ICTP, visited the Indian Institute of Astrophysics in October, 2006.

It is well established that more than 99% of the baryonic matter in the universe is in the plasma state. Most astrophysical systems could be approximated as conducting fluids in a gravitational field. It is the combined effect of these two that gives rise to the rich variety of configurations in the form of filaments, loops, jets and arches. The plasma structures that cannot last for more than a second or less in the laboratory remain intact on astronomical time and spatial scales. High energy radiation sources such as active galactic nuclei involve coherent plasma radiation processes for their exceptionally large output from regions of relatively small physical sizes. The generation of magnetic field, anomalous transport of angular momentum with decisive bearing on star formation processes, the ubiquitous MHD turbulence under

VI   Preface

conditions not reproducible in terrestrial laboratories are some of the generic issues still awaiting a concerted effort to be properly understood. Quantum plasmas, pair plasmas and pair-ion plasmas exist under extreme conditions in planetary interiors and exotic stars.

This monograph, consisting of 22 contributions, is organized in six parts dealing with astrophysical turbulence, dynamos, pulsar radiation mechanisms, quantum plasmas, accretion disks, and solar and space plasmas. The workshop brought together several international scientists and young researchers working in plasma astrophysics.

The workshop owes its success to the efforts of a large number of persons, including V. Krishan, the course director, K. E. Rangarajan, the convener, and R. T. Gangadhara, the coordinator. In addition, I am grateful to all the speakers for readily accepting to participate in the workshop and for a timely submission of their manuscripts. I am thankful to the scientific and administrative staff of the Indian Institute of Astrophysics at the Bangalore and Kodaikanal campuses for providing local support.

Bangalore
April 2008

S. S. Hasan

Contents

Part I Astrophysical Turbulence

Aspects of Hydrodynamic Turbulence in Classical and Quantum Systems
J.J. Niemela 3

Observations and Modeling of Turbulence in the Solar Wind
Melvyn L. Goldstein 21

Power Spectra of the Fluctuations in the Solar Wind
V. Krishan 35

Part II Astrophysical Dynamos

Alpha Effect in Partially Ionized Plasmas
V. Krishan and R. T. Gangadhara 55

Constraints on Dynamo Action
A. Mangalam 69

Planetary Dynamos
Vinod K. Gaur 85

Part III Pulsar Radiation Mechanism

Pulsars as Fantastic Objects and Probes
Jin Lin Han 99

VIII   Contents

Pulsar Radio Emission Geometry
R. T. Gangadhara 113

Millisecond Pulsar Emission Altitude from Relativistic Phase Shift: PSR J0437-4715
R. T. Gangadhara and R. M. C. Thomas 137

Magnetosphere Structure and the Annular Gap Model of Pulsars
G.J. Qiao, K.J. Lee, H.G. Wang, and R.X. Xu 147

Wave Modes in the Magnetospheres of Pulsars and Magnetars
C. Wang, D. Lai 169

Polarization of Coherent Curvature Radiation in Pulsars
R. M. C. Thomas and R.T. Gangadhara 177

Part IV Quantum Plasmas

Nonlinear Quantum Plasma Physics
Padma K. Shukla, Bengt Eliasson, Dastgeer Shaikh 191

Dust Plasma Interactions in Space and Laboratory
Padma K. Shukla, Bengt Eliasson, Dastgeer Shaikh 213

Part V Accretion Disks

Magnetorotational Instability In Accretion Disks
V. Krishan and S.M. Mahajan 233

Hybrid Viscosity and Magnetoviscous Instability in Hot, Collisionless Accretion Disks
Prasad Subramanian, Peter A. Becker, Menas Kafatos 249

Transonic Properties of Accretion Disk Around Compact Objects
Banibrata Mukhopadhyay 261

Maximum Brightness Temperature for an Incoherent Synchrotron Radio Source
Ashok K. Singal 273

Nonlinear Jeans Instability in an Uniformly Rotating Gas
Nikhil Chakrabarti, Barnana Pal and Vinod Krishan 281

Part VI Solar and Space Plasmas

An Overview of the Magnetosphere, Substorms and Geomagnetic Storms
G. S. Lakhina, S. Alex, R. Rawat 293

Monte Carlo Simulation of Scattering of Solar Radio Emissions
G. Thejappa, R. J. MacDowall 311

Evolution of Magnetic Helicity in NOAA 10923 Over Three Consecutive Solar Rotations
Sanjiv Kumar Tiwari, Jayant Joshi, Sanjay Gosain and P. Venkatakrishnan 329

Stability of Double Layer in Multi-Ion Plasmas
A.M. Ahadi, S. Sobhanian 337

List of Contributors

J.J. Niemela
The Abdus Salam International Center for Theoretical
Physics, Strada Costiera 11, 34014 Trieste, Italy
niemela@ictp.it

Melvyn L. Goldstein
NASA Goddard Space Flight Center, USA
melvyn.l.goldstein@nasa.gov

V. Krishan[1,2]
[1]Indian Institute of Astrophysics, Bangalore-560034, India
[2]Raman Research Institute, Bangalore-560080, India
vinod@iiap.res.in

R. T. Gangadhara
Indian Institute of Astrophysics, Bangalore-560034, India
ganga@iiap.res.in

R. M. C. Thomas
Indian Institute of Astrophysics, Bangalore-560034, India
mathew@iiap.res.in

A. Mangalam
Indian Institute of Astrophysics, Bangalore-560034, India
mangalam@iiap.res.in

Vinod K. Gaur
Indian Institute of Astrophysics, Bangalore-560034, India
vgaur@iiap.res.in

Jin Lin Han
National Astronomical Observatories, Chinese Academy of Sciences,
Jia-20 DaTun Road, Chaoyang District, Beijing 100012, China
hjl@bao.ac.cn

G.J. Qiao
Department of Astronomy, Peking University, Beijing
100871, China
gjn@pku.edu.cn

K.J. Lee
Department of Astronomy, Peking University, Beijing
100871, China
k.j.lee@water.pku.edu.cn

H.G. Wang
Center for astrophysics, Guangzhou University, Guangzhou 510006, China
hgwang@gzhu.edu.cn

XII   List of Contributors

R.X. Xu
Department of Astronomy, Peking
University, Beijing
100871, China
r.x.xu@pku.edu.cn

C. Wang[1,2]
[1]National Astronomical Observatories, Chinese Academy of
Sciences. A20 Datun Road, Chaoyang
District, Beijing 100012, China
[2]Center for Radiophysics and Space
Research, Department of Astronomy,
Cornell University. Ithaca, NY
14853, USA
wangchen@bao.ac.cn

Padma K. Shukla
Theoretische Physik IV, Ruhr-
Universität Bochum, D-44780
Bochum, Germany
ps@tp4.rub.de

Bengt Eliasson
Theoretische Physik IV, Ruhr-
Universität Bochum, D-44780
Bochum,
Germany
bengt@tp4.rub.de

Dastgeer Shaikh
Institute of Geophysics and
Planetary Physics,
University of Californina, Riverside,
CA 92521, USA
shaikh@ucr.edu

Menas Kafatos
College of Science, George Mason
University, Fairfax, VA 22030, USA

Prasad Subramanian
Indian Institute of Astrophysics,
Bangalore - 560034, India
psubrama@iiap.res.in

Peter A. Becker
College of Science, George Mason
University, Fairfax, VA 22030, USA

Banibrata Mukhopadhyay
Astronomy and Astrophysics
Programme, Department of Physics,
Indian Institute of Science,
Bangalore-560012, India
bm@physics.iisc.ernet.in

Ashok K. Singal
Astronomy & Astrophysics Division,
Physical Research
Laboratory, Navrangpura, Ahmed-
abad - 380009, India
asingal@prl.res.in

G. S. Lakhina
Indian Institute of Geomagnetism,
Plot no. 5, Sector-18,
Kalamboli Highway, Panvel (W),
Navi Mumbai-410 218, India
lakhina@iigs.iigm.res.in

S. Alex
Indian Institute of Geomagnetism,
Plot no. 5, Sector-18,
Kalamboli Highway, Panvel (W),
Navi Mumbai-410 218, India
salex@iigs.iigm.res.in

R. Rawat
Indian Institute of Geomagnetism,
Plot no. 5, Sector-18,
Kalamboli Highway, Panvel (W),
Navi Mumbai-410 218, India
rashmir@iigs.iigm.res.in

G. Thejappa
Department of Astronomy, University of Maryland, College
Park, MD 20742
thejappa@astro.umd.edu

R. J. MacDowall
NASA, Goddard Space Flight
Center, Greenbelt, MD 20771
Robert.MacDowall@nasa.gov

List of Contributors XIII

Sanjiv Kumar Tiwari
Udaipur Solar Observatory,
Physical Research Laboratory,
P. Box - 198, Dewali, Bari Road,
Udaipur - 313 001, Rajasthan, India
stiwari@prl.res.in

Jayant Joshi
Udaipur Solar Observatory,
Physical Research Laboratory,
P. Box - 198, Dewali, Bari Road,
Udaipur - 313 001, Rajasthan, India

Sanjay Gosain
Udaipur Solar Observatory,
Physical Research Laboratory,
P. Box - 198, Dewali, Bari Road,
Udaipur - 313 001, Rajasthan, India

P. Venkatakrishnan
Udaipur Solar Observatory,
Physical Research Laboratory,
P. Box - 198, Dewali, Bari Road,
Udaipur - 313 001, Rajasthan, India

S.M. Mahajan
Institute for Fusion Studies,
The University of Texas at Austin,
Austin, Texas 78712, USA

A.M. Ahadi
Physics Department, Shahid
Chamran University, Ahvaz, Iran
E.Mail: ahadi.am@gmail.com

S. Sobhanian
Faculty of Physics, Tabriz University,
Tabriz, Iran
E-Mail:
sobhanian@tabrizu.ac.ir

Nikhil Chakrabarti
Saha Institute of Nuclear Physics,
1/AF Bidhannagar, Kolkata - 700064
E-Mail:
nikhil.chakrabarti@saha.ac.in

Barnana Pal
Saha Institute of Nuclear Physics,
1/AF Bidhannagar, Kolkata - 700064

Part I

Astrophysical Turbulence

Aspects of Hydrodynamic Turbulence in Classical and Quantum Systems

J.J. Niemela

The Abdus Salam International Center for Theoretical Physics, Strada Costiera 11, 34014 Trieste, Italy niemela@ictp.it

Summary. Turbulence is a complex phenomenon, characterized by many interacting degrees of spatial and temporal freedom. It is widespread, and indeed nearly the rule, in the flow of classical fluids. The complexity of the underlying equations has impeded analytical progress and therefore direct numerical simulations of the equations, as well as experimental work, are key to making further progress. Here, special attention is given to the role of low temperature helium as a test fluid, including the superfluid phase which also exhibits turbulence.

Key words: Turbulence, helium, superfluid, convection, Reynolds number

1 Introduction

Turbulence is important in a wide variety of phenomena found in nature and in engineering [1]. In the oceans and atmosphere, the large scale circulations are turbulent and both weather and climate prediction depend on some knowledge of their dynamics. Turbulence can also be of great importance in the design of various means of transportation, whether ships, submarines, aircrafts or cars as well as in heat transfer applications where either natural or forced convection is operational. And it plays a role in nature not only in the part of the earth we see but beneath it where turbulent convection in the fluid outer core gives rise to the magnetic field [2]. It also can be found beyond Earth: in fact, one of the most ubiquitous fluid flows in the universe is the turbulent convection of heat from the outer radiative zone to surface of stars [3].

On a more fundamental level, fluid turbulence is characterized by having many degrees of freedom, both spatial and temporal, and is thus an example of the science of "complexity". It has potentially useful analogies to many diverse areas of interest, as disparate as stock market fluctuations [4]. Of course, the connection is not very direct: we rely on the fact that in the case of fluid turbulence we happen to know the exact equations of motion and for that reason it is a good model system. On the other hand, it must also be

4   J.J. Niemela

pointed out that the complexity of these equations has contributed to a very slow progress in analytical understanding of turbulence. For this reason there has been a particular emphasis on developing numerical and experimental approaches. With regard to the former, the computing power needed to solve large turbulent flows accurately is still years away [5], placing a still larger burden on experimental input which we will discuss in this article.

There have been several approaches to strengthening the experimental effort ranging from developing more powerful wind tunnels to developing better instrumentation. There has been another effort devoted to the identification and utilization of better test fluids, besides the more easily usable air and water.

While non-conventional test fluids may be desirable, their use may require considerable further development work in instrumentation, suitable apparatus, techniques, etc, and therefore their choice depends on the precise questions that one wishes to ask, and clearly on whether the new fluid offers a significant or unique solution. One such unconventional test fluid worth mentioning is helium at cryogenic temperatures, whether in gaseous or liquid phase. This lecture will briefly survey some of the applications of this test fluid to the study of turbulence and borrows from a previous review [6].

1.1 Fluid equations

The equations of motion are expressed in a coordinate system that is fixed in space. For a fluid "element" of density ρ and moving at a velocity $\mathbf{u}$, mass conservation is given by

$$\frac{\partial \rho}{\partial t} + \nabla \cdot (\rho \mathbf{u}) = \mathbf{0}. \tag{1}$$

It is usually the case that the density is, or can be considered to be constant in which case this reduces to

$$\nabla \cdot \mathbf{u} = \mathbf{0}. \tag{2}$$

For an ordinary Newtonian fluid momentum conservation reduces to

$$\frac{D\mathbf{u}}{Dt} = -\frac{1}{\rho}\nabla p + \nu \nabla^2 \mathbf{u} + \mathbf{F_{ext}}, \tag{3}$$

for a fluid having pressure p, and kinematic viscosity $\nu = \mu/\rho$, where μ is the shear viscosity. We have used the so-called substantive or convective derivative $\frac{D}{Dt} = \frac{\partial}{\partial t} + \nabla \cdot \mathbf{u}$ for convenience. The extra term $\mathbf{F}_{ext}$ represents a body force which could, e.g., be due to gravity, rotation, or magnetic field. In the case of thermal convection, which is of particularly common occurrence and interest (and a focus of this lecture), we have $F_{ext} = g\alpha\Delta T$, where g is the acceleration gravity, α is the isobaric coefficient of thermal expansion and ΔT is the vertical (in the direction of the gravity) temperature difference across a layer of fluid in the direction of the gravity. Assuming further that the density is not affected strongly by the thermal expansion (i.e., we consider only "thin layers"), these equations can then considered to be in the Boussinesq approximation [7]. The final equation in this approximation is that for energy conservation:

$$\frac{\partial T}{\partial t} = \kappa_f \nabla^2 T - \mathbf{u} \cdot \nabla T, \tag{4}$$

where κ_f is the thermal diffusivity of the fluid.

These partial differential equations do not suffice by themselves, of course; the complete description of the flow requires the specification of the initial and the boundary conditions.

The equations presented above can be made more useful by scaling all variables: using a characteristic velocity U, length L, time by L/U and pressure ρU^2. Then we have

$$\frac{\partial \mathbf{u}}{\partial t} + \mathbf{u} \cdot \nabla \mathbf{u} = -\nabla p + \frac{1}{Re} \nabla^2 \mathbf{u}, \tag{5}$$

where a Reynolds number $Re = UL/\nu$ now appears as a consequence of the scaling.

In its dimensionless form the momentum equation can account for many different phenomena, from microscopic scales to astrophysical scales. The sole difference in the description of diverse flows is the value of the Reynolds number Re that appears there, and of course the initial and the boundary conditions which supplement the equation. The Reynolds number in fact is also referred to as a similarity parameter in the context of aerodynamic model-testing, in which reliable information can be obtained from smaller scale models of aircraft, cars, etc. Of course, all this is true for simple pressure-driven flows that are also not too fast (to avoid approaching the speed of sound and therefore introducing the Mach number). An example of a simple flow is illustrated in Figure 1, which shows flow going from left to right past a vertical cylinder protruding above the fluid surface. In the left-most photo the flow corresponds to a Reynolds number of about 26. There we see a prominent back-eddy, which is somewhat regular, or "laminar" in appearance. In the right-hand photo the flow configuration is the same, but with $Re = 2000$. Note that in this case the flow behind the cylinder is turbulent with many length (and time) scales evident as opposed to the flow on the left for which there is only flow roughly at the scale of the cylinder. This is a signature of a turbulent flow and distinguished it from one which is laminar, with the Reynolds number being a quantitative measure of its intensity; i.e. turbulence exists for $Re \gg 1$.

Other similarity parameters can also be identified; for instance, in thermally generated flows, the Rayleigh number

$$Ra = g\alpha \Delta T L^3 / \nu \kappa_f, \tag{6}$$

results from making the equations with the appropriate body force term non-dimensional and physically represents the ratio of the rate of potential energy

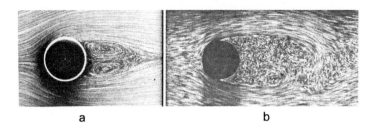

Fig. 1. Flow of a fluid past a circular cylinder. (a) $Re = 26$. (b) $Re = 2000$. After Van Dyke [8].

release due to buoyancy to the rate of its dissipation due to thermal and viscous diffusion [9] where L is the height of a horizontally oriented fluid layer having a vertical temperature difference ΔT across it. Another useful number is the Prandtl number

$$Pr = \nu/\kappa_f \tag{7}$$

which can be understood as the ratio of time scales due to thermal diffusion ($\tau_\theta = L^2/\kappa_f$) and momentum diffusion ($\tau_v = L^2/\nu$).

For a general discussion on turbulence, it is easier to consider only those flows which are driven by pressure gradients; i.e. when the principal control parameter is Re. When it is fully developed, the turbulence so generated can have fluctuations or "eddies" of many different scales. Clearly, the largest such scale is associated with the size of the system, which in the case of an experiment is the size of the confining box or apparatus.

The smallest scales are identified as those which are damped out quickly by viscosity, converting the energy that is transferred form the largest scales in near-scale interactions. This is the concept of a turbulent cascade of energy. The smallest scale, for which viscosity dominates, is also referred to as the dissipation (or Kolmogorov) scale, η.

In between the dissipation scale and the largest scales at which energy is injected, lies the so-called "inertial" range. It is in this range that the energy cascades from scale to scale and where viscosity is unimportant. The rate of energy dissipation per unit mass, ϵ in steady state is clearly the same as that of energy input; by dimensional analysis we can estimate it as proportional to u^3/ℓ, where u is a characteristic measure of the velocity fluctuation at the energy injection scale ℓ.

The main thing to note is that either decreasing the viscosity, or increasing the Reynolds number of the large scale extends the inertial range by making the dissipation scale ever smaller (the large scales are typically fixed by physical constraints). Dimensional analysis gives this smallest scale as

$$\eta = (\nu^3/\epsilon)^{1/4} = \ell Re_\ell^{-3/4}. \tag{8}$$

Here, $Re_\ell = u\ell/\nu$ is the Reynolds number of the large scale.

We can also define the energy cascade in terms of an energy spectrum $E(k)$, where k is the wavenumber and $E(k)dk$ is the kinetic energy contained in wavenumbers between k and $k + dk$. Using the cascade picture described above [10], we have

$$E(k) = C_k \epsilon^{2/3} k^{-5/3}. \tag{9}$$

The constant C_k is empirically determined [11, 12] to have a value near unity (about 1.5).

We expect generally that the turbulence can be nearly homogeneous and isotropic in the inertial range, following Kolmogorov's hypothesis of local isotropy [10]. Since the inertial range grows as a function of Reynolds number, if there are any universal statistical properties of turbulence we will probably find them in flows for which $Re \to \infty$, which must then be an aim for numerical and laboratory experiments.

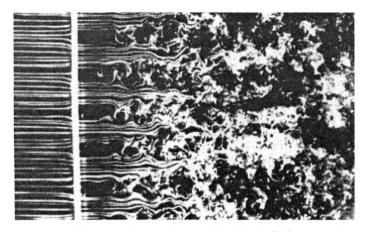

Fig. 2. Grid turbulence. After Frisch [13].

One of the most common methods for generating approximately homogeneous and isotropic turbulence in experiments is to place a grid of crossed tines perpendicular to the direction of flow as shown in Figure 2. Here wakes produced from each tine form small jets at the mesh scale, which then interact through the nonlinear terms in the Navier-Stokes equations to create the nearly homogeneous and isotropic turbulence further downstream. We shall consider this example later in the context of an experiment using a quantum fluid.

2 Low temperature helium as a test fluid

We will consider in some detail here the use of low temperature helium for the purpose of increasing the control parameters, Re and Ra, for fluid tur-

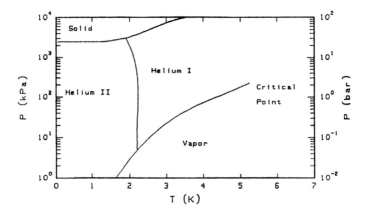

Fig. 3. Pressure-temperature phase diagram for helium.

bulence. In fact, cryogenic helium has the lowest value of kinematic viscosity of any known fluid. It also has some rather interesting quantum properties near absolute zero that are both interesting in themselves, and which perhaps could shed some light on classical flows.

Figure 3 shows the pressure-temperature phase diagram of helium. One aspect of the phase diagram that stands out is that helium remains a liquid even at the absolute zero of temperature. This results from a number of considerations. First, helium is a noble gas and hence the interaction between neighboring atoms is inherently weak. Because the atoms have little mass the quantum mechanical zero-point energy is sufficient to keep the atoms well separated so that the liquid does not solidify under normal pressure. In fact, one needs to pressurize helium to roughly 25 atmospheres at absolute zero in order to form the solid phase as can be seen from figure 3. Below about 2.177 kelvin at the saturated vapor pressure (SVP) the liquid phase becomes a superfluid, referred to as helium II, whereas above the transition point at SVP the liquid phase is completely classical and referred to as helium I.

Helium II is characterized by quantum effects: it is related to a form of Bose-Einstein condensation, in which some fraction of the atoms are condensed into the lowest quantum state. Because it is liquid, helium II is far from being a non-interacting ideal gas and so even at absolute zero temperature only a small fraction of the atoms are condensed into the ground state.

One can apply a hydrodynamic description of helium II where it is found to exhibit so-called two-fluid behavior. In this description the fluid is composed of two interpenetrating fluids: a normal fluid, which may be considered as composed of thermal excitations with a density ρ_n, and a superfluid with density ρ_s, with the total density given by $\rho = \rho_n + \rho_s$. In addition, each fluid has its own velocity field $\mathbf{v_n}$ and $\mathbf{v_s}$. The normal fluid, being in some sense a gas of thermal excitations, carries the thermal energy in the system and there

Aspects of Hydrodynamic Turbulence in Classical and Quantum Systems

is little reason not to consider it as a classical fluid. It has a shear viscosity given by η_n, while the superfluid component has zero viscosity; as we shall see there can be viscous dissipation at high enough flow rates due to the presence of vortices.

One of the most important facts is that the superfluid component of helium II is irrotational ($\mathrm{curl}\mathbf{v_s} = \mathbf{0}$). It can however have circulation, $\oint \mathbf{v}_s \cdot d\mathbf{r}$, in a multiply-connected geometry. We consider, then, the flow around a hollow vortex core where the superfluid velocity is given by

$$\mathbf{v}_s = \frac{\hbar}{m_4}\nabla S. \tag{10}$$

Here, the phase, S, of the wave function of the condensed atoms is a function of the spatial coordinates. Integrating the superfluid velocity around a closed curve (and requiring the condensate wave function to be single-valued) results in the circulation $\kappa = n2\pi\,\hbar/m_4 = nh/m_4$ where n is an integer. It turns out that the quantum number $n = 1$ so that the circulation around any quantized vortex is simply given by $\kappa = h/m_4$. Further details can be found in a recent review [14].

We shall consider below quantum turbulence generated in helium II, where the statistical properties of the turbulence agree with predictions for classical fluids despite the seemingly severe quantum restriction on the circulation.

2.1 Fluid properties

Returning to the general issue of producing turbulence at high values of $Re = UL/\nu$, we note that there are three ways in which laboratory experiments can affect this: by maximizing some combination of length scale L, velocity U or kinematic viscosity ν. First of all it is not always feasible to increase L arbitrarily due to available space and financial considerations (keeping in mind that Re for turbulent flows is measured by its logarithm). The velocity U cannot be made arbitrarily large compared to the speed of sound, c, in the fluid without introducing another similarity parameter, the Mach number $M = U/c$. Hence, the best method is to search for fluids with low kinematic viscosities. As we had mentioned above, helium has the lowest kinematic viscosity of any known fluid, but how does it compare to the more common test fluids such as water and air? This comparison is shown in Table 1.

Clearly helium has a substantial advantage in terms of viscosity over air and water; on the other hand, these latter fluids are easier to obtain and much easier to use. we note that helium II has a kinematic viscosity which is similar to that of helium I, defining it as the ratio of the normal shear viscosity to the total helium II density $(\rho_s + \rho_n)$. This remains true even to the absolute zero of temperature where an *effective* kinematic viscosity, proportional to

Table 1. Kinematic viscosities for some test fluids. The liquid phases of helium are evaluated at the saturated vapor pressure (SVP).

$Fluid$	$T(K)$	$P(Bar)$	$\nu(cm^2/s)$
Air	293	1	0.15
Water	293	1	0.01
Helium I	2.2	SVP	1.8×10^{-4}
Helium II	1.8	SVP	8.9×10^{-5}
Helium gas	5.5	2.8	3.2×10^{-4}

the quantum of circulation κ acts in turbulent flows even in the absence of a normal component [14].

Let us turn our attention briefly to thermal turbulence and examine other beneficial properties of helium. Recall that the relations 6 and 7 represent the two principal similarity parameters—the Rayleigh number Ra and the Prandtl number Pr— for systems in which buoyancy, derived from unstable heating, drives the flow. Ra is the primary control parameter and fully developed thermal turbulence occurs at very large values of Ra; again, we wish to produce flows in which Ra varies over a wide range. It is clear that the small kinematic viscosity of helium can make Ra also quite large. In addition, for helium gas, the thermal diffusivity $\kappa_f = k/\rho C_P$, where k is the thermal conductivity of the fluid, ρ its density, and C_P is the constant pressure specific heat, is of the same order of ν (except near the critical point, where, in fact, it becomes vanishingly small because C_P diverges there). The Prandtl number $Pr = \nu/\kappa_f$ has a nearly constant value near 0.7 sufficiently away from the critical point, as for air and other non-interacting gases, and increases rapidly only near the critical point.

The other fluid parameter which enters the definition of the Rayleigh number is the isobaric thermal expansion coefficient $\alpha \equiv -\rho^{-1}\partial\rho/\partial T$, where the derivative is evaluated at constant pressure. The sign of α determines the direction of the temperature gradient that is needed to induce buoyancy; since it is positive in most cases, the heating needs to be from below in order to destabilize the flow and set up turbulence. A notable exception is water for which the sign of α reverses below 4 degrees Celsius, as is also the case for liquid helium just above the lambda point.

For non-interacting gases, $\alpha = 1/T$, and so, low temperatures themselves have a particular advantage for buoyancy-driven flows. For helium near its critical point, α is thermodynamically related to the specific heat and also diverges. In summary, it is the combination $\alpha/\nu\kappa_f$ that determines the Rayleigh number, and we show it in Table 2 for helium, air and water.

One can use to advantage the enormously large values of the combination $\alpha/\nu\kappa_f$ near the critical point and generate large values of Ra. For illustration, if we consider a fluid layer some 10 meters tall (difficult and expensive, but well within the capacity of presently available cryogenic capability) and a

Aspects of Hydrodynamic Turbulence in Classical and Quantum Systems   11

Table 2. Values of the combination of fluid properties $\alpha/\nu\kappa_f$ for air, water and helium

$Fluid$	$T(K)$	$P(Bar)$	$\alpha/\nu\kappa_f$
Air	293	1	0.12
Water	293	1	14
Helium I	2.2	SVP	2.3×10^5
Helium II	1.8	SVP	$- - -$
Helium gas	5.25	2.36	6×10^9
Helium gas	4.4	2×10^{-4}	6×10^{-3}

reasonable temperature difference of $0.5K$, Rayleigh numbers of the order 10^{21} are possible. Such large values are characteristic of turbulent convection in the Sun [15].

2.2 Small "table-top" apparatus

Since the audience is presumably composed mostly of academic persons, we should re-emphasize that for any given value of Re, the helium experiment can be made to fit within a standard laboratory, yet yield world-class values of Re. This we can illustrate by taking the example of pipe flow, considering both laminar and fully turbulent conditions. The classical experiment is due to Nikuradse [16] who obtained turbulent flow conditions with Re of up to 3×10^6 in a water filled apparatus occupying two storeys of a building. Using liquid helium I this same range of Re has been duplicated in a standard dewar (13 cm diameter, 1.5 m tall) containing a pipe flow apparatus, where the pipe was only 5 mm in diameter [17, 18].

3 Alternatives

It is quite reasonable to ask why nature's "laboratories" such as the oceans or the atmosphere, which achieve large L without the added cost, cannot be instrumented and studied instead of investing in laboratory experiments using an unconventional fluid. The answer is that while this is, in fact, routinely done, the observed flows are not subject to well-controlled boundary conditions, and hence there are limitations as to the extent to which a deeper understanding of turbulence can be obtained; i.e., where small deviations from expected behavior may contain substantial new insights.

Of course, the reader is also aware that computers these days are quite powerful and are advancing their capabilities at a rapid rate. So it is also very reasonable to ask why we cannot simply run direct numerical simulations (DNS) in which the appropriate equations are solved on a computer without making any approximation. There are in fact many more numerical experiments than laboratory ones, but still this wonderful tool has a limited applicability for simulating practical flows at high Reynolds and Rayleigh numbers. Indeed, the DNS efforts are hampered by the large number of degrees of freedom in turbulence that grows as $Re^{9/4}$. This can be easily seen from 8, where it is clear that the range of scales needed to be well resolved, ℓ/η, grows like $Re^{3/4}$. Hence, if the flow is to be followed in a numerical simulation with a uniform grid, the minimum number of necessary grid points (in three dimensions) is proportional to $Re^{9/4}$. The state of the art in full DNS calculations of turbulent flows [13] is around $Re \sim 10^4$ whereas most atmospheric flows and many engineering flow of interest are for $Re > 10^7$.

Large eddy simulations (which compute only the large scales but model the small scales) and turbulence models (which compute only average information) are quite useful in getting to higher Re, and can produce useful information, but they are not satisfactory as universal recipes.

4 Examples of laboratory turbulence using helium

We now consider two types of flows where low temperature helium has been used effectively to push the limits of turbulence research.

4.1 Turbulent thermal convection

Turbulent thermal convection is a problem of much importance for a number of reasons: it plays a prominent role in the energy transport within stars, atmospheric and oceanic circulations, the generation of the earth's magnetic field, and countless engineering processes in which heat transport is important. Of course actual flows are often very complex and it is not reasonable to attempt to replicate in a laboratory every aspect related to complex boundary conditions, rotation, magnetic fields, ionization, chemical reactions, etc. For this reason, it is customary to consider a simpler problem that still contains the essential physics. In the case of buoyancy-driven turbulence, this model is Rayleigh-Bénard convection (RBC). In RBC, a thin fluid layer of infinite lateral extent is contained between two isothermal surfaces with the bottom surface maintained slightly hotter. When the expansion coefficient is positive (the usual case), an instability develops because the hot fluid from below rises to the top and the colder fluid from above sinks to the bottom. The applied driving force is measured in terms of a Rayleigh number, Ra, which is a non-dimensional measure of the imposed temperature difference across the fluid layer, see equation 6. The Prandtl number, Pr, defined in equation 7, is an additional dynamical parameter because it determines the state and thickness of the viscous and thermal boundary layers set up on the boundaries. With increasing Ra the dynamical state of RBC goes from a uniform and parallel roll pattern at the onset ($Ra \sim 10^3$) to turbulent state at $Ra \sim 10^7$–10^8.

Here, the quantity roughly corresponding to drag in an isothermal flow is the heat transport, given non-dimensionally in terms of the Nusselt number Nu

$$Nu = \frac{q}{q_{cond}} = \frac{qH}{k\Delta T}, \tag{11}$$

where q is the total heat flux, q_{cond} is the heat flux in the absence of convection, given by Fourier's law. Nu represents the ratio of the effective turbulent thermal conductivity of the fluid to its molecular value and can reach values of over 10^4 in helium experiments [19] thus demonstrating the enormous enhancement of transport possible.

The ability to accurately predict Nu for the Ra corresponding to many natural and engineering flows is out of reach of laboratory experiments using conventional fluids, since extrapolation in Ra by many orders of magnitude is required. The use of helium has extended the range of Ra by at least 6 orders of magnitude compared with the best water experiments, reaching values corresponding to many natural flows, and substantially reducing the needed extrapolation for others.

We expect $Nu = f(Ra, Pr, \Gamma, ...)$ to attain some limiting power-law form in fully developed turbulence as $Ra \to \infty$. Thus, we may hope to extrapolate results from experiments at the highest Ra if we have, indeed, exceeded the Ra beyond which the functional relation ceases to change.

In a simplistic view, it is possible to see two limiting cases for the scaling of Nu. In the first, we imagine that the global flux of heat is determined by processes occurring in the two thermal boundary layers at the top and bottom of the heated fluid layer. These layers, defined as the extent over which the molecular conduction remains important, are always present at the solid boundary where a heat flux is imposed, although they become thin in turbulent convection. Their thickness λ can be estimated as the distance over which the imposed temperature gradient would produce a state of marginal stability for convection; that is, when the Rayleigh number based on their height exceeds something of order 10^3. Assuming that the temperature outside the thermal boundary layer is uniform (because of turbulent mixing), one can easily show that $\lambda = H/2Nu$. In helium experiments λ can be of the order of hundreds of micrometers in a sample cell of height one meter [19]. The resulting large gradient in temperature acts as a source of stress, resulting in bursts of hot or cold fluid into the interior, commonly referred to as "plumes" or "thermals" [20, 21].

Now, in the sense that the bulk fluid outside the boundary layers is fully turbulent and "randomized", it acts more or less as a thermal short circuit. Therefore its precise nature or spatial extent is immaterial to the flux of heat and from the definition of Nu given in (14) we can then immediately write down the asymptotic power law [22] as $Nu \sim Ra^{1/3}$, where we assume that the heat flux q appearing in Nu has no implicit height dependence (no internal energy sources). This simple argument has been made more precise by others [23] and rigorously shown [24] to hold when Pr and Ra are both very large.

There is another limit, one can think of, in which molecular properties lose relevance in determining heat transport—for instance, when boundary layers are mixed by a vigorous turbulent flow so that molecular conduction is supplanted entirely by small scale turbulent motion. Here, an exponent of 1/2 (modulo logarithmic corrections) has been proposed; the phenomenological theory has been developed by Kraichnan [25].

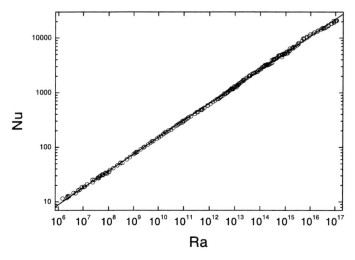

Fig. 4. Log-log plot of the Nusselt number versus Rayleigh number. The line through the data is a least-square fit over the entire Ra range, and represents a $dlogNu/dlogRa$ slope of 0.32.

Figure 4, plotted using the data from Niemela *et al.* [19], illustrates the enormous range of Ra and Nu possible in low temperature experiments of quite modest size. The average slope over 11 decades of Ra is 0.32, rather close to 1/3. These data have been corrected for some minor effects due to the finite conductivity of the horizontal boundaries (although the conductivity of pure metals at low temperature compared to the fluid is much larger than for corresponding room temperature experiments) and another effect due to heat being carried up the sidewalls and then flowing laterally into the fluid, which can drive the large eddies which then feed back on the sidewall conduction. The models developed for this are rather rudimentary [26, 27, 28, 29] and a better knowledge of this effect might slightly alter the exponent. However, it appears as though an exponent of 1/3 is much more characteristic of high-Ra turbulence than 1/2, and the use of helium has been crucially important to making this statement.

4.2 Superfluid grid turbulence

As we discussed earlier, the canonical method of creating a nearly isotropic and homogeneous turbulent flow is to force a fluid through a grid of crossed bars (see 2). A series of measurements have also been made in which a grid is towed through a stationary sample of helium II, contained within a square channel of 1 cm^2 cross-section [30, 31, 32, 33]. We should point out that whether it is the grid or the fluid which moves is, of course, irrelevant.

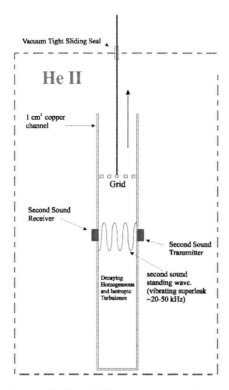

Fig. 5. Schematic of superfluid grid flow apparatus. Attenuation of a standing second sound wave is used to detect the decay of average quantized vortex line density.

The apparatus is shown schematically in figure 5. The mesh Reynolds number defined as $Re_M = v_g M \rho / \mu$, where μ is the dynamic viscosity of the normal fluid, ρ the total density of the fluid (helium II), M the mesh size, and v_g the velocity at which the grid was pulled, varied between 5000 and 200,000. These mesh Reynolds numbers must be considered quite large given the compactness of the apparatus. There is some ambiguity as to which density must be used in defining the Reynolds number in helium II. The present choice seems plausible because, when the superfluid vortex tangle becomes polarized and the superfluid and normal fluid velocity fields are correlated, helium II should behave as a single fluid with the combined density ρ for scales large compared with the average inter-vortex line spacing [34].

On the opposing walls of the channel shown in figure 5 two second-sound transducers were situated, using flush-mounted vibrating superleaks. In this case the channel acted as a second sound resonator. The length of quantized vortex line per unit volume, L, can be deduced from second sound measurements through the relation

$$L = \frac{16\Delta_0}{B\kappa_c}\left(\frac{A_0}{A} - 1\right), \qquad (12)$$

where A and A_0 are, respectively, the amplitudes of the second sound standing wave resonance with and without the vortices present, Δ_0 is the full width at half maximum of the second sound resonance peak in the absence of vortices, B is the (frequency-dependent) mutual friction constant [35]. Throughout this section, we keep the conventional symbol L for the line density, distinguishing it from the length scale used previously, e.g., in equation 6, by context.

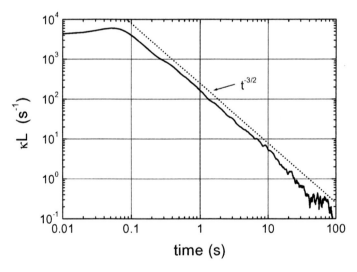

Fig. 6. The log-log plot of the decaying vortex line density versus time in helium II. The "classical" exponent $-3/2$ is shown by the displaced dashed line.

An example of decaying line density, after the grid is pulled through a column of helium II, is shown in figure 6. The slope of $-3/2$ is precisely what one might expect from a classical description of the superfluid turbulence, which we will now attempt to show.

For classical homogeneous turbulence, the turbulent energy dissipation rate per unit mass, ϵ is exactly equal to kinematic viscosity times the mean squared vorticity. For decaying grid turbulence in helium II, we may *assume* [36, 14] that the total energy dissipation per unit mass in the turbulent fluid is analogously given by

$$\epsilon = \nu' \kappa_c^2 L^2, \tag{13}$$

where ν' is a proportionality constant with units of kinematic viscosity.

Second, the energy dissipation rate in fully developed turbulence is thought to be independent of viscosity, so that

$$\epsilon = C_\epsilon u^3 / \ell, \tag{14}$$

where C_ϵ is a constant for sufficiently large Reynolds number. Here, u can be taken as the root-mean-square value of the turbulent velocity and $C_\epsilon = 0.5$, dependent on the definition used for ℓ [37]. In the following we assume that the same type of relation holds for helium turbulence as well.

At late stages of decay following a grid pull, ℓ will become comparable to the channel width d and will cease to grow further. Then the above relation becomes

$$u^2 = \left[\frac{\epsilon d}{C_\epsilon} \right]^{2/3}. \tag{15}$$

Noting further that $d/dt(\frac{3}{2}u^2) = -\epsilon$, we have

$$-\epsilon = \left(\frac{d}{C_\epsilon} \right)^{2/3} \epsilon^{-1/3} \frac{d\epsilon}{dt}. \tag{16}$$

Integrating we have

$$\kappa_c L = \left[\frac{27}{\nu'} \right]^{1/2} \frac{d}{C_\epsilon} t^{-3/2}. \tag{17}$$

This -3/2 decay is just that observed in the superfluid data (see figure 6) and shown there by the dashed line, indirectly showing that approximately homogeneous and isotropic turbulence in the quantum fluid is similar to that which we would expect for a classical fluid, even though the quantum restrictions on the circulation about any vortex line are severe. We can furthermore solve equation 17 for the effective viscosity of quantum turbulence. This has been done and shown [14] to be in excellent agreement with theoretical results assuming an effective kinematic viscosity proportional to the quantum of circulation κ_c.

Finally, we point out that the highest values of mesh Re obtained in these experiments are comparable to those obtained in large wind tunnels, while the relatively small size of the channel (1 cm by 1 cm) makes it quite accessible to academic facilities.

18   J.J. Niemela

5 Summary

In summary, we have demonstrated that large values of the turbulence control parameters, Re and Ra, can be obtained under well-controlled laboratory conditions using low temperature helium as the test fluid. In the case of thermal turbulence this brings us closer to identifying the asymptotic scaling relation between Ra and the dimensionless heat transport Nu.

The observation that turbulence below the superfluid transition is amenable to a quasi-classical description is interesting and may lead to a better understanding of the nature of turbulence above the superfluid transition, which remains one of the major challenging problems of classical physics.

References

1. D.J. Tritton, *Physical Fluid Dynamics*. Clarendon Press, Oxford (1988).
2. G.A. Glatzmaier, R.C. Coe, L. Hongre, P.H. Roberts, *Nature* **401**, 885 (1999).
3. E.M. Spiegel, *Ann. Rev. Astron. Astrophys.* **9**, 323 (1971).
4. B.B. Mandelbrot, *Scientific American* **280**, 50 (1999).
5. National Research Council Report: *Condensed Matter and Materials Physics: Basic Research for Tomorrow's Technology*, 308 pages, National Academy Press (1999).
6. J.J. Niemela and K.R. Sreenivasan, *J. Low Temp. Phys.* **146**, 499-510 (2007).
7. A. Oberbeck, *Annalen der Physik und Chemie* **7**, 271 (1879).
8. M. Van Dyke, *An Album of Fluid Motion* The Parabolic Press 1982.
9. F.H. Busse, *Rep. Prog. Phys.* **41**, 1929 (1978).
10. A.N. Kolmogorov, *Dokl. Akad. Nauk SSSR* **30**, 9 (1941).
11. A.S. Monin, A.M. Yaglom, *Statistical Fluid Mechanics, vol. 2*, MIT Press, Cambridge, USA (1975).
12. K.R. Sreenivasan, *Phys. Fluids* **7**, 2778 (1995).
13. U. Frisch, *Turbulence* (Cambridge University Press, 1995).
14. W.F. Vinen, J.J. Niemela, *J. Low Temp. Phys.* **128**, 167 (2002).
15. K.R. Sreenivasan, R.J. Donnelly, *Adv. Appl. Mech.* **37**, 239-276 (2001).
16. J. Nikuradse, NASA TT F-10359 (1966). Translated from the German original *Forsch. Arb. Ing.-Wes.* No. 356 (1932).
17. C.J. Swanson, B. Julian, G.G. Ihas, R.J. Donnelly, *J. Fluid. Mech.* **461**, 51 (2002).
18. B.J. McKeon, C.J. Swanson, M.V. Zagarola, R.J. Donnelly, A.J. Smits, *J. Fluid Mech.* **511**, 41 (2004).
19. J.J. Niemela, L. Skrbek, K.R. Sreenivasan, R.J. Donnelly, *Nature* **404**, 837 (2000).
20. E.M. Sparrow, R.B. Husar, R.J. Goldstein, *J. Fluid Mech.* **41**, 793 (1970).
21. S.A. Theerthan, J.H. Arakeri, *J. FLuid Mech.* **373**, 221 (1998).
22. W.V.R. Malkus, *Proc. R. Soc. Lond. A* **225**, 196 (1954).
23. L.N. Howard, in *Proc. 11th Intern. Cong. Appl. Mech.* (ed. H. Gortler), Springer, Berlin, p. 1109 (1966).
24. P. Constantin, C.R. Doering, *J. Stat. Phys.* **94**, 159 (1999).
25. R.H. Kraichnan, *Phys. Fluids* **5**, 1374 (1962).

Aspects of Hydrodynamic Turbulence in Classical and Quantum Systems   19

26. P. Roche, B. Castaing, B. Chabaud, B. Hebral, J. Sommeria, *Euro. Phys. J.* **24**, 405 (2001).
27. G. Ahlers, *Phys. Rev. E.* **63**, art. no. 015303 (2001).
28. J.J. Niemela, K.R. Sreenivasan, *J. Fluid. Mech.* **481**, 355 (2003).
29. R. Verzicco, *J. Fluid Mech.* **473**, 201 (2002).
30. M.R. Smith, Ph.D. thesis, University of Oregon, Eugene (1992).
31. M.R. Smith, R.J. Donnelly, N. Goldenfeld, W.F. Vinen, *Phys. Rev. Lett.* **71**, 2583 (1993).
32. S.R. Stalp, Ph.D. thesis, University of Oregon, Eugene (1998).
33. J.J. Niemela, R.J. Donnelly, K.R. Sreenivasan, *J. Low Temp. Physics* **138** 537 (2005).
34. C.F. Barenghi, S. Hulton, D.C. Samuels, *Phys Rev. Lett.* **89** 275301 (2002).
35. C.F. Barenghi, R.J. Donnelly, W.F. Vinen, *J. Low Temp. Phys.* **52** 189 (1983).
36. S.R. Stalp, L. Skrbek, R.J. Donnelly, *Phys. Rev. Lett.* **82**, 4831 (1999).
37. K.R. Sreenivasan, *Phys. Fluids* **27**, 1048 (1984).

Observations and Modeling of Turbulence in the Solar Wind

Melvyn L. Goldstein

NASA Goddard Space Flight Center, USA
E-mail: melvyn.l.goldstein@nasa.gov

Summary. Alfvénic fluctuations are a ubiquitous component of the solar wind. Evidence from many spacecrafts indicate that the fluctuations are convected out of the solar corona with relatively flat power spectra and constitute a source of free energy for a turbulent cascade of magnetic and kinetic energy to high wave numbers. Observations and simulations support the conclusion that the cascade evolves most rapidly in the vicinity of velocity shears and current sheets. Numerical solutions of the magnetohydrodynamic equations have elucidated the role of expansion on the evolution of the turbulence. Such studies are clarifying not only how a turbulent cascade develops, but also the nature of the symmetries of the turbulence.

Key words: Magnetohydrodynamics-solar wind: Turbulence-simulations

1 Observed Properties of the Solar Wind

The solar wind is the continuous outflow of ionized gas from the solar corona [1]. It is composed primarily of supersonic protons. Alpha particles comprise a few percent of the total and there are traces of heavier ions. The ion flow is generally about Mach 10. In contrast, the electrons are subsonic with a Mach number of ≈ 0.2. Although the temperature of the corona is typically $1 - 2$ MK, the temperature of electron and protons near Earth orbit is only about a tenth of that. The flow becomes supersonic within a few solar radii. Because the solar wind plasma is highly conductive, the "frozen-in" coronal magnetic field is pulled into the interplanetary medium by the flow and solar rotation winds it into an "Archimedian" spiral pattern [2]. The wind rapidly attains a nearly constant terminal speed after which its density decreases as the square of the radial distance from the sun.

Solar wind flow patterns

Solar wind flow patterns vary with solar cycle. The most well-organized patterns occur during times of minimum solar activity when there are regions of

22   Melvyn L. Goldstein

steady fast wind that originate from regions of open magnetic field lines where the corona is cooler than in surrounding loops. These regions appear darker in EUV and X-ray wavelengths and are referred to as *coronal holes*. Fast solar wind has speeds exceeding 600 km/s and can reach $\approx$ 800 km/s. Slow solar wind, which is rather unsteady, has speeds less than about 500 km/s and originates near closed loops and coronal streamers. This region is also associated with coronal mass ejections (CMEs) that generate highly transient flows that include large-scale magnetic clouds.

1.1 Alfvén Waves

Soon after the discovery of the solar wind it was realized [4, 5, 6, 11, 43] that the fluctuations in the magnetic field and velocity were often so highly correlated that they appeared as nearly perfect Alfvén waves [3]. This phenomenon was most pronounced in fast solar wind, especially in the trailing edges of the fast streams. However, highly "Alfvénic" flows have been observed in slow wind as well.

If one normalizes the fluctuating magnetic field, $\mathbf{b}$, by dividing by $\sqrt{\mu_0 \rho}$ (where ρ is the mass density), Alfvén waves are defined by $\mathbf{v} \equiv \pm\mathbf{b}$, where $\mathbf{v}$ is the fluctuating component of the velocity. The $\pm$ sign represents wave propagation antiparallel $(+)$ and parallel $(-)$ to the ambient magnetic field. It is often convenient to use new variables, referred to as Elsässer variables that are defined by $\mathbf{z}^\pm = \mathbf{v} \pm \mathbf{b}$. The predominance of outward propagation suggests that the Alfvénic fluctuations are generated below the height where the flow becomes super-Alfvénic. Thus, only outward propagating fluctuations are convected into the heliosphere.

An example of an Alfvénic interval is shown in Figure 1 from data taken by Ulysses during a polar pass. The data are displayed as component fluctuations in Sun-Earth-Ecliptic coordinates (radial, normal, tangential, i.e., RTN) of the magnetic field (thin lines) and velocity vector (thick lines). The high correlations, also referred to as high *Alfvénicity*, are obvious, and the positive sense of correlation for sunward-pointing magnetic field implies that the fluctuations are propagating outward.

The very high Alfvénicity of solar wind fluctuations might lead one to wonder whether or not the medium is actively evolving because *pure* Alfvén waves are an exact solution of the incompressible, ideal, equations of magnetohydrodynamics. However, the solar wind also exhibits the contrasting quality, first noted by [12], viz., that the power spectra of the magnetic (and velocity) fields resemble that of fully developed fluid turbulence [24] in that the spectral index of the power spectra is $\approx -5/3$. Analyses of Helios data show, however, that while the power spectra of fluctuations observed in slow wind and in fast wind at 1 AU (and beyond) typically have an "inertial range" spectrum with a slope of $-5/3$, at 0.3 AU in fast wind, only a very limited inertial range exits and the spectrum is dominated by an extended energy-containing range with an spectral index of -1.

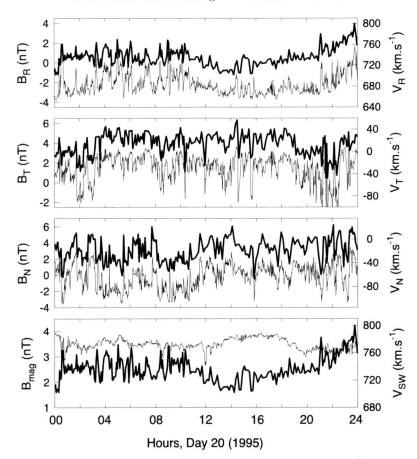

Fig. 1. RTN components and magnitude of the solar wind magnetic field (thin line, left hand scale) and those of the velocity of the solar wind (thick line, right hand scale) measured by Ulysses in the fast solar wind.

Relationship of Solar Wind Turbulence to Fluid Turbulence

In Figure 2 we show the classic example of fully developed fluid turbulence [20], and, for comparison, in Figure 3, an example computed from Voyager magnetometer data obtained near 1 AU [26]. Figure 4 shows the difference between fast and slow wind at 0.3 AU from [42]. The Helios data have been plotted using Elsässer variables, $z^{\pm}$.

As is clear from these spectra, when turbulence has had the opportunity to develop an inertial range, the slope is essentially $-5/3$, the Kolmogorov value for fully developed, incompressible Navier-Stokes fluid turbulence. The remarkable result is still not fully understood, because the solar wind is neither a Navier-Stokes fluid nor is it incompressible.

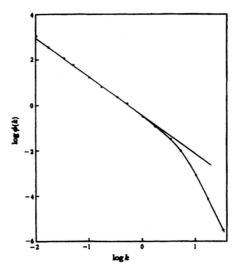

Fig. 2. A plot of the one-dimensional spectrum of tidal channel data collected on March 10, 1959. The straight line has a slope of $-5/3$. From [20].

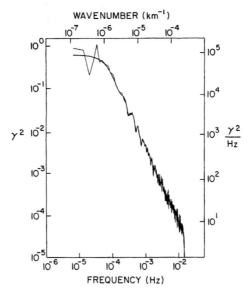

Fig. 3. The magnetic energy spectrum of data collected at 1 AU in units of nT^2. The abscissa is given in both frequency and wave number. Spectra constructed using the Blackman-Tukey algorithm [8] and the fast Fourier transform technique are shown. To an excellent approximation, the slope of the power spectrum is $-5/3$.

[12] noted the strong association of a well-defined −5/3 inertial range with (slow) flows that were found in the vicinity of the heliospheric current sheet where velocity gradients tend to be large and where the magnetic field changes sign as one crosses the current sheet. [12] argued that turbulence in the solar wind was driven by the energy in velocity shears. [36], in an extensive analysis of Helios data, confirmed this conjecture.

Properties of Solar Wind Turbulence

The variation in the degree of Alfvénicity of the magnetic and velocity fluctuation with both distance and with latitude have been an intense area of study. In a series of papers, [36, 38] examined the evolution of the solar wind with distance using data from both Helios and Voyager. That analysis showed clearly that in the outer heliosphere the Alfvénicity tended to decrease—higher wave numbers retained their Alfvénic nature to larger heliospheric distances. (There are exceptions to this as discussed in [38]).

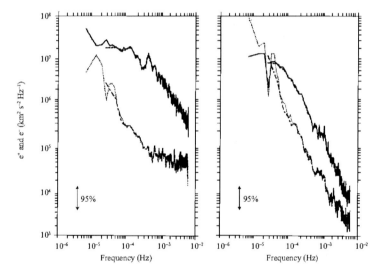

Fig. 4. Power spectra of the Elsässer variables for fast wind (left panel) and slow wind (right panel) from data obtained by Helios at 0.3 AU. Adapted from [42].

Data from Ulysses gives another view of how Alfvénic fluctuations evolve in the heliosphere. Evolution should be quite slow at the high heliospheric latitudes probed by Ulysses because there velocity gradients are small; and indeed, the fluctuations remain Alfvénic even near 4 AU at those high latitudes, as is illustrated in Figure 5 [21, 23].

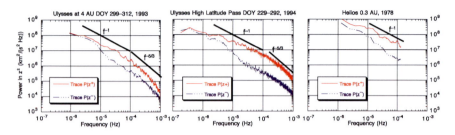

Fig. 5. The spectrum from Ulysses at 4 AU (left) are not very Alfvénic, while the spectrum from the high latitude pass (center) shows little evolution with latitude. The Helios spectrum (right) also suggests that the source spectrum of solar wind fluctuations is both highly Alfvénic and has a spectral slope of −1 (cf. Figure 4, above). Adapted from [17].

1.2 Radial and Temporal Evolution of Solar Wind Turbulence

Another measure of Alfvénicity is the cross helicity, defined by

$$H_c = \tfrac{1}{2} \int dx^3 \mathbf{v} \cdot \mathbf{b}.$$

A useful related quantity is the normalized cross helicity, defined by $\sigma_c \equiv 2H_c^r(k)/E^r(k)$, where $E^r(k)$ is the reduced spectral energy (magnetic plus kinetic). These quantities are reduced in the sense that, due to the super-Alfvénic nature of the solar wind flow, the one-dimensional power spectra are equivalent to three-dimensional spectra that have been integrated over perpendicular wave numbers [26].

A clear indication that velocity shear is the primary driver of solar wind turbulence and the cause of the reduction with heliocentric distance of cross helicity is shown in Figure 6. In the figure, the normalized cross helicity, σ_c, which is generally close to unity in fast wind, fluctuates around zero where B_R changes sign and where the velocity gradients are large. This situation has been simulated using both two-dimensional incompressible and compressible codes in Cartesian (periodic) geometry as well as in three dimensions in spherical geometry [18]. The two-dimensional results are shown in Figure 7. The top panel shows results at $T = 4$ eddy-turnover times from an incompressible MHD simulation, where $\omega \equiv \nabla \times \mathbf{v}$ is the vorticity. The bottom panel shows similar results at $T = 2$ eddy-turn-over times from a solution of the compressible equations.

1.3 The Symmetry Properties of the Magnetic Fluctuations

The magnetic and velocity field fluctuations in the solar wind are intrinsically three-dimensional. However, as mentioned above, the supersonic/super-Alfvénic nature of the flow makes it difficult with only single point measurements to determine more than the reduced, one-dimensional, properties of the

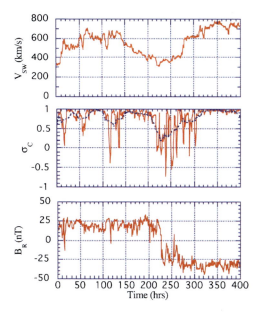

Fig. 6. A current sheet crossing observed by Helios 2 near 0.3 AU (days 93 − 110 of 1976). The data are hour averages. The blue curve of σ_c is a 25−hour running mean. The minimum values of σ_c are found in the low speed region where B_R changes sign and where the velocity gradients are large. Adapted from [34] and [37]

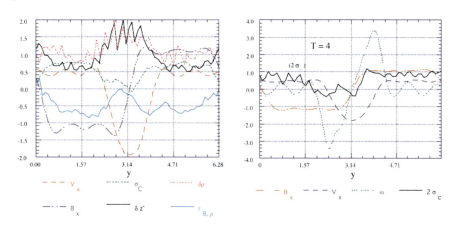

Fig. 7. Left: Results at $T = 2$ eddy-turn-over times from a solution of the compressible equations. The panel includes density fluctuations, ρ, δz_-, and the correlation between density and magnetic field magnitude, r_B ($\delta \rho$ and δz_- are normalized to twice their maximum values of 0.042 and 0.075, respectively). Right: The solution at $T = 4$ eddy-turnover times from an incompressible simulation (ω is the vorticity). From [35]

fluctuations. It is possible, however, even with a single spacecraft (and some additional assumptions) to explore at least two-dimensional symmetries. [40] analyzed the relatively few intervals during which the background magnetic field **B** was either parallel or transverse to the solar wind velocity. Because of a paucity of examples they could not reach any definitive conclusions, but suggested that the magnetic fluctuations could not be isotropic, although for periods when **B** was perpendicular to the solar wind velocity, the isotropic model was a reasonable approximation and parallel propagating Alfvén waves offered a poor approximation that was only approximately valid during the radial **B** flow intervals. They suggested that a combination of magnetosonic and Alfvén waves with approximately one quarter as much power in magnetosonic waves as in Alfvén waves fit both data intervals.

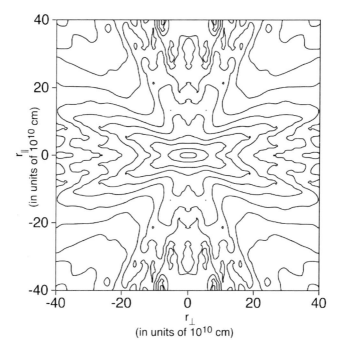

Fig. 8. Contour plot of the two-dimensional correlation function of solar wind fluctuations as a function of distance parallel and perpendicular to the mean magnetic field. The four quadrant plot was produced by reflecting the data across the axes from the first quadrant (from [29]).

The first comprehensive effort to deduce at least the two-dimensional structure of the magnetic fluctuations was carried out using approximately 16 months of nearly continuous data from ISEE 3 by [29]. The results suggested that the solar wind fluctuations were made of two components: parallel propagating planar (slab) Alfvén waves and fluctuations with wave vectors nearly transverse to the mean magnetic field. The analysis was done on stationary data intervals [27, 28, 31, 33] and assumed that all the data intervals (463 15−min averages) were members of the same statistical ensemble. The two-dimensional correlation function from that analysis is shown in Figure 8. More recently, [13] has carried out a similar analysis, but the intervals were sorted by whether the flows were fast or slow. They concluded that the fast and slow wind belonged to separate ensembles and that fast wind was more dominated by slab-like fluctuations and slow wind contained more quasi-two-dimensional fluctuations.

[7] suggested that typically 80% of the magnetic fluctuation energy resided in the quasi-two-dimensional component and only 20% was in the "slab" component. Subsequently, [41], using Ulysses data, showed that between 1991 and 1995, a more typical value was 50%. There was no obvious way to tell, however, whether the "quasi-2D component" reflected a nearly-incompressible population of fluctuations or merely one that contained nearly perpendicular wavevectors.

[39] used a three-dimensional spherical simulation to demonstrate that small velocity shears would produce rapid "phase mixing" of planar Alfvén waves. [19] later generalized the results of [15, 16] to three dimensions, showing that small velocity shears superposed on an initial wave packet of planar Alfvén waves could evolve with distance to produce two-dimensional correlation functions reminiscent of the ISEE-3 analysis.

Consequently, it appears that the $50 - 80\%$ of the fluctuation energy that is quasi-two-dimensional, does not reflect a nearly-incompressible component of the turbulence that arose in the corona, but rather indicates either local generation or phase mixing. Ascertaining the nature of these fluctuations is important because the distribution of power between parallel and perpendicular wave numbers determines the spatial diffusion of solar and galactic energetic particles since only power in $k_{\parallel}$ resonantly scatters energetic charged particles.

2 Simulating Solar Wind Turbulence

The evolution of turbulence in a spherically expanding magnetofluid such as the solar wind requires solving the MHD equations using a finite difference algorithm. One approach has been to use fourth-order flux corrected transport (FCT) [9, 14, 44, 18], although other approaches and algorithms have been used. In the approach used by [18], the magnetic field is determined using Faraday's law, which ensures that $\nabla \cdot \mathbf{B} = 0$ is preserved to within numerical round-off errors. The dependent variables are density, ρ, momentum $\rho\mathbf{v}$, total energy (internal (ρe) + kinetic $(\rho v^2/2)$ + magnetic $(B^2/2\mu_0)$), and the magnetic field $\mathbf{B}$. The equation of state is that of an ideal gas $(p = \rho RT)$, where R is the gas constant and T is temperature.

The turbulence solutions are initiated with supersonic/super-Alfvénic time-dependent inflow boundary conditions that are imposed at some radial distance above the critical point, typically taken to be between $r = 0.2 - 0.5$ AU. The solutions include a tilted rotating current sheet, incoming Alfvénic or two-dimensional modes with finite transverse correlation lengths, microstreams, and pressure-balanced structures.

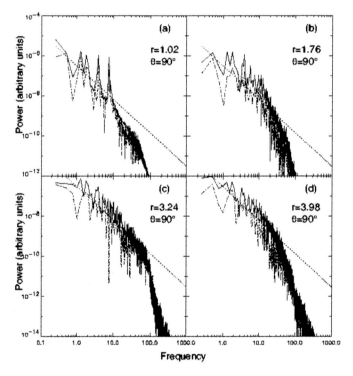

Fig. 9. Power spectra deduced from the time series of the magnetic field components and $|B|$ (dashed line) along a radius in slow speed flow at $\theta = 90°$ and four radii ranging from $r = 1.02 - 3.98$ AU. Also shown on each plot is a line with slope of $-5/3$. From [18].

2.1 Simulations of a Turbulent Cascade

Using algorithms that are sufficiently dissipationless, one can study the development of a cascade of magnetic energy from the input energy-containing scales toward the dissipation range. Eventually, the residual numerical dissipation does steepen the spectrum and damp fluctuations. In Figure 9, a series of power spectra are plotted that were computed from a time series of magnetic field values taken at four sites in the simulation domain. Simulation output is for "slow" flow near $\theta = 90°$ at four radial points ranging from $1.02 - 3.98$ AU. This region near the current sheet, stirred as it is by strong velocity gradients and weak magnetic fields, is where the most rapid evolution takes place. The turbulent mixing generates about a decade of inertial range. The straight line on the plots has a slope of $-5/3$.

3 Summary

This brief review has omitted many topics in solar wind turbulence studies. Many questions remain unanswered. For example, we do not know if the corona is turbulent. Indications from Helios observations suggest that it is not, but rather that fluctuations are continuously pumping the medium via a log-normal process [30] that results in a k^{-1} power spectrum at the transition to the solar wind. A curious aspect of the Alfvénic turbulence in the solar wind is that their minimum variance direction tends to lie along the direction of the local magnetic field (see, for example, [22]). This simple observation has yet to be satisfactorily explained and simulations have been unsuccessful in replicating the observations. Recently [32], magnetic field data obtained by the four Cluster spacecraft when their separation reached 10,000 km has been used to compute wavenumber spectra parallel and perpendicular to the local magnetic field. Those preliminary results suggest that, indeed, quasi-two-dimensional fluctuations predominate over slab Alfvénic fluctuations. For a review of the theory of the solar wind, see [25]. An extensive review of solar wind observations, including topics such as intermittency that have not be mentioned here, can be found in the Living Reviews article by [10].

References

1. Parker, E.N.: ApJ vol. 128, p 664 (1958)
2. Parker, E.: Interplanetary Dynamical Processes. Wiley, New York (1963)
3. Alfvén, H.: Cosmical Electrodynamics. Clarendon, Oxford (1950)
4. Barnes, A.: Rev. Geophys. vol. 17, p 596 (1979a)
5. Barnes, A.: Hydromagnetic waves and turbulence in the solar wind. In: Parker, E.N., Kennel, C.F., Lanzerotti, L.J. (eds.) Solar System Plasma Physics, vol. 1 p 249 North-Holland, Amsterdam (1979b)
6. Belcher, J.W., Davis, L.: J. Geophys. Res. vol. 76, p 3534 (1971)
7. Bieber, J.W., Wanner, W., Matthaeus, W.H.: J. Geophys. Res. vol. 101(A2), p 2511 (1996)
8. Blackman, R., Tukey, J.: Measurement of Power Spectra. Dover, New York (1958)
9. Boris, J.P., Book, D.L.: J. Comput. Phys. vol. 11, p 38 (1973)

10. Bruno, R., Carbone, V.: Living Reviews in Solar Physics **2**(4) (2005). http://www.livingreviews.org/lrsp-2005-4
11. Coleman, P.J.: Phys. Rev. Lett. vol. 17, p 207 (1966)
12. Coleman, P.J.: ApJ vol. 153, p 371 (1968)
13. Dasso, S., Milano, L.J., Matthaeus, W.H., Smith, C.W.: ApJ vol. 635(2), p L181 (2005)
14. DeVore, C.R.: J. Comput. Phys. vol. 92, p 142 (1991)
15. Ghosh, S., Matthaeus, W.H., Roberts, D.A., Goldstein, M.L.: J. Geophys. Res. vol. 103(A10), p 23691 (1998a)
16. Ghosh, S., Matthaeus, W.H., Roberts, D.A., Goldstein, M.L.: J. Geophys. Res. vol. 103(A10), p 23705 (1998b)
17. Goldstein, B.E., Smith, E.J., Balogh, A., Horbury, T.S., Goldstein, M.L., Roberts, D.A.: Geophys. Res. Lett. vol. 22(23), p 3393 (1995)
18. Goldstein, M.L., Roberts, D.A., Deane, A.E., Ghosh, S., Wong, H.K.: J. Geophys. Res. vol. 104(A7), p 14437 (1999)
19. Goldstein, M.L., Roberts, D., Deane, A.: In: Velli, M., Bruno, R., Malara, F. (eds.) Solar Wind 10. AIP Conf. Proceedings, vol. 679, 405. Amer. Inst. Phys. (2003)
20. Grant, H.L., Stewart, R.W., Moilliet, A.: J. Fluid Mech. vol. 12, p 241 (1962)
21. Horbury, T.S., Balogh, A.: J. Geophys. Res. (Space) vol. 106(A8), p 15929 (2001)
22. Horbury, T.S., Balogh, A., Forsyth, R.J., Smith, E.J.: Geophysical Research Letters vol. 22, p 3405 (1995)
23. Horbury, T.S., Balogh, A., Forsyth, R.J., Smith, E.J.: Astron. Astrophys. vol. 316(2), p 333 (1996)
24. Kolmogorov, A.N.: Dokl. Akad. Nauk SSSR vol. 30, p 299 (1941)
25. Marsch, E., Axford, W.I., McKenzie, J.M.: In: Dwivedi, B.N. (ed.) Dynamic Sun, 374. Cambridge Univ. P. (2003)
26. Matthaeus, W.H., Goldstein, M.L.: J. Geophys. Res. vol. 87, p 6011 (1982)
27. Matthaeus, W.H., Goldstein, M.L.: J. Geophys. Res. vol. 87, p 10347 (1982)
28. Matthaeus, W.H., Goldstein, M.L., King, J.H.: J. Geophys. Res. vol. 91, p 59 (1986)
29. Matthaeus, W.H., Goldstein, M.L., Roberts, D.A.: J. Geophys. Res. vol. 95, p 20673 (1990)
30. Matthaeus, W., Goldstein, M.L.: Phys. Rev. Lett. vol. 57, p 495 (1986)
31. Monin, A.S., Yaglom, A.M.: Statistical Fluid Mechanics: Mechanics of Turbulence. vol. 2. MIT Press, Cambridge, Mass. (1975)
32. Narita, Y., Glassmeier, K.H., Goldstein, M.L., Treumann, R.A.: In: Shaikh, D., Zank, G.P. (eds.) Turbulence and Nonlinear Processes in Astrophysical Plasmas; 6th Annual International Astrophysics Conference, APS, p 215. American Physical Society, Oahu, Hawaii (USA) (2007)
33. Panchev, S.: Random Functions and Turbulence. Pergamon, New York (1971)
34. Roberts, D.A., Ghosh, S., Goldstein, M.L., Matthaeus, W.H.: Phys. Rev. Lett. vol. 67, p 3741 (1991)
35. Roberts, D.A., Goldstein, M.L.: Turbulence and waves in the solar wind. In: Shea, M.A. (ed.) U.S. National Report to International Union of Geodesy and Geophysics, p 932. Amer. Geophys. Union, Washington DC (1991)
36. Roberts, D.A., Goldstein, M.L., Klein, L.W., Matthaeus, W.H.: J. Geophys. Res. vol. 92, p 12023 (1987a)

37. Roberts, D.A., Goldstein, M.L., Matthaeus, W.H., Ghosh, S.: J. Geophys. Res. vol. 97, p 17115 (1992)
38. Roberts, D.A., Klein, L.W., Goldstein, M.L., Matthaeus, W.H.: J. Geophys. Res. vol. 92, p 11021 (1987b)
39. Ruderman, M.S., Goldstein, M.L., Roberts, D.A., Deane, A., Ofman, L.: J. Geophys. Res. vol. 104(A8), p 17057 (1999)
40. Sari, J.W., Valley, G.C.: J. Geophys. Res. vol. 81, p 5489 (1976)
41. Smith, C.W.: In: Velli, M., Bruno, R., Malara, F. (eds.) Solar Wind 10. AIP Conf. Proceedings, vol. 679, 413. Amer. Inst. Phys. (2003)
42. Tu, C.Y., Marsch, E., Thieme, K.M.: J. Geophys. Res. vol. 94(A9), p 11739 (1989)
43. Unti, T.W., Neugebauer, M.: Phys. Fluids vol. 11, p 563 (1968)
44. Zalesak, S.T.: J. Comp. Phys. vol. 31, p 335 (1979)

Power Spectra of the Fluctuations in the Solar Wind

V. Krishan[1,2]

[1]Indian Institute of Astrophysics, Bangalore-560034, India
[2]Raman Research Institute, Bangalore-560080, India
e-mail: vinod@iiap.res.in

Summary. The power spectra of the velocity, the magnetic field and the density fluctuations and their inter-relationships are investigated in the turbulent solar wind using the dimensional approach of the Kolmogorovic type. While the velocity and the magnetic field fluctuations are dynamically related within the framework of the magnetohydrodynamic (MHD) turbulence, the density fluctuations could behave as a passive scalar and be simply convected by the velocity or the magnetic field fluctuations or they could dynamically participate in the joint production mechanism of all the fluctuations. The spectrum of the density fluctuations can distinguish between these two possibilities. Further the inclusion of the Hall effect, arising from the two fluid treatment, near the ion- inertial scale generates different spectra for the velocity and the magnetic fluctuations adding steeper branches to the ideal MHD spectra. Which spectrum would the density fluctuations, behaving as a passive scalar, follow in such a case? The answer leads to the interesting consequence that the electron density fluctuations and the ion density fluctuations have different spectra at spatial scales equal to and smaller than the ion-inertial scale. This result clearly demonstrates the two fluid picture brought in by the Hall effect.

Key words: solar wind, turbulent spectra, Hall effect

1 Introduction

The solar wind is a continuous, radial, supersonic outflow of plasma from the solar atmosphere and extends to the farthest reaches of the solar system to merge eventually with the interstellar medium. At the earth's orbit, the electron- proton plasma, with some trace elements, has a density of 5-10 particles/cc, a temperature of $\sim 100,000$ K and a velocity $V \sim 300$ Km/s. Supporting small deviations of ~ 10 Km/s, from the radial flow, it is collisionless, inhomogeneous and turbulent on several time and spatial scales. The source of energy for the hot plasma is the million degree solar corona; solar gravity being inadequate to hold the corona in static equilibrium. Fluctuations in the density, the velocity and the magnetic fields exist on three major

36   V. Krishan

scales : (1) 11 year solar cycle related variations and fast stream - slow stream interactions due to solar rotation; (2) transient disturbances originating on the sun and propagating out such as those associated with Solar flare caused blast wave with energy $\sim 10^{32}$ erg and (3) on hours or less scales are the waves and turbulence in the plasma. For example the Alfven waves fluctuations have power law spectra K^{-p}, $p \sim 5/3$ along with other values of p. Key Observations of the Solar Wind Turbulence are: 1. velocity, density, magnetic field and temperatures vary in time; 2. magnetohydrodynamics accounted for fluctuations reasonably well, particularly the shear Alfven waves, indicating that the magnetosonic waves are damped by kinetic effects; 3. Alfvén waves are found always propagating outwards from the Sun; 4. similarity between the power spectra of the magnetic fluctuations and the velocity fluctuations for an isotropic magnetofluid as well as fluid turbulence; 5. turbulence is driven by stream - shear Instabilities.The Alfvén waves being the exact solutions lack evolution. However, the solar wind turbulence shows evolution. Perhaps it is not all Alfvénic; 6. Voyager and Helios missions provided observations from 0.3 to > 30 AU; 7. Reduced Spectra are obtained by averaging over the two directions perpendicular to the solar wind velocity (V_s). The study of the solar wind turbulence is of immence importance on several counts. The solar wind is a distant plasma system that is accessible to direct observations of nonlinear processes such as the wave-particle and wave-wave interactions which have essential bearing on the propagation of the cosmic rays. The coupling of the solar wind and the magnetosphere defines the solar-terrestrial relationship. The turbulence modifies the transport processes with consequences for the space weather.

The reduced Spectra of the fluctuations are obtained by averaging over the two directions perpendicular to the solar wind velocity V_s. The spectra are a function of the wavenumber along V_s. The spectral energy distributions of the velocity and the magnetic field fluctuations in the solar wind are now known in a wide frequency range, starting from much below the proton cyclotron frequency (0.1 - 1 Hz) to hundreds of Hz. The inferred power spectrum [1] of magnetic fluctuations (Fig.1) consists of multiple segments- a Kolmogorov like branch ($\propto k^{-5/3}$) flanked, on the low frequency end by a flatter branch($\propto k^{-1}$) and, on the high frequency end, by a much steeper branch ($\propto k^{-\alpha_1}, \alpha_1 \simeq$ 3-4). Attributing the Kolmogorov branch ($\propto k^{-5/3}$) to the standard inertial range cascade, initial explanations invoked dissipation[2] processes, in particular, the collisionless damping of Alfvén and magnetosonic waves , to explain the steeper branch ($\propto f^{-\alpha_1}$, $\alpha_1 \simeq$ 3-4). However, a recent critical study has concluded that damping of the linear Alfvén waves via the proton cyclotron resonance and of the magnetosonic waves by the Landau resonance, being strongly k (wave vector) dependent, is quite incapable of producing a power-law spectral distribution of magnetic fluctuations, the damping mechanisms lead, instead, to a sharp cutoff in the power spectrum[3]. Cranmer and Ballogoeijen[4] have however, demonstrated a weaker than an exponential dependence of damping on the wave vector by including kinetic ef-

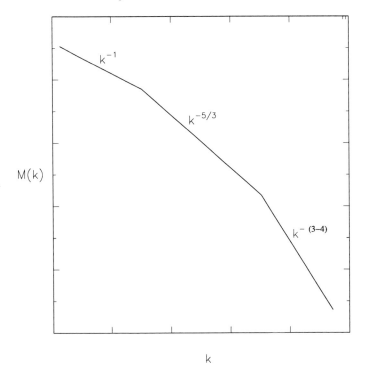

Fig. 1. Schematic representation of the observed magnetic energy spectrum in the solar wind.

fects. However it is still steeper than that required for explaining the observed spectrum.

An alternative possibility, suggested by Ghosh[5]et al., links the spectral break and subsequent steepening to a change in the controlling invariants of the system in the appropriate frequency range. Stawicki[6]et al. have invoked the short wavelength dispersive properties of the magnetosonic/whistler waves to account for the steepened spectrum and christened it as the spectrum in the dispersion range. Krishan and Mahajan[7],[8], however, invoke the Hall effect to model the steepened part of the spectrum, and this should be correct since the steepening begins at a scale close to the ion-inertial scale, a hallmark of the Hall effect. The incompressible turbulence has no associated density fluctuations. The density fluctuations, produced by an independent mechanism, could be convected by the velocity and the magnetic field fluctuations as a passive scalar or they may be concomitantly generated in compressible turbulence along with the other fluctuations. I recapitulate the power spectra of the fluctuations in Alfvénic turbulence in the next section. The relation between the velocity and the magnetic field fluctuations, including the Hall effect, is derived in section 3. The power spectra of fluctuations within the Hall- MHD invoking the new invariant, the generalized helicity, are derived in section 4.

38 V. Krishan

The dilemma of the density fluctuations, whether being convected by the velocity or the magnetic fluctuations at spatial scales near the ion-inertial scale, is discussed in section five and the paper ends with a section on conclusion.

2 Power spectra of fluctuations in Alfvénic turbulence

The solar wind turbulence is modeled in terms of the magnetohydrodynamic fluctuations. An assumed input spectrum k^{-1} of the Alfvén waves is believed to decay to generate the Kolmogorov $k^{-5/3}$ spectrum. since the Alfvénic fluctuations are characterized by velocity fluctuation $\boldsymbol{V} = \pm\boldsymbol{B}$, the velocity and the magnetic fluctuations have identical spectra. The question of the origin of the k^{-1} spectrum has as yet no satisfactory answer although such a spectrum is observed under different and disparate circumstances related to self-organized criticality. One alternative is to derive the spectra using the dimensional arguments of the Kolomogorov hypotheses. The incompressible magnetohydrodynamic turbulence represented by the Alfvénic fluctuations supports two global invariants: the total energy E and the magnetic helicity H_M defined as:

$$\text{Total Energy} \quad E = \frac{1}{2}\int(V^2 + B^2)d^3x \quad = \frac{1}{2}\sum_k |V_k|^2 + |B_k|^2 \quad (1)$$

$$\text{Magnetic Helicity } H_M = \int \boldsymbol{A}\cdot\boldsymbol{B}d^3x \quad = \sum_k \frac{i}{k^2}(\boldsymbol{k}\times\boldsymbol{B}_k)\cdot\boldsymbol{B}_{-k} \quad (2)$$

where $\boldsymbol{A}$ is the vector potential. In order to derive the spectral energy distributions we resort to the Kolmogorov hypotheses according to which the spectral cascades proceed at a constant rate governed by the eddy turn over time $(k\boldsymbol{V}_k)^{-1}$. For ε_E denoting the constant cascading rate of the total energy E, we get the dimensional equality

$$(kV_k)V_k^2 = \varepsilon_E. \quad (3)$$

The omnidirectional spectral distribution function $W_E(k)$ (kinetic energy per gram per unit wave vector, $\frac{V_k^2}{2k}$), then, takes the form

$$W_E(k) = \frac{(\varepsilon_E)^{\frac{2}{3}}}{2}k^{-\frac{5}{3}}. \quad (4)$$

The equipartition between the kinetic and the magnetic energy in Alfvénic turbulence yields:
$$M_E(k) = W_E(k). \quad (5)$$

where $M_E(k) = (B_k^2/2k)$ is the similarly defined omnidirectional spectral distribution function of the magnetic energy density.

The cascading of the magnetic helicity H_M (ε_H being the cascading rate for the magnetic helicity) produces a different dimensional equality

$$(kV_k)\left(\frac{B_k^2}{k}\right) = \varepsilon_H \qquad (6)$$

resulting in the following kinetic and magnetic spectral energy distributions:

$$W_H(k) = 2^{-1}(\varepsilon_H)^{\frac{2}{3}} k^{-1} \qquad (7)$$

$$M_H(k) = W_H(k). \qquad (8)$$

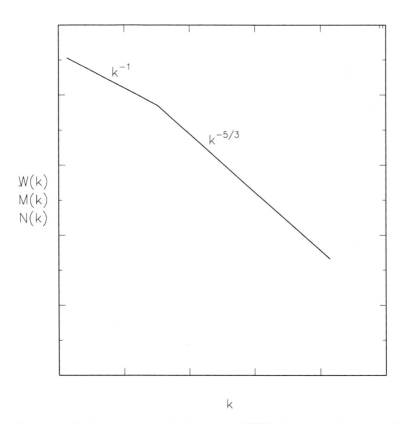

Fig. 2. Spectral distribution of kinetic energy $W(k)$, the magnetic energy $M(k)$ and the density $N(k)$ in Alfvénic turbulence.

The observed solar wind magnetic spectrum, at spatial scales much larger than the ion-inertial scale can be generated if we could string together the two branches $M_E(k) \propto k^{-5/3}$, and $M_H(k) \propto k^{-1}$. The rationale as well as

40   V. Krishan

the modality for stringing different branches originates in the hypothesis of selective dissipation. It was, first, invoked in the studies of two-dimensional hydrodynamic turbulence[9]. The idea is that in a given k range, the particular invariant which suffers the strongest dissipation, controls the spectral behavior (determined, in turn, by arguments a la Kolmogorov). Thus if the k ranges associated with different invariants are distinct and separate, we have a straightforward recipe for constructing the entire k-spectrum in the extended inertial range. In 2-D hydrodynamic turbulence[9], for instance, the enstrophy invariant, because of its stronger k dependence (and hence larger dissipation) compared to the energy invariant, dictates the large k spectral behavior. Therefore, the entire inertial range spectrum has two segments- the energy dominated low k, and the enstrophy dominated high k ($\propto k^{-3}$). The procedure amounts to placing the spectrum with the highest negative exponent at the highest k-end, and the one with the lowest negative exponent of k at the lowest k-end. This procedure generates the low k-end spectral branches k^{-1} and $k^{-5/3}$ for the kinetic energy and the magnetic energy spectrum. Now, the density fluctuations if convected as a passive scalar would have the spectral distributions, $N_E(k)$ and $N_H(k)$, defined as $n_k^2 = kN(k)$, same as the kinetic and the magnetic energy as exhibited in Fig.(2). However, if the density fluctuations are generated by the same mechanism as the velocity and the magnetic field fluctuations as for example in magnetosonic modes, the three spectra would be determined from their dynamical relationships. In a magnetosonic mode propagating perpendicular ($k_z = 0$) to the uniform ambient magnetic field $\boldsymbol{B}_0$, say in the z direction, one has:

$$V_\perp = n \left(1 + \beta\right)^{1/2}, \boldsymbol{B} = B_z = n \tag{9}$$

in dimensionless form. Here n is the density fluctuation and β is the plasma β. Since the relationships amongst the three fluctuations are independent of $k_\perp$, the spectral distributions are same as in the case of the incompressible turbulence. However, the relative magnitudes of the power in the three spectra are now given by equation (10). One notices that the magnetic fluctuations and the density fluctuations have identical spectra but the velocity fluctuations can be much larger than the density or the magnetic fluctuations depending on the value of the plasma β. For oblique waves, the corresponding spectra would be anisotropic.

3 Relation between the velocity and the magnetic field fluctuations in Hall-MHD

In the HALL-MHD (HMHD) comprising of the two fluid Model, the electron fluid equation is given by

$$m_e n_e \left[\frac{\partial \boldsymbol{V}_e}{\partial t} + (\boldsymbol{V}_e . \nabla) \boldsymbol{V}_e \right] = -\nabla p_e - e n_e \left[\boldsymbol{E} + \frac{1}{c} \boldsymbol{V}_e \times \boldsymbol{B} \right] \tag{10}$$

Assuming inertialess electrons ($m_e \leftarrow 0$), the electric field is found to be:

$$E = -\frac{1}{c}V_e \times B - \frac{1}{n_e e}\nabla p_e \tag{11}$$

The Ion fluid equation is :

$$m_i n_i [\frac{\partial V_i}{\partial t} + (V_i.\nabla)V_i] = -\nabla p_i + e n_i [E + \frac{1}{c}V_i \times B] \tag{12}$$

Substitution for E from the inertialess electron Eq. begets:

$$m_i n_i [\frac{\partial V_i}{\partial t} + (V_i.\nabla)V_i] = -\nabla(p_i + p_e) + \frac{1}{c}J \times B \tag{13}$$

The magnetic Induction equation becomes:

$$\frac{\partial B}{\partial t} = -c\nabla \times E = \nabla \times (V_e \times B) \tag{14}$$

where B is seen to be frozen to electrons. Substituting for $V_e = V_i - J/en_e$, one gets :

$$\frac{\partial B}{\partial t} = \nabla \times \left[\left(V_i - \frac{J}{en_e}\right) \times B\right] \tag{15}$$

We see that B is not frozen to the ions, $n_e = n_i$.

The Hall term dominates for $(n_e ec)^{-1}J \times B \geq V_i \times B/c$ or the length scale $L \leq M_A c/\omega_{pi}$ and the time scale $T \geq \omega_{ci}^{-1}$ where M is the Alfvenic Mach number and ω_{pi} is the ion plasma frequency.

That is the Hall term decouples electron and ion motion on ion inertial length scales and ion cyclotron times. The Hall effect does not affect mass and momentum transport but it does affect the energy and magnetic field transport.

The importance of the nonlinear Alfvénic state for MHD prompts one to speculate if a similar kind of an exact solution exists for Hall MHD[9],[10], a system which encompasses MHD. We write the equations in a dimensionless form. The magnetic and the velocity fields are respectively normalized by the uniform ambient field B_0 and the Alfvén speed $V_{Ai} = B_0/\sqrt{4\pi\rho_i}$, where ρ_i is the uniform ion mass density. The time and the space variables are normalized, respectively, with the Alfvén travel time $t_A = L/V_{Ai}$, and a scale length L. In these units the following dimensionless equations

$$\frac{\partial B}{\partial t} = \nabla \times [(V_i - \epsilon\nabla \times B) \times B] \tag{16}$$

$$\frac{\partial(\nabla \times V_i)}{\partial t} = \nabla \times [V_i \times (\nabla \times V_i) - B \times (\nabla \times B)] \tag{17}$$

constitute the Hall-MHD in the incompressible limit. Here $\epsilon = \lambda_i/L = c/\omega_{pi}L = V_{Ai}/L\omega_{ci}$ and $\omega_{pi} = (4\pi e^2 n_i/m_i)^{1/2}$ is the ion plasma frequency, λ_i

42   V. Krishan

is the ion inertial length. Equation (18) has been obtained by taking the curl of the equation of motion of the ion fluid (Equation (14)) . We split the fields into their ambient and the fluctuating parts:

$$\boldsymbol{B} = \widehat{e}_z + \boldsymbol{b}; \qquad \boldsymbol{V}_i = \boldsymbol{V}_0 + \boldsymbol{v} \tag{18}$$

where $\boldsymbol{V}_0 = 0$ and substitute in Eqs. (17) and (18) to get :

$$\frac{\partial \boldsymbol{b}}{\partial t} = \nabla \times \left[(\boldsymbol{v} - \epsilon \nabla \times \boldsymbol{b}) \times \widehat{e}_z + (\boldsymbol{v} - \epsilon \nabla \times \boldsymbol{b}) \times \boldsymbol{b} \right], \tag{19}$$

$$\frac{\partial}{\partial t}(\nabla \times \boldsymbol{v}) = \nabla \times \left[\boldsymbol{v} \times (\nabla \times \boldsymbol{v}) \right.$$
$$\left. + (\nabla \times \boldsymbol{b}) \times \widehat{e}_z + (\nabla \times \boldsymbol{b}) \times \boldsymbol{b} \right]. \tag{20}$$

Assuming a plane wave form $(\boldsymbol{V}_k, \boldsymbol{B}_k) exp(ikz - i\omega t)$ we get the linear relations:

$$\boldsymbol{V}_k - \epsilon \nabla \times \boldsymbol{B}_k = -\frac{\omega}{k}\boldsymbol{B}_k \tag{21}$$

$$\nabla \times \boldsymbol{B}_k = -\frac{\omega}{k}\nabla \times \boldsymbol{V}_k \tag{22}$$

The solution of equqtions (22, 23) furnishes:

$$\nabla \times \boldsymbol{B}_k = \lambda \boldsymbol{B}_k, \ \lambda^2 = k^2, \ \boldsymbol{B}_k = \alpha(k)\boldsymbol{V}_k, \ \alpha = -\frac{\omega}{k} \tag{23}$$

and the dispersion relation is:

$$\alpha = -\frac{\epsilon k}{2} \pm \left(\frac{\epsilon^2 k^2}{4} + 1 \right)^{\frac{1}{2}} \tag{24}$$

One notices that the $\boldsymbol{V}_k$, $\boldsymbol{B}_k$ relation of the waves is now k dependent, the waves are dispersive and that they are nonlinear since for $\boldsymbol{b}$ given by eqs.(22) and (23), the nonlinear terms in eqs.(20 and 21) vanish. For $k \ll 1$,

$$\alpha \rightarrow \pm 1, \omega \rightarrow \mp k \tag{25}$$

reproducing the k independent MHD Alfvénic relationship for both the co- and the counter propagating waves. For $k \gg 1$, it is easy to recognize, in analogy with the linear theory, that the α_+ wave is the shear-cyclotron branch, while the α_- represents the compressional-whistler mode. The frequency of the α_+ wave approaches the ion gyro frequency. The $\boldsymbol{V}_k$, $\boldsymbol{B}_k$ relation would now give different spectral distributions for the kinetic and the magnetic energy. The Hall-MHD also supports an additional invariant called the generalized helicity. The additional invariant and its cascading characteristics along with the new spectral relations arising from the new $\boldsymbol{V}_k$, $\boldsymbol{B}_k$ relation are discussed in the next section.

4 Generalized helicity and power spectra in Hall-MHD

The additional invariant of the Hall-MHD system, the generalized helicity[12] is defined as:

$$H_G = \int d^3x \left(\boldsymbol{A} + \epsilon \boldsymbol{V}_i \right) \cdot \left(\boldsymbol{B} + \epsilon \nabla \times \boldsymbol{V}_i \right)$$
$$= \sum \left(i \frac{\boldsymbol{k} \times \boldsymbol{B}_k}{k^2} + \epsilon \boldsymbol{V}_k \right) \cdot \left(\boldsymbol{B}_k + i\epsilon \boldsymbol{k} \times \boldsymbol{V}_k \right) \tag{26}$$

where $\boldsymbol{A}$ is the vector potential. Notice that $H_G - H_M$ is a combination of the kinetic and the cross helicities. The cascading of the generalized helicity with a constant rate ε_G, using the relation $B_k = \alpha(k)V_k$ gives

$$(kV_k)\left[k^{-1}g(k)V_k^2 \right] = \varepsilon_G, \tag{27}$$
$$g(k) = (\alpha + \epsilon k)^2$$

leading to the spectral energy distributions:

$$W_G(k) = (\varepsilon_G)^{\frac{2}{3}} \left[g(k) \right]^{-\frac{2}{3}} k^{-1}, \tag{28}$$

and

$$M_G(k) = (\alpha)^2 W_G(k).$$

The spectral distributions $W_G(k)$ and $M_G(k)$ are shown in figures (3,4), respectively, for the two roots α_+ and α_- for different values of the Hall parameter ϵ. The generalized helicity reduces to the magnetic helicity for $\epsilon = 0$ and $\alpha_\pm = \pm 1$. The steepening of the spectra towards large values of k is evident for nonzero values of ϵ.

The spectral distributions obtained from the cascading of the total energy are accordingly modified to:

$$W_E(k) = 2^{-1/3} \left(\varepsilon_E \right)^{\frac{2}{3}} \left[1 + (\alpha)^2 \right]^{-\frac{2}{3}} k^{-\frac{5}{3}}. \tag{29}$$

and

$$M_E(k) = (\alpha)^2 W_E(k). \tag{30}$$

The spectral distributions $W_E(k)$ and $M_E(k)$ are shown in figures (5,6) respectively, for the two roots α_+ and α_- for different values of the Hall parameter ϵ. The steepening of the spectra towards large values of k is evident for nonzero values of ϵ.

The spectral distributions obtained from the cascading of the magnetic helicity are accordingly modified to:

$$W_H(k) = 2^{-1} \left(\varepsilon_H \right)^{\frac{2}{3}} (\alpha)^{\frac{-4}{3}} k^{-1}, \tag{31}$$

and

$$M_H(k) = (\alpha)^2 W_H(k). \tag{32}$$

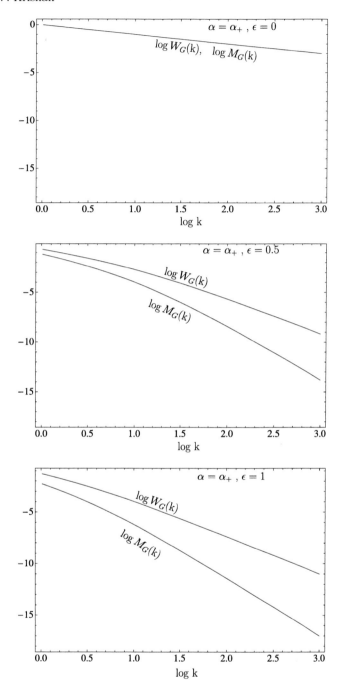

Fig. 3. Power spectra of the kinetic energy density $W_G(k)$ and the magnetic energy density $M_G(k)$ derived from the generalized helicity invariant H_G with Hall effect for α_+ root.

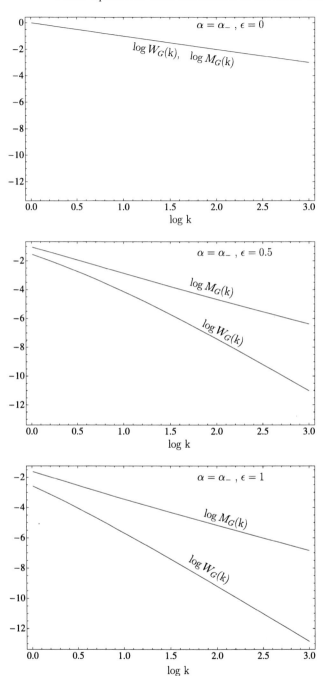

Fig. 4. Power spectra of the kinetic energy density $W_G(k)$ and the magnetic energy density $M_G(k)$ derived from the generalized helicity invariant H_G with Hall effect for α_- root.

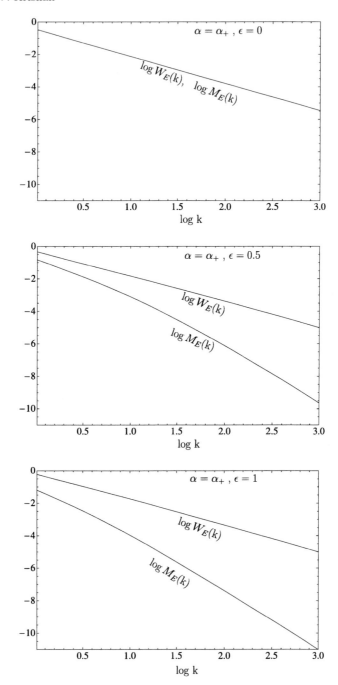

Fig. 5. Power spectra of the kinetic energy density $W_E(k)$ and the magnetic energy density $M_E(k)$ derived from the total energy invariant E with Hall effect for the α_+ root.

Power Spectra of the Fluctuations in the Solar Wind 47

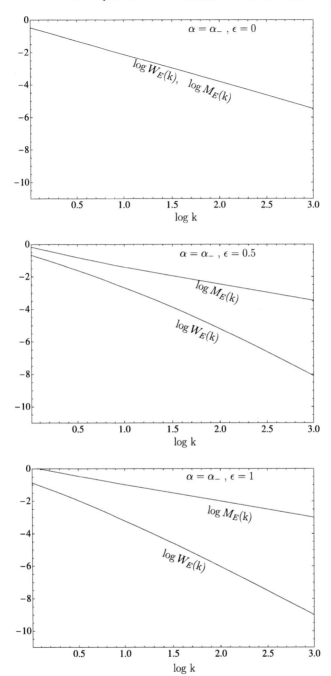

Fig. 6. Power spectra of the kinetic energy density $W_E(k)$ and the magnetic energy density $M_E(k)$ derived from the total energy invariant E with Hall effect for the α_- root.

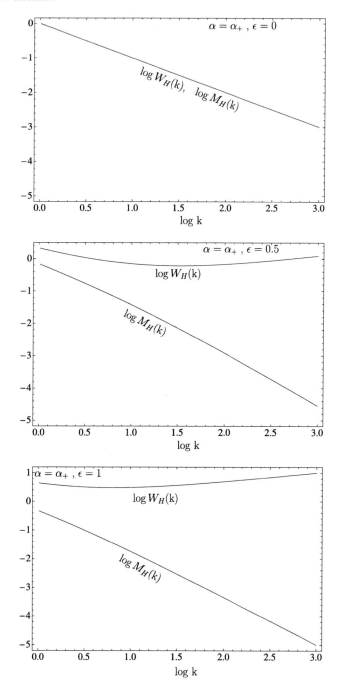

Fig. 7. Power spectra of the kinetic energy $W_H(k)$ and the magnetic energy $M_H(k)$ derived from the mahnetic helicity invariant H_M with Hall effect for α_+ root.

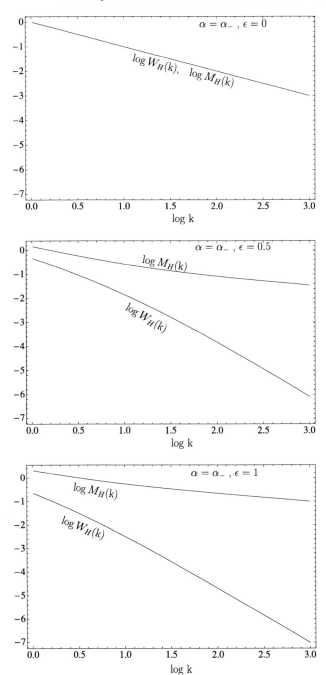

Fig. 8. Power spectra of the kinetic energy $W_H(k)$ and the magnetic energy $M_H(k)$ derived from the mahnetic helicity invariant H_M with Hall effect for α_- root.

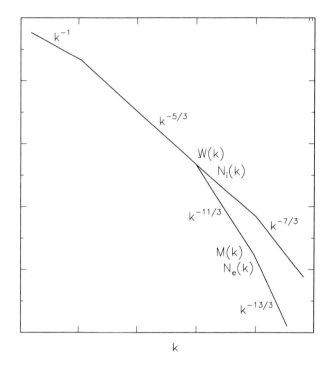

Fig. 9. Power spectra of the kinetic and the magnetic energy density fluctuations including Hall effect.

The spectral distributions $W_H(k)$ and $M_H(k)$ are shown in figures (7 & 8) for the two roots α_+ and α_- for different values of the Hall parameter ϵ. The steepening of the spectra towards large values of k is evident for nonzero values of ϵ.

It is clear that the spectral distributions (W_E, M_E, W_H, M_H) reduce to the ones obtained in the MHD case for $\alpha = 1$. For large k, α reduces to the two values: ϵk and $\epsilon^{-1} k^{-1}$. For $\alpha = \epsilon^{-1} k^{-1}$, corresponding to the shear Alfven wave, one can determine the kinetic and the magnetic energy spectra using again the receipe for stringing together the various spectral branches. This spectra is shown in fig.(9) where the low k end is represented by the MHD spectra and the high k end by the Hall-MHD spectra.

This is in accordance with the observed magnetic spectra with the steepened part, now, being proposed to be in the inertial range, in contrast to the other proposals which have unsucessfully put it in the dissipation range. In the HMHD regime the kinetic and the magnetic enery spectra are different as the two fluids now have their own dynamics.

5 Power spectra of density fluctuations in Hall-MHD

If the spectra of the kinetic and the magnetic fluctuations are different in HMHD then which would carry the passive density fluctuations? In order to answer this question let us look at the induction eq.(15) which shows that the magnetic field is frozen to the electrons. Thus one would conclude that the electron density fluctuations would be frozen to the magnetic field fluctuations and would have the same spectrum as the magnetic field. An inspection of the induction equation in eq.(16) shows that the magnetic field is not frozen to the ions. Thus one would conclude that the ion density fluctuations would be carried by the velocity field fluctuations and would have the same spectrum as the velocity field. So, the electron and the ion density fluctuations have different spectral distributions without causing any violation of the quasineutrality. The entire picture is presented in fig.(10).

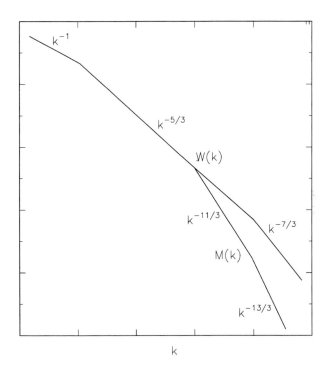

Fig. 10. Power spectra of the electron and the ion density fluctuations including Hall effect.

6 Conclusion

It has been shown that the observed spectral distributions of the velocity, the magnetic and the density fluctuations in the solar wind can be modeled within the framework of the Hall magnetohydrodynamics using the dimensional arguments of the Kolmogorov hypotheses. The Hall effect is particularly needed to account for the high k end of the spectra. The k dependent relation between the velocity fluctuations and the magnetic field fluctuations arising from the Hall-MHD waves results in different spectra for the fluctuations. Additionally, the spectra for the electron and the ion density fluctuations, treated as passive scalars, also differ at the high k end, again a consequence of the two fluid treatment which brings in the Hall effect. The spectrum of the electron density fluctuations steeper than the Kolmogorov spectrum at the ion inertial scale has been inferred from the interplanetary scintillation studies[13]. The spectrum of ion density fluctuations could be inferred from the plasma wave insitu measurements. Several other consequences of the foray into the two-fluid treatment such as the inclusion of compressibility and the ensuing wave modes and anisotrpies need to be investigated.

References

1. M.L. Goldstein, D.A. Roberts and W.H. Matthaeus, Annual Rev. Astron. Astrophys., 33, 283, (1995).
2. S.P. Gary, J. Geophys. Res., 104, 6759, (1999).
3. H. Li, P. Gary and O. Stawicki, Geophys. Res. Lett, 28, 1347, (2001).
4. S.R. Cranmer, and A.A. von Ballagooijen, Astrophys. J., 594, 573, (2003).
5. S. Ghosh, E. Siregar, D.A. Roberts, and M.L. Goldstein, J. Geophys. Res., 101, 2493, (1996).
6. O. Stawicki, P.S. Gary and H. Li, J. Geophys. Res., 106, 8273, (2001).
7. V. Krishan and S.M. Mahajan, J.G.R., 109, A11105, (2005).
8. V. Krishan and S.M. Mahajan, Solar Physics., 220, 29, (2004).
9. A. Hasegawa, Advances in Phys., 34, 1, (1985).
10. S.M. Mahajan and V. Krishan, MNRAS 359, L27, (2005).
11. V. Krishan and B.A. Varghese, Solar Physics, (2007).
12. Z. Yoshida and S.M. Mahajan, Phys. Rev. Lett., 88, 095001, (2002).
13. P.K. Manoharan, M. Kojima, and H. Misawa, J. Geophys. Res., 99, 23411, (1994).

Part II

Astrophysical Dynamos

Alpha Effect in Partially Ionized Plasmas

V. Krishan[1,2] and R. T. Gangadhara[1]

[1]Indian Institute of astrophysics, Bangalore-560034, India
[2]Raman Research Institute, Bangalore 560080, India

Summary. There are several astrophysical situations where one needs to study the dynamics of magnetic flux in partially ionized turbulent plasmas. In a partially ionized plasma the magnetic induction is subjected to the ambipolar diffusion and the Hall effect in addition to the usual resistive dissipation. In this paper we initiate the study of the kinematic dynamo in a partially ionized turbulent plasma. The Hall effect arises from the treatment of the electrons and the ions as two separate fluids and the ambipolar diffusion due to the inclusion of neutrals as the third fluid. It is shown that these nonideal effects modify the so called α effect and the turbulent diffusion coefficient β in a rather substantial way. The Hall effect may enhance or quench the dynamo action altogether. The ambipolar diffusion brings in an α which depends on the mean magnetic field. The new correlations embodying the coupling of the charged fluids and the neutral fluid appear in a decisive manner. The turbulence is necessarily magnetohydrodynamic with new spatial and time scales. The nature of the new correlations is demonstrated by taking the Alfvénic turbulence as an example.

Key words: Partially ionized plasma, Dynamo, Hall effect, Ambipolar diffusion.

1 Introduction

The kinematic dynamo has revealed many an essential workings of an astrophysical dynamo for the generation of magnetic field in objects varying from stars to molecular clouds to accretion disks. The kinematic dynamo ([1] – [4]) is based on the possible generation of an electromotive force parallel to the mean magnetic field in a reflexionally asymmetric turbulence, the so called α effect. Here α is a measure of the net kinetic helicity. The corresponding turbulent diffusion coefficient β becomes a function of the mean turbulent kinetic energy. The scale separation is an integral part of the kinematic dynamo. A weakly ionized plasma is defined by the condition [5] that the electron-neutral collision frequency $\nu_{en} \sim 10^{-15} n_n \sqrt{8K_B T/(\pi m_{en})}$ is much larger than the electron-ion collision frequency $\nu_{ei} \sim 6 \times 10^{-24} n_i \Lambda Z^2 (K_B T)^{-3/2}$. This translates into the ionization fraction $n_e/n_n < 5 \times 10^{-11} T^2$ [5] where n's are the particle densities and T is the temperature in Kelvin. Although the ideal magnetohydrodynamics (MHD) is often used as a starting point of an astrophysical investigation, there are many systems with a rather low degree of ionization dominated by the charged particle-neutral collisions and the neutral particle dynamics. A major part of the solar photosphere ([6], [7]), the protoplanetary disks [8] and the molecular clouds [9] are some of the examples of weakly ionized astrophysical plasmas. The dynamo action in such a plasma would be affected by the multifluid interactions. The issue of possible disconnection between the sub-surface and the surface solar magnetic field, recently emphasized by [10], may have some bearing on the neglect of the neutral fluid-plasma coupling in the flux transport on the solar photosphere. Zweibel ([11]) studied the dynamo process in a partially ionized plasma within a single fluid description. Including only the ambipolar diffusion, she determined the velocity, the density and the magnetic field fluctuations self consistently in the form of magnetohydrodynamic waves and thus went beyond the kinematic dynamo. Here we recapitulate the essentials of the mean filed dynamo in a single fluid in section one. We develop a three fluid framework for a kinematic dynamo including the Hall effect and the ambipolar diffusion in section three. The α effect of the kinematic dynamo is formulated in section four. The new correlations arising due to the coupling amongst different fluids are understood by taking the Alfvénic turbulence as an example and we end the paper with a section on discussion and conclusion.

2 Mean Field Dynamo in Single Fluid

Following Krause and Rädler [12], we begin with the basic equations of non-relativistic magnetohydrodynamics:

$$\nabla \times \boldsymbol{E} = -\frac{1}{c}\frac{\partial \boldsymbol{B}}{\partial t}, \quad \nabla \times \boldsymbol{B} = \frac{4\pi}{c}\boldsymbol{J}, \quad \nabla \cdot \boldsymbol{B} = 0, \tag{1}$$

$$\boldsymbol{J} = \sigma(\boldsymbol{E} + \frac{\boldsymbol{u} \times \boldsymbol{B}}{c}), \tag{2}$$

where $\boldsymbol{B}$ is the magnetic flux density, $\boldsymbol{E}$ the electric field, $\boldsymbol{J}$ the electric current density, σ the electrical conductivity, and $\boldsymbol{u}$ the fluid velocity.

$$\rho\left(\frac{\partial \boldsymbol{u}}{\partial t} + (\boldsymbol{u} \cdot \nabla)\boldsymbol{u}\right) = \frac{\boldsymbol{J} \times \boldsymbol{B}}{c} - \nabla p \tag{3}$$

Equation (3) describes the momentum conservation of the fluid of mass density ρ and pressure p. For a given $\boldsymbol{u}$, these equations determine $\boldsymbol{B}$, $\boldsymbol{E}$ and $\boldsymbol{J}$. They show that

$$\frac{\partial \boldsymbol{B}}{\partial t} = \nabla \times (\boldsymbol{u} \times \boldsymbol{B}) + \eta \nabla^2 \boldsymbol{B}, \tag{4}$$

where $\eta = 1/\sigma$ is the electrical resistivity.

2.1 Averaging operations

In a turbulent medium, all the fields vary irregularly in space and time. Let $\boldsymbol{F}$ be such a fluctuating field, considered as a random function of space and time. The corresponding mean field, $\overline{\boldsymbol{F}}$, is then defined to be the expected value of $\boldsymbol{F}$ in an ensemble of identical systems, and $\boldsymbol{F}'$ shall be used to denote the difference, $(\boldsymbol{F} - \overline{\boldsymbol{F}})$, between $\boldsymbol{F}$ and $\overline{\boldsymbol{F}}$. We can thus write

$$\boldsymbol{F} = \overline{\boldsymbol{F}} + \boldsymbol{F}', \quad \overline{\boldsymbol{F}'} = 0. \tag{5}$$

The averaging operator commutes with the differentiation and integration operators in both space and time.

2.2 The equations for the mean fields

The problem of the mean-field electrodynamics is that of determining $\overline{\boldsymbol{B}}$, $\overline{\boldsymbol{E}}$ and $\overline{\boldsymbol{J}}$ when $\overline{\boldsymbol{u}}$ and some other properties of $\boldsymbol{u}'$ are known. Writing $\boldsymbol{B} = \overline{\boldsymbol{B}} + \boldsymbol{B}'$, $\boldsymbol{E} = \overline{\boldsymbol{E}} + \boldsymbol{E}'$ and $\boldsymbol{u} = \overline{\boldsymbol{u}} + \boldsymbol{u}'$, and substituting in Eq. (1), and averaging, we get

$$\nabla \times \overline{\boldsymbol{E}} = -\frac{\partial \overline{\boldsymbol{B}}}{\partial t}, \quad \nabla \times \overline{\boldsymbol{B}} = \frac{4\pi}{c} \overline{\boldsymbol{J}}, \quad \nabla \cdot \overline{\boldsymbol{B}} = 0, \tag{6}$$

$$\overline{\boldsymbol{J}} = \sigma(\overline{\boldsymbol{E}} + \overline{\boldsymbol{u}} \times \overline{\boldsymbol{B}} + \overline{\boldsymbol{u}' \times \boldsymbol{B}'}). \tag{7}$$

These equations would be identical with Eqs. (1) and (2), were it not the single new term arising from a non-linearity, which gives rise to an additional electromotive force in Ohm's law for the mean fields. The new term called the "turbulent electromotive force" is denoted as:

$$\mathcal{E} = \overline{\boldsymbol{u}' \times \boldsymbol{B}'}. \tag{8}$$

In order to determine $\overline{\boldsymbol{B}}$, $\overline{\boldsymbol{E}}$ and $\overline{\boldsymbol{J}}$, we must know not only $\overline{\boldsymbol{u}}$ but also this electromotive force $\mathcal{E}$.

Let us now consider the electromotive force $\mathcal{E}$ in more detail. Substituting $\boldsymbol{B} = \overline{\boldsymbol{B}} + \boldsymbol{B}'$ and $\boldsymbol{u} = \overline{\boldsymbol{u}} + \boldsymbol{u}'$ in the induction Eq. (4), we obtain

$$\frac{\partial \overline{\boldsymbol{B}}}{\partial t} + \frac{\partial \boldsymbol{B}'}{\partial t} = \nabla \times (\overline{\boldsymbol{u}} \times \overline{\boldsymbol{B}}) + \nabla \times (\overline{\boldsymbol{u}} \times \boldsymbol{B}') + \nabla \times (\boldsymbol{u}' \times \overline{\boldsymbol{B}}) +$$
$$\nabla \times (\boldsymbol{u}' \times \boldsymbol{B}') + \eta \nabla^2 (\overline{\boldsymbol{B}} + \boldsymbol{B}'). \tag{9}$$

These equations determine $\boldsymbol{B}'$, if $\overline{\boldsymbol{B}}$, $\overline{\boldsymbol{u}}$ and $\boldsymbol{u}'$ are known. We may therefore consider $\mathcal{E}$ as a functional of $\overline{\boldsymbol{B}}$, $\overline{\boldsymbol{u}}$ and the average statistical properties of $\boldsymbol{u}'$. The solution of Eq. (9) depends on the initial and the boundary conditions.

For known $\overline{\boldsymbol{u}}$ and $\boldsymbol{u}'$, it is possible to determine $\overline{\boldsymbol{B}}$, $\overline{\boldsymbol{E}}$ and $\overline{\boldsymbol{J}}$. Consequently, the determination of that functional dependence is the crucial point in the development of mean-field electrodynamics.

By taking average of Eq. (9), we obtain

$$\frac{\partial \overline{\boldsymbol{B}}}{\partial t} = \nabla \times \left(\overline{\boldsymbol{u}} \times \overline{\boldsymbol{B}} \right) + \nabla \times \left(\overline{\boldsymbol{u}' \times \boldsymbol{B}'} \right) + \eta \nabla^2 \overline{\boldsymbol{B}} . \tag{10}$$

Next, by substituting back Eq. (10) into Eq. (9), we obtain

$$\frac{\partial \boldsymbol{B}'}{\partial t} = \nabla \times \boldsymbol{\mathcal{E}} + \nabla \times (\boldsymbol{u}' \times \boldsymbol{B}') + \nabla \times \left(\boldsymbol{u}' \times \overline{\boldsymbol{B}} \right) +$$
$$\nabla \times \left(\overline{\boldsymbol{u}} \times \boldsymbol{B}' \right) + \eta \nabla^2 \boldsymbol{B}' . \tag{11}$$

In the limit of high-conductivity, Eq. (11) can be solved by retaining only the linear terms in fluctuating quantities and assuming $\overline{\boldsymbol{u}} = 0$:

$$\boldsymbol{B}'(\boldsymbol{x}, t) = \boldsymbol{B}'(\boldsymbol{x}, t_0) + \int_{t_0}^{t} \nabla \times \left(\boldsymbol{u}'(\boldsymbol{x}, t') \times \overline{\boldsymbol{B}}(\boldsymbol{x}, t') \right) dt', \tag{12}$$

where t_0 is arbitrary, and $t - t_0$ is to be chosen so large that no correlation between $\boldsymbol{B}'(\boldsymbol{x}, t_0)$ and $\boldsymbol{u}'(\boldsymbol{x}, t)$ exists. Thus we assume

$$t - t_0 \gg (\lambda_{cor}^2/\eta) \gg \tau_{cor}, \tag{13}$$

where λ_{cor}^2 and τ_{cor} are the correlation length and time, respectively. The appropriate solution takes the form

$$\boldsymbol{B}'(\boldsymbol{x}, t) = \int_{t_0}^{t} \nabla \times \left(\boldsymbol{u}'(\boldsymbol{x}, t') \times \overline{\boldsymbol{B}}(\boldsymbol{x}, t') \right) dt'. \tag{14}$$

We shall now derive the turbulent electromotive force $\boldsymbol{\mathcal{E}}$ by taking the cross product of Eq. (12) with $\boldsymbol{u}'(\boldsymbol{x}, t')$ and taking the average. The resulting expression does not depend on t_0 as far as Eq. (13) holds. Hence we can put $t_0 = -\infty$ and obtain

$$\boldsymbol{\mathcal{E}}(\boldsymbol{x}, t) = \int_{-\infty}^{t} \overline{\boldsymbol{u}(\boldsymbol{x}, t')' \times \nabla \times \left(\boldsymbol{u}'(\boldsymbol{x}, t') \times \overline{\boldsymbol{B}}(\boldsymbol{x}, t') \right)} dt' . \tag{15}$$

The integrand is significantly different from zero only if $t - t' < \tau_{cor}$. If we replace t' by $t - \tau$ we derive

$$\boldsymbol{\mathcal{E}}(\boldsymbol{x}, t) = \int_{0}^{\infty} \overline{\boldsymbol{u}'(\boldsymbol{x}, t) \times \nabla \times \left(\boldsymbol{u}'(\boldsymbol{x}, t - \tau) \times \overline{\boldsymbol{B}}(\boldsymbol{x}, t) \right)} d\tau . \tag{16}$$

We now write Eq. (16) in terms of its component equations in right-handed Cartesian coordinate system (x_1, x_2, x_3).

$$\mathcal{E}_1 = -(w_{312} - w_{213})\overline{B_1} - w_{22}\frac{\partial \overline{B}_3}{\partial x_2} + w_{33}\frac{\partial \overline{B}_2}{\partial x_3},$$

$$\mathcal{E}_2 = -(w_{123} - w_{321})\overline{B_2} - w_{33}\frac{\partial \overline{B}_1}{\partial x_3} + w_{11}\frac{\partial \overline{B}_3}{\partial x_1}, \tag{17}$$

$$\mathcal{E}_3 = -(w_{231} - w_{132})\overline{B_3} - w_{11}\frac{\partial \overline{B}_2}{\partial x_1} + w_{22}\frac{\partial \overline{B}_1}{\partial x_2},$$

where the notations

$$w_{ijk} = \int_0^\infty \overline{u_i'(\boldsymbol{x}, t)\frac{\partial u_k'(\boldsymbol{x}, t - \tau)}{\partial x_j}}\, d\tau, \tag{18}$$

$$w_{ij} = \int_0^\infty \overline{u_i'(\boldsymbol{x}, t)u_j'(\boldsymbol{x}, t - \tau)}\, d\tau, \tag{19}$$

have been used. Comparing Eq. (17) with

$$\boldsymbol{\mathcal{E}} = \overline{\boldsymbol{u}' \times \boldsymbol{B}'} = \alpha\overline{\boldsymbol{B}} - \beta\nabla \times \overline{\boldsymbol{B}} \tag{20}$$

we immediately find that

$$\alpha = -(w_{312} - w_{213}) = -(w_{123} - w_{321}) = -(w_{231} - w_{132})$$
$$= -\frac{1}{3}(w_{123} + w_{231} + w_{312} - w_{132} - w_{321} - w_{213}), \tag{21}$$

and

$$\beta = w_{12} = w_{22} = w_{33} = \frac{1}{3}(w_{11} + w_{22} + w_{33}). \tag{22}$$

Inserting Eqs. (18) and (19) into Eqs. (21) and (22) we find finally

$$\alpha = -\frac{1}{3}\int_0^\infty \overline{\boldsymbol{u}'(\boldsymbol{x}, t) \cdot \nabla \times (\boldsymbol{u}'(\boldsymbol{x}, t - \tau))}\, d\tau\ , \tag{23}$$

and

$$\beta = \frac{1}{3}\int_0^\infty \overline{\boldsymbol{u}'(\boldsymbol{x}, t) \cdot \boldsymbol{u}'(\boldsymbol{x}, t - \tau)}\, d\tau\ . \tag{24}$$

For an estimate of the order of magnitude we evaluate the integrals (23) and (24) such that

$$\alpha = -\frac{1}{3}\overline{\boldsymbol{u}' \cdot \nabla \times \boldsymbol{u}'}\, \tau_{cor}\ , \tag{25}$$

and

$$\beta = \frac{1}{3}\overline{\boldsymbol{u}'^2}\tau_{cor} \ . \tag{26}$$

The dynamo equation, now, becomes

$$\frac{\partial \overline{\boldsymbol{B}}}{\partial t} = \nabla \times \left(\alpha \overline{\boldsymbol{B}}\right) + (\beta + \eta)\nabla^2 \overline{\boldsymbol{B}} \ . \tag{27}$$

The first term on the right hand side determines the growth rate of the large scale magnetic field $\overline{\boldsymbol{B}}$.

3 Three-Component Magnetofluid

We begin with the three component partially ionized plasma consisting of electrons (e), ions (i) of uniform mass density ρ_i and neutral particles (n) of uniform mass density ρ_n. The equation of motion of the electrons can be written as:

$$m_e n_e \left[\frac{\partial \boldsymbol{V}_e}{\partial t} + (\boldsymbol{V}_e \cdot \nabla)\boldsymbol{V}_e\right] = -\nabla p_e -$$
$$e n_e \left[\boldsymbol{E} + \frac{\boldsymbol{V}_e \times \boldsymbol{B}}{c}\right] - m_e n_e \nu_{en}(\boldsymbol{V}_e - \boldsymbol{V}_n) \ , \tag{28}$$

where the electron-ion collisions have been neglected since the ionized component is of low density. On neglecting the electron inertial force, the electric field $\boldsymbol{E}$ is found to be:

$$\boldsymbol{E} = -\frac{\boldsymbol{V}_e \times \boldsymbol{B}}{c} - \frac{\nabla p_e}{e n_e} - \frac{m_e}{e}\nu_{en}(\boldsymbol{V}_e - \boldsymbol{V}_n) \ . \tag{29}$$

This gives us Ohm's law. For $\delta = (\rho_i/\rho_n) \ll 1$ the ion dynamics can be ignored. The ion force balance then becomes:

$$0 = -\nabla p_i + e n_i \left[\boldsymbol{E} + \frac{\boldsymbol{V}_i \times \boldsymbol{B}}{c}\right] - \nu_{in}\rho_i(\boldsymbol{V}_i - \boldsymbol{V}_n) \ , \tag{30}$$

where ν_{in} is the ion-neutral collision frequency, and the ion–electron collisions have been neglected for the low density ionized component. Substituting for $\boldsymbol{E}$ from Eq. (29) we find the relative velocity between the ions and the neutrals:

$$\boldsymbol{V}_n - \boldsymbol{V}_i = \frac{\nabla(p_i + p_e)}{\nu_{in}\rho_i} - \frac{\boldsymbol{J} \times \boldsymbol{B}}{c\nu_{in}\rho_i}, \tag{31}$$

where

$$\boldsymbol{J} = e n_e(\boldsymbol{V}_i - \boldsymbol{V}_e) \ . \tag{32}$$

The equation of motion of the neutral fluid is:

$$\rho_{\mathrm{n}} \left[\frac{\partial V_{\mathrm{n}}}{\partial t} + (V_{\mathrm{n}} \cdot \nabla) V_{\mathrm{n}} \right] = -\nabla p_{\mathrm{n}} - \nu_{\mathrm{ni}} \rho_{\mathrm{n}} (V_{\mathrm{n}} - V_{\mathrm{i}}) -$$
$$\nu_{\mathrm{ne}} \rho_{\mathrm{n}} (V_{\mathrm{n}} - V_{\mathrm{e}}) , \tag{33}$$

where the viscosity of the neutral fluid has been neglected. Substituting for $V_{\mathrm{n}} - V_{\mathrm{i}}$ from Eq. (31), and using $\nu_{\mathrm{in}} \rho_{\mathrm{i}} = \nu_{\mathrm{ni}} \rho_{\mathrm{n}}$ we find:

$$\rho_{\mathrm{n}} \left[\frac{\partial V_{\mathrm{n}}}{\partial t} + (V_{\mathrm{n}} \cdot \nabla) V_{\mathrm{n}} \right] = -\nabla p + \frac{J \times B}{c} , \tag{34}$$

where $p = p_{\mathrm{n}} + p_{\mathrm{i}} + p_{\mathrm{e}}$. Observe that the neutral fluid is subjected to the Lorentz force as a result of the strong ion-neutral coupling due to their collisions.

Consider Faraday's law of induction:

$$\frac{\partial B}{\partial t} = -c \nabla \times E \tag{35}$$

By substituting for the electric field from Eq. (2), we get

$$\frac{\partial B}{\partial t} = \nabla \times (V_{\mathrm{e}} \times B) + \eta \nabla^2 B , \tag{36}$$

where the pressure gradient terms have been dropped for the incompressible case with constant temperature. Here $\eta = m_{\mathrm{e}} \nu_{\mathrm{en}} c^2 / (4\pi e^2 n_{\mathrm{e}})$ is the electrical resistivity predominantly due to electron-neutral collisions. Using the construction

$$V_{\mathrm{e}} \times B = [V_{\mathrm{n}} - (V_{\mathrm{n}} - V_{\mathrm{i}}) - (V_{\mathrm{i}} - V_{\mathrm{e}})] \times B , \tag{37}$$

and substituting for the relative velocity of the ion and the neutral fluid from Eq. (31), Eq. (36) becomes:

$$\frac{\partial B}{\partial t} = \nabla \times \left[\left(V_{\mathrm{n}} - \frac{J}{e n_{\mathrm{e}}} + \frac{J \times B}{c \nu_{\mathrm{in}} \rho_{\mathrm{i}}} \right) \times B \right] + \eta \nabla^2 B \tag{38}$$

One can easily identify the Hall term ($J/e n_{\mathrm{e}}$), and the ambipolar diffusion term ($J \times B$) [13]. The Hall term is much larger than the ambipolar term for large neutral particle densities or for $\nu_{\mathrm{in}} \gg \omega_{\mathrm{ci}}$ where ω_{ci} is the ion cyclotron frequency. In this system the magnetic field is not frozen to any of the fluids. Equations (34) and (38) along with the mass conservation

$$\nabla \cdot V_{\mathrm{n}} = 0 \tag{39}$$

form the basis of our investigation.

4 The Alpha Effect in Three-Component Magnetofluid

The α effect along with its several variants is the key concept in the generation of large scale magnetic fields from small scale velocity and magnetic fields in

62   V. Krishan and R. T. Gangadhara

the kinematic dynamo process [12]. The magnetic induction equation (38) is written as:

$$\frac{\partial \boldsymbol{B}}{\partial t} = \nabla \times [\boldsymbol{V}_E \times \boldsymbol{B}] + \eta \nabla^2 \boldsymbol{B} \ , \tag{40}$$

where

$$\boldsymbol{V}_E = \boldsymbol{V}_n + \boldsymbol{V}_H + \boldsymbol{V}_{Am} \tag{41}$$

with

$$\boldsymbol{V}_H = -\frac{\boldsymbol{J}}{en_e} \tag{42}$$

as the Hall velocity and

$$\boldsymbol{V}_{Am} = \frac{\boldsymbol{J} \times \boldsymbol{B}}{c\nu_{in}\rho_i} \tag{43}$$

could be called the ambipolar velocity. Following the standard procedure [12] the velocity $\boldsymbol{V}_E$ and the magnetic field $\boldsymbol{B}$ are split into their average large scale parts and the fluctuating small scale parts as:

$$\boldsymbol{V}_E = \overline{\boldsymbol{V}_E} + \boldsymbol{V}'_E, \tag{44}$$
$$\boldsymbol{B} = \overline{\boldsymbol{B}} + \boldsymbol{B}' \tag{45}$$

such that

$$\overline{\boldsymbol{V}'_E} = 0, \qquad \overline{\boldsymbol{B}'} = 0. \tag{46}$$

In the kinematic dynamo the magnetic induction equation is solved for large and small scale fields. Substituting Eqs. (44) and (45) into the induction equation (38), we find, in the first order smoothing approximation,

$$\boldsymbol{V}'_E = \boldsymbol{V}'_n - \frac{\boldsymbol{J}'}{en_e} + \frac{\boldsymbol{J}' \times \overline{\boldsymbol{B}}}{c\nu_{in}\rho_i} + \frac{\overline{\boldsymbol{J}} \times \boldsymbol{B}'}{c\nu_{in}\rho_i} \tag{47}$$

and the mean flow is found to be:

$$\overline{\boldsymbol{V}_E} = \overline{\boldsymbol{V}_n} - \frac{\overline{\boldsymbol{J}}}{en_e} + \frac{\overline{\boldsymbol{J}} \times \overline{\boldsymbol{B}}}{c\nu_{in}\rho_i} \ . \tag{48}$$

The turbulent electromotive force $\boldsymbol{\mathcal{E}}$ is defined as $\boldsymbol{\mathcal{E}} = \overline{\boldsymbol{V}'_E \times \boldsymbol{B}'}$ and is a function of the mean magnetic induction $\overline{\boldsymbol{B}}$ and mean quantities formed from the fluctuations. The fluctuations in turbulence have, generally, a correlation in spatial scale L_{cor} and time scale τ_{cor}. In a two-scale turbulence, $L_{cor} \ll \overline{L}$ and $\tau_{cor} \ll \overline{\tau}$, where $\overline{L}$ and $\overline{\tau}$ represent the scales of the large scale quantities. Thus the fluctuations need to be determined in the immediate vicinity of the point at which the large scale quantity is to be found. This enables us to employ Taylor's expansion for $\overline{B}$ and express the turbulent electromotive force as (Eq. 5.4 of [12], retaining only the first order spatial derivatives and omitting all time derivatives,

$$\mathcal{E}_i = \left(\overline{V'_E \times B'}\right)_i = a_{ij}\overline{B}_j + b_{ijk}\frac{\partial \overline{B}_j}{\partial x_k} \tag{49}$$

For a zero mean flow ($\overline{V}_E = 0$), homogeneous, isotropic, steady and non-mirror symmetric turbulent velocity field V'_E, the coefficients a_{ij} and b_{ijk} become isotropic, and can be expressed as:

$$a_{ij} = \alpha\delta_{ij}; \tag{50}$$
$$b_{ijk} = \beta\epsilon_{ijk} \tag{51}$$

and the electromotive force can be expressed as:

$$\mathcal{E} = \alpha\overline{B} - \beta\nabla \times \overline{B} , \tag{52}$$

The coefficients a_{ij} or α and b_{ijk} or β become functions of $\overline{B}$ for large $\overline{B}$ as we find here from the contribution of the ambipolar diffusion. The quantity α, a pseudoscalar, turns out to be the kinetic helicity of the turbulence and is defined as:

$$\alpha = -\frac{\tau_{cor}}{3}\overline{V'_E \cdot (\nabla \times V'_E)}$$
$$= \alpha_v + \alpha_H + \alpha_{Am} . \tag{53}$$

Here, retaining terms quadratic in fluctuations,

$$\alpha_v = -\frac{\tau_{cor}}{3}\overline{V'_n \cdot \Omega'_n} \tag{54}$$

is the measure of the average kinetic helicity of the neutral fluid in the turbulence possessing correlations over time τ_{cor} and

$$\alpha_H = \frac{2\tau_{cor}}{3en_e}\overline{J' \cdot \Omega'_n} \tag{55}$$

represents the contribution of the Hall effect. The coupling of the charged components with the neutral fluid is clearly manifest through the possible correlation between the current density fluctuations and the vorticity fluctuations of the neutral fluid $\Omega'_n = \nabla \times V'_n$. The ambipolar term gives rise to

$$\alpha_{Am} = \boldsymbol{\alpha}_A \cdot \overline{B} , \tag{56}$$

with

$$\boldsymbol{\alpha}_A = \frac{2\tau_{cor}}{3c\rho_i\nu_{in}}\overline{J' \times \Omega'_n} , \tag{57}$$

as the contribution from the ambipolar diffusion with its essential nonlinear character manifest through its dependence on the average magnetic induction.

One also observes that the Hall alpha (Eq. 55) requires a component of the fluctuating current density along the fluctuating vorticity of the neutral fluid whereas the ambipolar effect (Eq. 56) thrives on the component of the fluctuating current density perpendicular to the fluctuating vorticity. The turbulent dissipation is given by

$$\beta = \frac{\tau_{\text{cor}}}{3}\overline{V_{\text{E}}'^2} = \beta_{\text{v}} + \beta_{\text{H}} + \beta_{\text{Am}} \tag{58}$$

with

$$\beta_{\text{v}} = \frac{\tau_{\text{cor}}}{3}\overline{V_{\text{n}}'^2} \tag{59}$$

as the measure of the average turbulent kinetic energy of the neutral fluid in the turbulence possessing correlations over time τ_{cor} and

$$\beta_{\text{H}} = \frac{2\tau_{\text{cor}}}{3en_e}\overline{\boldsymbol{J}' \cdot \boldsymbol{V}_{\text{n}}'} \tag{60}$$

represents the contribution of the Hall effect. The coupling of the charged components with the neutral fluid is clearly manifest through the possible correlation between the current density fluctuations and the velocity fluctuations of the neutral fluid. The ambipolar term furnishes

$$\beta_{\text{Am}} = \boldsymbol{\beta}_A \cdot \overline{\boldsymbol{B}} , \tag{61}$$

$$\boldsymbol{\beta}_A = \frac{2\tau_{\text{cor}}}{3c\rho_i\nu_{\text{in}}}\overline{\boldsymbol{J}' \times \boldsymbol{V}_{\text{n}}'} \tag{62}$$

with its essential nonlinear character manifest through its dependence on the average magnetic induction. One also observes that the Hall β_{H} requires a component of the current density fluctuations along the velocity fluctuations of the neutral fluid whereas the ambipolar effect thrives on the component of the current density fluctuations perpendicular to the velocity fluctuations. We have used rigid or perfectly conducting boundary conditions (all surface contributions vanish) while determining the averages. Here we consider what is known as the α^2 dynamo and take the mean flow $\overline{V_{\text{E}}} = 0$. This actually determines the relative mean flow amongst the three fluids. The dynamo equation reduces to

$$\frac{\partial \boldsymbol{B}}{\partial t} = \nabla \times [\alpha\boldsymbol{B} - \beta\nabla \times \boldsymbol{B}] + \eta\nabla^2\boldsymbol{B} . \tag{63}$$

From now onwards we omit the bar on the large scale magnetic field. Assuming one dimensional space dependence, we assume the magnetic induction $\boldsymbol{B} = (0, B, \partial A/\partial x)$ in Cartesian coordinates (x, y, z). In the corresponding spherical configuration, one identifies the Cartesian coordinates (x, y, z) with the polar coordinates (θ, ϕ, r). Thus B and $\partial A/\partial x$ represent the toroidal and the poloidal parts of the field, respectively [4]. The boundary conditions then turn out to be the vanishing of B and A at the endpoints of a finite

Alpha Effect in Partially Ionized Plasmas   65

x-interval, say $x = 0$ and $x = \pi R$ corresponding to the poles of the sphere. It is convenient to put the induction equation in a dimensionless form using a normalizing magnetic field B_0, a spatial scale R, a time scale R^2/η_1 and writing $A = A'R$. It begets:

$$\frac{\partial B}{\partial t} = -R_\alpha \frac{\partial^2 A\prime}{\partial x^2} + \frac{\partial^2 B}{\partial x^2} - R_{\alpha A} \frac{\partial}{\partial x} \left[\left(B + a\frac{\partial A'}{\partial x} \right) \frac{\partial A'}{\partial x} \right] +$$

$$R_{\beta A} \frac{\partial}{\partial x} \left[\left(B + b\frac{\partial A'}{\partial x} \right) \frac{\partial B}{\partial x} \right] \tag{64}$$

and

$$\frac{\partial A'}{\partial t} = R_\alpha B + \frac{\partial^2 A'}{\partial x^2} + R_{\alpha A} \left(B + a\frac{\partial A'}{\partial x} \right) B +$$

$$R_{\beta A} \left(B + b\frac{\partial A'}{\partial x} \right) \frac{\partial^2 A'}{\partial x^2} , \tag{65}$$

where

$$R_\alpha = \frac{\alpha_1 R}{\eta_1}, \tag{66}$$

$$R_{\alpha A} = \frac{\alpha_{Ay} R B_0}{\eta_1}, \tag{67}$$

$$R_{\beta A} = \frac{\beta_{Ay} B_0}{\eta_1}, \tag{68}$$

$$a = \frac{\alpha_{Az}}{\alpha_{Ay}}, \tag{69}$$

$$b = \frac{\beta_{Az}}{\beta_{Ay}}, \tag{70}$$

$$\eta_1 = \eta + \beta_v + \beta_H, \tag{71}$$

$$\alpha_1 = \alpha_v + \alpha_H . \tag{72}$$

Here one observes that since the Hall effect contributes linearly, it can be combined with the standard α_v effect. The ambipolar effect is nonlinear and appears separately in the induction equation. The Hall effect can completely quench or enhance the standard α_v contribution to the dynamo for $V'_n = \pm J'/(en_e)$. In the absence of the ambipolar effect one recovers the standard α^2 effect with an exponential growth rate of the magnetic induction.

It is instructive to examine the new correlations for the case of say Alfvénic turbulence. Now in the weakly ionized case the relation between the velocity and the magnetic field fluctuations for the Alfvén waves is given by $V'_n = \pm \delta B'/\sqrt{4\pi\rho_i}$ [7] with $\delta = \rho_i/\rho_n$. Substituting these results in the expression for α_H we find:

$$\alpha_H = \pm \frac{2\lambda_H}{\delta\lambda_n}\alpha_v \tag{73}$$

where $\lambda_H = c/\omega_{\text{pi}}$ is the ion inertial scale and $\omega_{\text{pi}} = \sqrt{4\pi n_i e^2/m_i}$ is the ion plasma frequency and

$$\lambda_n = \frac{\overline{\Omega'_n \cdot V'_n}}{\overline{\Omega'^2_n}} \tag{74}$$

is the ratio of the average kinetic helicity and the average enstrophy of the neutral fluid turbulence. It is interesting to note that the ambipolar α effect vanishes for the Alfvénic turbulence. In the next section we discuss the results of the numerical solutions of Eqs. (64) and (65).

5 Discussion and Conclusion

We solve the field Eqs. (64) and (65) demonstrating the linear and the nonlinear α effect for the initial conditions given by [4]: $A'(x, 0) = 0$, $B(x, 0) = \sin x$. In the absence of the ambipolar contribution ($R_{\alpha A} = 0, R_{\beta A} = 0$). The Eqs. (64) and (65) become linear with a solution of the form $\exp[i(kx + \gamma t)]$, where $\gamma = -k^2 \pm kR_\alpha$ and k is the dimensionless wavenumber. Then the field grows for $k < R_\alpha$ i.e., for large values of the effective α at large spatial scale.

With the inclusion of the ambipolar contributions equations become nonlin-

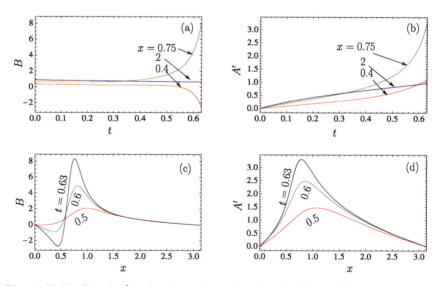

Fig. 1. Fields B and A' as functions of t and x. Panels (a) and (b) show their time variations at fixed x values, while the panels (c) and (d) show the spatial variations at some fixed t values. Chosen $R_\alpha = 1.6$, $R_{\alpha A} = 1.7$, $R_{\beta A} = 0.1$, $a = 1$ and $b = 0.7$.

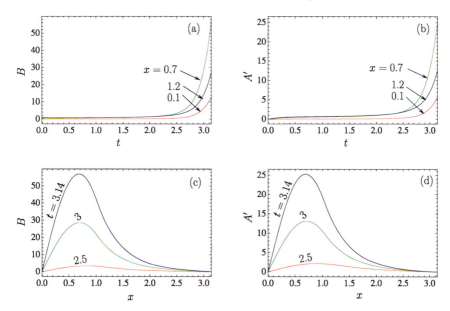

Fig. 2. Fields B and A' as functions of t and x. Panels (a) and (b) show their time variations at fixed x values, while the panels (c) and (d) show the spatial variations at some fixed t values. Chosen $R_\alpha = 0.2$, $R_{\alpha A} = 3.5$, $R_{\beta A} = 1.5$, $a = 1$ and $b = 1$.

ear. In Fig. 1, we present a case with the Hall and the ambipolar contributions. The comparable values of the coefficients of the linear and the nonlinear α terms with $R_\alpha = 1.6$, $R_{\alpha A} = 1.7$, $R_{\beta A} = 0.1$, $a = 1$ and $b = 0.7$ lead to a near constant toroidal field near $x = 2$, fast decaying solution at $x = 0.4$ and growing solution at $x = 0.75$ beyond $t \sim 0.5$. The poloidal field, however, grows at all values of x. The panels c and d demonstrate the expected formation of spatially sharp magnetic structures due to the nonlinearity of the ambipolar diffusion [9]. The toroidal field in addition undergoes a reversal at $x \sim 0.5$. Figure 2 shows the dominant effect of the ambipolar term with $R_\alpha = 0.2$, $R_{\alpha A} = 3.5$, $R_{\beta A} = 1.5$, $a = 1$ and $b = 1$. Both the components of the magnetic field, after an initial near steady state, grow rather fast and again the formation of small spatial scale structures is evident. Thus the inclusion of the Hall and the ambipolar effects opens up a range of possible profiles of the magnetic field.

In this first attempt, a framework and some instructive examples of the dynamo solutions in a 3-component magnetofluid have been given. In subsequent work we plan to investigate the role of the Hall and the ambipolar terms in some realistic situations such as the Solar surface, molecular clouds and the accretion disks. In order to deal with these systems, the differential rotation in the objects must be included. In the three component system, one would have to specify the rotation profile of all the components since the system can

68 V. Krishan and R. T. Gangadhara

afford to carry a net current density. The inclusion of the ion-neutral collisions introduces an additional time scale with which the turbulence correlation time needs to be contrasted. The inclusion of the Hall effect brings in the physics at the ion-inertial spatial scale and ion gyration time scale. The possibilities are many and varied and should be explored in a system specific manner.

References

1. Parker E.N., 1955, ApJ, 122, 293
2. Steenbeck M., Krause F., Rädler K.-H., 1966, ZNat, 21a, 369
3. Moffatt H.K., 1970, JFM, 41, 435
4. Stix M., 1972, A&A, 20, 9
5. Alfvén H., Fälthammer C.-G., 1962, Cosmic Electrodynamics, Clarendon Press, Oxford
6. Leake J.E., Arber T.D., 2006, A&A, 450, 805
7. Krishan, V., Varghese, A.B., 2007, Solar Physics (in Press)
8. Krishan V., Yoshida Z., 2006, Phys. Plasmas, 13, 092303
9. Brandenburg A., Zweibel E.G., 1994, ApJ, 427, L91
10. Schüssler M., 2005, Astronomical Notes, Astronomische Nachrichten, 326, 194
11. Zweibel E.G., 1988, ApJ, 329, 384
12. Krause F., Rädler K.-H., 1980, Mean Field Magnetohydrodynamics and Dynamo Theory, Pergman Press
13. Chitre S.M., Krishan V., 2001, MNRAS, 323, L23

Constraints on Dynamo Action

A. Mangalam

Indian Institute of Astrophysics, Koramangala, Bangalore 560034
`mangalam@iiap.res.in`

Summary. We discuss first geometrical constraints on dynamo action. Magnetic field decay is unavoidable under certain conditions of symmetry. These known results are encompassed by a calculation for flows described in terms of a generalized poloidal-toroidal representation of the magnetic field with respect to an arbitrary two dimensional surface. We show that when the velocity field is two dimensional, the dynamo growth, if any, that results, is linear in one of the projections of the field while the other projections remain constant. We discuss criteria for the existence of and classification into two and three dimensional velocity results which are satisfied by a restricted set of geometries. Secondly, helicity conservation imposes constraints on dynamo action. The calculation of helicity is technically complicated because of open boundaries and the usual form for the MHD invariant is modified to take this into account. Here we present a simple application of the quasi linear model to the galactic dynamo.

1 Geometrical Constraints

The classical problem of the magnetic dynamo concerns the question of the amplification or maintenance of the magnetic field in cases where the induction equation is valid. The equations yield only decaying solutions when the velocity and magnetic fields are both axisymmetric [1, 2] or if the geometry has planar symmetry [3]. In both situations, the velocity and magnetic field can be three dimensional but do not depend on at least one of the coordinates. In situations where the velocity is two dimensional, but the magnetic field is three dimensional, the impossibility of dynamo action has been proven if the flow is planar [4, 5, 6] or spherical [7, 8].

In the following, we use a generalized toroidal-poloidal representation of the magnetic field, $\mathbf{B}$, with respect to an arbitrary two dimensional surface and derive two scalar equations for the poloidal and the toroidal potentials from the induction equation. We prove a result, which is an extension of the one in [9], that incompressible two-dimensional velocity flows in situations other than in the above mentioned antidynamo theorems (where the field decays),

70   A. Mangalam

lead to linear growth in one of the field components and are otherwise slow. Here, the fluid velocity is two dimensional in the sense that it lies entirely on surfaces which can be described by $\chi(\mathbf{r}) = $ constant. The approach taken here also lends itself to a unified and simpler exposition of the previously cited results [10].

1.1 Normal projection of the induction equation

The starting point of all the above investigations is the well known induction equation in MHD

$$\partial_t \mathbf{B} + \nabla \times (\mathbf{B} \times \mathbf{v}) = -\nabla \times (\eta \nabla \times \mathbf{B}), \tag{1}$$

with the constraint

$$\nabla \cdot \mathbf{B} = 0, \tag{2}$$

where $\eta = c^2/4\pi\sigma$ is the magnetic viscosity and σ is the conductivity. As in the above cases, the fluid is assumed to be incompressible ($\nabla \cdot \mathbf{v} = 0$). We make the additional simplification of taking η to be uniform. Later, we discuss the effects of relaxing the latter simplification.

The RHS of the equation (1) represents resistive dissipation whereas the LHS contains a term which represents stirring of the $\mathbf{B}$ field by fluid motions. In analogy with the heat conduction equation, this shows that in a static fluid the fields decay, while stirring may induce field generation.

Consider the fluid velocity and the magnetic field to be described in terms of components perpendicular and parallel to the surfaces defined by $\chi(\mathbf{r}) = $ constant. The quantities, $B_\chi \equiv \mathbf{B} \cdot \nabla\chi$ and $v_\chi \equiv \mathbf{v} \cdot \nabla\chi$, satisfy the reduced form of the induction equation projected normal to the surface

$$d_t B_\chi - \mathbf{B} \cdot \nabla v_\chi - \eta(\nabla^2 \mathbf{B})_\chi = 0, \tag{3}$$

where $d_t \equiv \partial_t + \mathbf{v} \cdot \nabla$, and the constraint (2) and incompressibility condition was taken into account. One can expand the third term of (3) and obtain

$$d_t B_\chi = \eta(\nabla^2 B_\chi - \nabla \cdot [(\nabla\chi \cdot \nabla)\mathbf{B} + (\mathbf{B} \cdot \nabla)\nabla\chi]) + \mathbf{B} \cdot \nabla v_\chi, \tag{4}$$

where the last term, $\mathbf{B} \cdot \nabla v_\chi$, equals zero when the velocity fields lie on the surfaces, $\chi(\mathbf{r}) = $ const. This can be further reduced to the dissipation theorem [9],

$$\frac{1}{2} d_t \int_V B_\chi^2 d^3 r = -\eta \int_V \{(\nabla B_\chi)^2 + B_\chi \Theta(\mathbf{B}, \chi)\} d^3 r, \tag{5}$$

where a surface integral over $\nabla \cdot (B_\chi \nabla B_\chi)$ obtained from Gauss's theorem vanishes at large distances. Here we have introduced a useful quantity

$$\Theta(\mathbf{Y}, \chi) \equiv \nabla \cdot [(\nabla\chi \cdot \nabla)\mathbf{Y} + (\mathbf{Y} \cdot \nabla)\nabla\chi]$$
$$= [2\partial_k(Y_i) + Y_k \partial_i]\partial_i \partial_k \chi + \nabla\chi \cdot \nabla(\nabla \cdot \mathbf{Y}). \tag{6}$$

It can be seen that the last term in equation (6) vanishes if $\mathbf{Y}$ is solenoidal or if $\nabla \cdot \mathbf{Y}$ is independent of a coordinate directed along $\nabla\chi$. The only positive contribution, leading to growth, can come from the tensor term in Θ on the RHS of (6). Clearly, if the surface is planar ($\chi(\mathbf{r}) = z$) or spherical ($\chi(\mathbf{r}) = r^2/2$) the term becomes zero ($\partial_i\partial_k$ equals 0 or δ_{ik}, respectively, and in the latter case the condition, (2) needs to be further applied), implying that B_χ decays. This point was made by [9, 11]. Hereafter, we suppress the notation $\Theta(\mathbf{Y}, \chi)$ to $\Theta(\mathbf{Y})$ unless χ is specified. In the following, where we keep the treatment general (by keeping the velocity and magnetic fields three dimensional), we find the same term occuring in the surface projection of the induction equation. This enables us to expand upon these conditions for field decay and generalize the antidynamo results cited earlier.

1.2 Parallel projection of the induction equation

It is convenient to express the magnetic field in terms of the local "poloidal" and "toroidal" components

$$\mathbf{B} = \mathbf{B}_P + \mathbf{B}_T = \nabla \times \nabla \times (\psi\nabla\chi) + \nabla \times (\Phi\nabla\chi), \tag{7}$$

where ψ and Φ are the generalized poloidal and toroidal flux functions. This is analogous to the description of magnetic field given by [12, 4] for spherical geometry. We express the field in a coordinate system in which a special coordinate, q, is normal to the surfaces and is given by

$$\chi(\mathbf{r}) = f(q). \tag{8}$$

The connection to the corresponding formulae in spherical geometry lies in the fact that any smooth surface has a local radius of curvature. Therefore, we implicity demand that the surface be smooth (have first derivatives defined). Next, it can be seen that

$$\begin{aligned}
B_\chi &= \nabla \times (\nabla\psi \times \nabla\chi) \cdot \nabla\chi \\
&= \nabla \cdot [(\nabla\psi \times \nabla\chi) \times \nabla\chi] \\
&= \nabla \cdot [(\nabla\psi \cdot \nabla\chi)\nabla\chi - (\nabla\chi)^2\nabla\psi] \\
&= -(\nabla\chi \times \nabla)^2\psi \equiv -\nabla_\parallel^2\psi.
\end{aligned} \tag{9}$$

In order to examine the local components parallel to the surface, we take a normal projection of the curl of the induction equation (1),

$$\partial_t[(\nabla \times \mathbf{B}) \cdot \nabla\chi)] + \nabla\chi \cdot [\nabla \times \nabla \times (\mathbf{B} \times \mathbf{v})] = -\nabla\chi \cdot [\nabla \times \nabla \times (\eta\nabla \times \mathbf{B})], \tag{10}$$

After expanding $\mathbf{B}$, the operand of the time derivative in the resulting equation can be reduced in a fashion similar to Eq. (9)

$$(\nabla \times \mathbf{B}) \cdot \nabla\chi = -\nabla_\parallel^2\Phi + \Theta(\nabla \times [\psi\nabla\chi]). \tag{11}$$

72   A. Mangalam

Defining, $\mathcal{C} \equiv \nabla \times \mathbf{B}$, and the term on the RHS of Eq. (10) yields,

$$
\begin{aligned}
- \eta(\mathrm{curl}^3 \mathbf{B}) \cdot \nabla \chi &= \eta(\nabla^2 \mathcal{C}) \cdot \nabla \chi \\
&= \eta \left(\nabla^2 \mathcal{C}_\chi - \nabla \cdot [(\nabla \chi \cdot \nabla)\mathcal{C} + (\mathcal{C} \cdot \nabla)\nabla \chi] \right) \\
&= -\eta \nabla^2 \nabla_\|^2 \Phi - \eta \Theta(\mathcal{C}) + \eta \nabla^2 \Theta(\nabla \times (\psi \nabla \chi)), \quad (12)
\end{aligned}
$$

after performing manipulations identical to those required in obtaining Eq. (6). The second term of equation (10) is

$$
- \nabla \chi \cdot \{\nabla \times \nabla \times [\mathbf{v} \times (\nabla \times \nabla \times (\psi \nabla \chi)) + \mathbf{v} \times (\nabla \Phi \times \nabla \chi)]\}. \quad (13)
$$

The second term may be evaluated in steps as follows

$$
\begin{aligned}
\mathbf{v} \times (\nabla \Phi \times \nabla \chi) &= v_\chi \nabla \Phi - (\mathbf{v} \cdot \nabla \Phi)\nabla \chi \\
\nabla \times [\mathbf{v} \times (\nabla \Phi \times \nabla \chi)] &= \nabla v_\chi \times \nabla \Phi - \nabla(\mathbf{v} \cdot \nabla \Phi) \times \nabla \chi \\
-\nabla \chi \cdot \nabla \times \{\nabla \times [\mathbf{v} \times (\nabla \Phi \times \nabla \chi)]\} &= -\nabla \cdot [\nabla \Phi(\nabla \chi \cdot \nabla v_\chi) - \nabla v_\chi (\nabla \chi \cdot \nabla \Phi)] \\
&\quad - \nabla_\|^2 (\mathbf{v} \cdot \nabla \Phi).
\end{aligned}
$$

Defining, $\mathcal{D} \equiv \mathbf{v} \times \mathbf{B}_\mathrm{P}$, $\mathcal{D}_\chi$ reduces to

$$
\mathcal{D}_\chi = \mathbf{v} \cdot (\mathbf{B}_\mathrm{P} \times \nabla \chi). \quad (14)
$$

The first term in Eq. (13) will reduce to

$$
- \nabla \chi \cdot [\nabla \times \nabla \times \mathcal{D}] = -\Theta(\mathcal{D}) + \nabla^2 \mathcal{D}_\chi \quad (15)
$$

where the properties in the calculation (12) are used.

1.3 Results

Now one can write the equation describing the evolution of the parallel components of the field by including all the terms simplified to the forms given in Eqs. (9)–(14), and rearranging terms, as

$$
\begin{aligned}
\nabla_\|^2 (d_t \Phi) - \eta \nabla^2 \nabla_\|^2 \Phi - \eta \Theta(\mathcal{C}) &= \nabla \cdot [\nabla v_\chi (\nabla \chi \cdot \nabla \Phi) - \nabla \Phi(\nabla \chi \cdot \nabla v_\chi)] \quad (16) \\
&\quad + (\partial_t - \eta \nabla^2)\Theta(\nabla \times [\psi \nabla \chi]) - \Theta(\mathcal{D}) + \nabla^2 \mathcal{D}_\chi.
\end{aligned}
$$

The full form of the corresponding equation for the normal component is

$$
[d_t - \eta \nabla^2]\nabla_\|^2 \psi - \eta \Theta(\mathbf{B}) = -(\mathbf{B} \cdot \nabla)v_\chi, \quad (17)
$$

where the second term involves the normal component of velocity.

Non-diffusive flows ($\eta = 0$)

An exclusion theorem proved by [9] that states that two dimensional non-diffusive flows with a stationary velocity field and the property $\mathbf{v} \cdot \mathbf{a} = 0$ (the flux helicity density) under the gauge condition, $\nabla \cdot \mathbf{a} = 0$, where $\mathbf{a}$ is the vector potential for the velocity, will lead to a conservation of $\mathbf{B} \cdot \mathbf{a}$.

We now generalize this result by not imposing any limitation on the gauge transformation- as it is impossible to always simultaneously satisfy the conditions $\mathbf{v} \cdot \mathbf{a} = 0$ and $\nabla \cdot \mathbf{a} = 0$. When the velocity is two dimensional ($\mathbf{v}_\chi = 0$) and $\eta \to 0$, it is clear from (17) that B_χ is an integral of motion and cannot grow with time. Further, if one takes a stationary velocity field with $\mathbf{v} \cdot \mathbf{a} = 0$ and $\nabla \cdot \mathbf{a} = 0$, this implies that the flow lines are along the intersection of two surfaces; i.e. $\mathbf{v} = \nabla \xi \times \nabla \chi$. As a result, B_ξ is another constant of motion ($\mathbf{v}_\xi = 0$). So the induction, $\mathbf{v} \times \mathbf{B} = B_\xi \nabla \xi - B_\chi \nabla \chi$, is independent of time and the growth of the remaining component of $\mathbf{B}$, (along $\mathbf{v}$), can utmost be linear. Hence, we can conclude the intuitive result that flows with zero linkage (and $\nabla \cdot \mathbf{a} = 0$) cannot lead to an exponential growth of the field, but utmost to a linear growth of the field in the direction of the flow.

Antidynamo theorems

Based on the equations (17) and (16), we can immediately divide the antidynamo results for two and three dimensional velocity fields. For the 3D results, we take both the velocity and the magnetic field to be three dimensional but invariant along the special coordinate q ($\partial_q = 0$). In the case of 2D results, the velocity field is two dimensional, $\mathbf{v}_\chi = 0$, and the magnetic field is three dimensional. Under either of these conditions the first term on the RHS of (16) vanishes. The advection-diffusion operator ($d_t - \eta \nabla^2$) can only manipulate the field and cannot cause growth (cf. (5)).

Here, an antidynamo case is defined as a situation in which the flux functions, ψ and $\Phi \to 0$ everywhere (as $t \to \infty$) with the boundary conditions that these flux functions vanish at remote surfaces which enclose the volume of fluid. So the strategy in finding antidynamo situations is to identify the conditions when one equation completely decouples from the other and the decay of the corresponding flux function kills the source terms in the other. It is natural to consider spherical and planar geometries first, since $\Theta(\mathbf{B})$ and $\Theta(\mathcal{C})$ would then be zero (cf. (6)). It is to be noted that $\Theta(\nabla \times [\psi \nabla \chi])$ is zero for spherical, planar and axisymmetry.

The cases of spherical and planar geometries

For spherical geometry ($\chi = \frac{1}{2} r^2$) or planar geometry ($\chi = z$), the equations (16)–(17) reduce to

$$[d_t - \eta \nabla^2]\Phi = \mathcal{Q}(\psi, \mathbf{v}) \tag{18}$$

74   A. Mangalam

$$[d_t - \eta\nabla^2]\nabla_\parallel^2\psi = -\mathbf{B}\cdot\nabla v_\chi \tag{19}$$

where

$$\mathcal{Q}(\psi,\mathbf{v}) \equiv [\nabla_\parallel^2]^{-1}\{\nabla^2\mathcal{D}_\chi - \Theta(\mathcal{D}) - \nabla\cdot[\nabla\Phi(\nabla\chi\cdot\nabla v_\chi)]\}, \tag{20}$$

and $\nabla_\parallel^2$ represents L^2, the angular momentum operator in the spherical case or $\partial_x^2 + \partial_y^2$ in the planar case. In the two dimensional case ($v_\chi = 0$), we have a source term only in the toroidal equation, (18), while ψ decays. According to the definition of $\mathcal{D}$, as $\psi \to 0$, $\mathcal{D} \to 0$ and the RHS of (20) vanishes. Therefore $\mathcal{Q} \to 0$ and Φ will decay when $t \to \infty$. This follows from the arguments after (5), and from potential theory which demands that the mean value of $[\nabla_\parallel^2]^{-1}(0)$ is zero in the volume enclosed by a surface on which it vanishes. Physically, the normal component diffuses out and the field is confined to two dimensions; as a result, the field is transported like a scalar, and hence decays.

Now, in the three dimensional planar case ($\chi = z, \partial_z = 0$) there is a source term only in the poloidal equation, (19). It can be easily seen from (6) and (14), that the toroidal source terms involving $\mathcal{D}$ and $\nabla\chi\cdot\nabla = \partial_z$ are zero. Subsequently, as $\Phi \to 0$, $\mathbf{B}\cdot\nabla \to B_z\partial_z(= 0)$ and ψ decays. It is interesting that in the three dimensional spherical case none of the source terms in the above pair of equations are zero to begin with, and hence dynamo action occurs.

1.4 The case of axisymmetry

In axisymmetry ($\chi = \phi, \nabla\chi\cdot\nabla = \varpi^{-2}\partial_\phi = 0$) and $\mathcal{D}$ is poloidal. As a result, $\mathcal{Q} = 0$, as the term involving $\mathcal{D}$ is zero [cf. (15)], while $\Theta(\mathbf{A},\phi) = -(2/\varpi)\partial_\varpi(A_\phi/\varpi)$, expressed in the cylindrical coordinates, (ϖ, ϕ, z). Then the equations, (16)–(17), simplify to

$$D^2[d_t - \eta D^2]\Phi = 0, \tag{21}$$

$$[d_t - \eta\varpi^{-2}D^2\varpi^2](B_\phi/\varpi) = -\mathbf{B}\cdot\nabla(v_\phi/\varpi), \tag{22}$$

where $D^2 \equiv \nabla^2 - \frac{2}{\varpi}\partial_\varpi$, and is known as the Stokes operator, and $\nabla_\parallel^2$ represents $[\varpi^{-1}\hat{\phi}\times\nabla]^2 = \varpi^{-2}D^2$. Since the D^2 term can be reduced to a divergence term and subsequently to a surface integral by Gauss's theorem which vanishes due to the dipole behavior of the field at large radius, it does not contribute to growth; see for example, [4](p. 114). Therefore $\Phi \to 0$, $\mathbf{B}\cdot\nabla \to B_\phi\partial_\phi(= 0)$ and ψ decays according to similar arguments in [4] (p. 115).

Effects of spatial variation of density and resistivity

It is easy to see from the form of the continuity equation in steady state or under the anelastic approximation when the density is dependent only on the special coordinate q,

$$\rho \nabla \cdot \mathbf{v} + v_q \nabla_q \rho = 0, \tag{23}$$

that $\nabla \cdot \mathbf{v} = 0$ is valid when $v_q = 0$. Therefore only the above 2D results still hold. Non-uniform resistivity introduces a term $(\nabla \eta \times \nabla \chi) \cdot (\nabla \times \mathbf{B})$ in the equation (17) for ψ. This term is zero if η is a function only of q and hence does not alter the decay of ψ. This was commented upon by [6] for the cases of spherical and planar geometries. We now show that Φ also decays in some special cases. The spatial variation of η in q, however introduces a nonvanishing term $(\partial_z \eta)(\partial_z \Phi)$ for planar geometry or $(1/r)(\partial_r \eta)(\partial_r[\Phi r])$ for spherical geometry in equation (18). The 3D planar case follows trivially. Now when $v_\chi = 0$, one can invoke a theorem on the resulting elliptic equation [13, 5], which states that only the constant solution ($\Phi \equiv 0$) is possible under the condition of Φ vanishing at large distances. Similarly for axisymmetry, $\eta(\phi)$ introduces the term $(\partial_\phi \eta)\mathcal{C}_\varpi$ in (21). When there is no differential rotation, $B_\phi \to 0$ and $\mathcal{C}_\varpi$ which depends solely on B_ϕ, vanishes and Φ decays as before. Therefore, the above 3D planar and 2D planar, spherical, and axisymmetric results are still valid if η is a function only of q. Also, if there is no differential rotation, axisymmetric fields cannot be maintained if η depends on ϕ. It shown that in axisymmetry [14], poloidal fields cannot be maintained even by a compressible fluid with η as a function of space and time.

1.5 Concluding remarks on geometrical constraints

In the analysis above, we have cast the induction equation (16)–(17) in a geometry given by the surfaces $\chi(\mathbf{r}) = $ constant. This was useful in extending a previous result for incompressible two dimensional flows while unifying, classifying, and simplifying the proofs of the known antidyanamo results, and thereby providing some new insights into the structure of the induction equation. In order to deduce the general conditions of decay, we can consider the order of decay of the flux functions, ψ and Φ. If Φ is to decay first, then all the potential source terms which are on the RHS of (16) containing ψ and $\mathcal{D}$ should be zero. This includes the condition that

$$\nabla^2 \mathcal{D}_\chi - \Theta(\mathcal{D}) = 0. \tag{24}$$

The above equation automatically ensures that $\Theta(\nabla \times [\psi \nabla \chi]) = \nabla \chi \cdot [\nabla \times \nabla \times \nabla \times (\psi \nabla \chi)]$ is zero (c.f . (15)). The equation, (24) is true for planar and axisymmetry. Consequently, the 3D results of planar and axisymmetry follow, as the first term on the RHS of (16) which involves $\nabla \chi \cdot \nabla$ is zero. On the other hand, if ψ is to decay first then the RHS of (17) should be zero demanding the 2D condition, $v_\chi = 0$. Further,

$$\int_V [\Theta(\mathbf{B}) - \nabla^2 \mathbf{B}_\chi] \mathrm{d}^3 \mathbf{r} \geq 0, \tag{25}$$

must hold to ensure that ψ decays as per (5) and the boundary conditions that the flux functions vanish at the remote boundaries. This takes care of the

76   A. Mangalam

decay of Φ since the remaining source terms, which involve ψ, in (16) vanish. The condition, (25), is true for planar, spherical and axisymmetric geometries. The unifying aspect of treatment of the boundary conditions used here is that $[\nabla_\parallel^2]^{-1}(0)$ is zero with vanishing flux at the boundaries.

We have been able show the impossibility of dynamo action (as defined in §4) with $\eta = $ const and $\nabla \cdot \mathbf{v} = 0$, exists only for a restricted group of geometries as allowed by (24) and (25) for the case of 3D and 2D results respectively. In addition, the above results need further qualifications that $\mathbf{v}$ is bounded for all time and that "spiky" time dependent behavior is excluded [15].

As suggested in a classic paper [16], using a different approach, that the impossibility of a steady dynamo for a given geometry depends on the existence of an arbitrary current, $\mathbf{j}'$, such that $\int \mathbf{j} \cdot \mathbf{j}' d^3\mathbf{r} = 0$, and that only a restricted set satisfies this equation. Here we have derived specific conditions that determine these geometries. However, a more rigorous analysis of the above two conditions is needed to find the set of all possible χ that satisfies the above criteria.

Including incompressibility in the case of axisymmetry has been done by ([14, 17]). Simulations of [18] reveal hints of a larger theorem involving non axisymmetric resistivity.

2 Constraints from magnetic helicity on turbulent dynamos

Two questions about the solar magnetic field might be answered together once their connection is identified. The first is important for large scale dynamo theory: What prevents the magnetic back-reaction forces from shutting down the dynamo cycle? The second question is: what determines the handedness of twist and writhe in magnetized coronal ejecta?

Magnetic helicity conservation is important for answering both questions. Conservation implies that dynamo generation of large scale writhed structures is accompanied by the oppositely signed twist along these structures. The latter is associated with the back-reaction force. It has been suggested that coronal mass ejections (CMEs) (eg. [19]) simultaneously liberate small scale twist and large scale writhe of opposite sign, helping to prevent the cycle from quenching and enabling a net magnetic flux change in each hemisphere. Observations and helicity spectrum measurements from a numerical simulation of a rising flux ribbon in the presence of rotation support this idea.

Keeping these ideas in mind, we aim to develop a detailed frame work to calculate the helicity dynamics including relative helicity in a quasi-linear back reaction model which is more appropriate in situation where there are flows or magnetic flux crossing the boundaries. The is an extension of a model by [20]; a more careful treatment using the concept of relative helicity is attempted here. The model is applicable to various astrophysical situations such as solar or galactic dynamos.

2.1 Magnetic helicity conservation and the dynamo

The magnetic helicity associated with a field $\mathbf{B} = \nabla \times \mathbf{A}$ is defined as $H = \int \mathbf{A}.\mathbf{B} \ \mathrm{d}^3 x$, where $\mathbf{A}$ is the vector potential ([4, 21]). Note that this definition of helicity is only gauge invariant (and hence meaningful) if the domain of integration is periodic, infinite or has a boundary where the normal component of the field vanishes. In this case, under a gauge transformation $\mathbf{A} \rightarrow \mathbf{A} - \nabla \psi$, the additional term in the helicity, $\int \nabla \psi.\mathbf{B} = \int \psi \mathbf{B}.d\mathbf{S} - \int \psi \nabla.\mathbf{B} \ \mathrm{d}^3 x = 0$. H measures the linkages and twists in the magnetic field. From the induction equation one can easily derive the helicity conservation equation,

$$\frac{\mathrm{d}H}{\mathrm{d}t} = -2\eta \int \mathbf{J}.\mathbf{B} \ \mathrm{d}^3 x. \tag{26}$$

So in ideal MHD with $\eta = 0$, magnetic helicity is strictly conserved. Under many astrophysical conditions where R_m is large (η small), the magnetic helicity H, is almost independent of time. In situations where the magnetic flux penetrates the boundaries the concept of relative helicity is used [21]. We incorporate this using gauge conditions where the form remains the same (see §2.2).

2.2 Calculation of Relative Helcity in C gauge

Here we present a shorter derivation of the relative helicity in "C gauge" of [22] and associated gauge conditions from first principles. The relative helicity should have the properties of gauge independence and in addition a measure of the helicity in the volume in question (say V_a).If the volume external to this is V_b, then we can define relative helicity defined over the entire volume as

$$\Delta H = \int_{V_a + V_b} (\mathbf{A}_1 \cdot \mathbf{B}_1 - \mathbf{A}_2 \cdot \mathbf{B}_2) \ \mathrm{d}^3 x, \tag{27}$$

where 1 and 2 describe two configurations where one is the given configuration and in 2, the external field is taken to be a reference field (usually a potential field). This can be written as a sum of integrals of $(\mathbf{A}_2 \cdot \mathbf{B}_1 - \mathbf{A}_1 \cdot \mathbf{B}_2) + (\mathbf{A}_1 + \mathbf{A}_2) \cdot (\mathbf{A}_2 - \mathbf{B}_2)$. The former is expected to be zero as both quantities in the term are the same, i.e. physically it is the linkage between field configurations 1 and 2. Now the latter term reduces to

$$\Delta H = \int_{V_a} (\mathbf{A}_1 + \mathbf{A}_2) \cdot (\mathbf{A}_2 - \mathbf{B}_2) \ \mathrm{d}^3 x. \tag{28}$$

To get the gauge of [23], choose the external reference field to be a potential field, $\mathbf{B}_2 = \mathbf{P}, \mathbf{A}_2 = \mathbf{A}_P$. To preserve gauge independence, this requires, $\mathbf{P} = \mathbf{B}_n$.

78   A. Mangalam

Further we write the integrand in the Finn-Antonsen gauge as

$$(\boldsymbol{A} + \boldsymbol{A}_P) \cdot (\boldsymbol{B} - \boldsymbol{P}) = \boldsymbol{A} \cdot \boldsymbol{B} - \boldsymbol{A}_P \cdot \boldsymbol{P} + (\boldsymbol{A}_P \cdot \boldsymbol{B} - \boldsymbol{A} \cdot \boldsymbol{P}) \qquad (29)$$

The last term can be reduced to a surface integral of $\boldsymbol{A} \times \boldsymbol{A}_P$. If we take this as another gauge condition which is equivalent to

$$\boldsymbol{A} \times \hat{n}]_s = \boldsymbol{A}_P \times \hat{n}]_s \qquad (30)$$

on the surface, and we obtain the gauge of [24],

$$\Delta H_{JC} = \int_{V_a} (\boldsymbol{A} \cdot \boldsymbol{B} - \boldsymbol{A}_P \cdot \boldsymbol{P}) \, \mathrm{d}^3 x \qquad (31)$$

Now if we take the coulomb gauge for A_P and the condition that $\boldsymbol{A}_P . \hat{n}|_s = 0$, then we obtain the C gauge,

$$H_B = \int_{V_a} \boldsymbol{C} \cdot \boldsymbol{B} \, \mathrm{d}^3 x, \qquad (32)$$

where $\boldsymbol{C}$ is determined by the conditions, $\boldsymbol{P} = \boldsymbol{B}|_s$ and $\boldsymbol{A}_P \cdot \hat{n}|_s = 0$. Further, we remove all gauge freedom by taking the coulomb gauge for $\boldsymbol{A}_P$ and $\boldsymbol{C}$.

In summary the full set of gauge boundary conditions for the $\boldsymbol{C}$ gauge are

$$\nabla \times \boldsymbol{A}_P = \boldsymbol{P}, \quad \nabla \times \boldsymbol{C} = \boldsymbol{B}, \quad B_n = P_n, \quad \nabla \cdot \boldsymbol{A}_P = 0, \quad \nabla \cdot \boldsymbol{C} = 0,$$
$$\boldsymbol{A}_P . \hat{n}|_s = 0; \quad \boldsymbol{A} \times \hat{n}]_s = \boldsymbol{A}_P \times \hat{n}]_s$$

The key point here is that as long the above boundary and gauge conditions are taken into account, the helicty may calculated using $H_R = H = \boldsymbol{C}_0 \cdot \boldsymbol{B}_0$, i.e. that relative helicity is the usual helicity.

2.3 Helicity Dynamics

We write the mean-field dynamo equation for $\mathbf{B}_0$ (eg. [4]) ,

$$\frac{\partial \mathbf{B}_0}{\partial t} = \nabla \times (\mathbf{V}_0 \times \mathbf{B}_0 + \varepsilon_T - \eta \nabla \times \mathbf{B}_0) . \qquad (33)$$

$$\varepsilon_T = \langle \mathbf{v}_T \times \mathbf{b} \rangle = \alpha_0 \mathbf{B}_0 - \eta_T \nabla \times \mathbf{B}_0. \qquad (34)$$

Here ε_T is the turbulent emf, $\alpha_0 = -(\tau/3)\langle \mathbf{b}_T \cdot (\nabla \times \mathbf{b}_T) \rangle$, is the dynamo α-effect, proportional to the kinetic helicity and $\eta_T = (\tau/3)\langle \mathbf{b}_T^2 \rangle$, is the turbulent magnetic diffusivity proportional to the kinetic energy of the turbulence. This kinematic mean-field dynamo equation, has exponentially growing solutions, provided a dimensionless dynamo number has magnitude above a critical value. While the α-effect is crucial for regeneration of poloidal from toroidal fields, the turbulent diffusion turns out to be also essential for allowing changes in the mean field flux.

The kinematic mean-field equation neglects the the back-reaction on the velocity due to the Lorentz forces. This rapidly becomes a bad approximation, due to the more rapid build up of magnetic noise compared to the mean field [25]. Both direct numerical simulations of the non-linear dynamo [26] and semi-analytic modeling of the non-linear effects point to the crucial role played by magnetic helicity conservation in limiting mean field growth.

The operation of the mean field dynamo automatically leads to the growth of linkages between the toroidal and poloidal mean fields. Such linkages measure the helicity associated with the mean field. To see how the mean field helicity arises, we need to split the helicity conservation equation into evolution equations of the sub-helicities associated with the mean field, say $H_0 = \int \mathbf{A_0}.\mathbf{B_0}\, \mathrm{d}^3 x$ and the fluctuating field $h = \int \langle \mathbf{a}.\mathbf{b} \rangle\, \mathrm{d}^3 x = \langle \mathbf{a}.\mathbf{b} \rangle V$.

The evolution equations for H_0 and h (the details will presented in a paper in preparation) are

$$\frac{\mathrm{d}H_0}{\mathrm{d}t} = 2 \int \boldsymbol{\varepsilon}_T \cdot \boldsymbol{B_0}\, \mathrm{d}^3 x - 2\eta \int \mathbf{J_0} \cdot \boldsymbol{B_0}\, \mathrm{d}^3 x + \oint (\eta \mathbf{J_0} - \boldsymbol{\varepsilon}_T) \times \boldsymbol{A} \cdot \hat{n}\, \mathrm{d}A$$

$$+ \oint [(\boldsymbol{A} \cdot \boldsymbol{B})\mathbf{b}_n - (\boldsymbol{A} \cdot \mathbf{b})B_n]\, \mathrm{d}A \tag{35}$$

$$\frac{\mathrm{d}h}{\mathrm{d}t} = -2 \int \boldsymbol{\varepsilon}_T \cdot \boldsymbol{B_0}\, \mathrm{d}^3 x - 2\eta \int \langle \mathbf{j} \cdot \mathbf{b} \rangle\, \mathrm{d}^3 x$$

$$+ \oint (-\mathbf{b} \times \boldsymbol{B} + \eta \mathbf{j} - \boldsymbol{\varepsilon}_T) \times \boldsymbol{a} \cdot \hat{n}\, \mathrm{d}A \tag{36}$$

The surface terms include the Poynting flux of the small and large scale helicity due to expulsion and twisting and writhing at the surface. Here we assume that the surface terms can either be neglected or are zero for simplicity and defer the inclusion for a detailed treatment later. We see that the turbulent emf ε transfers helicity between large and small scales; it puts equal and opposite amounts of helicity into the mean field and the small-scale field, conserving the total helicity $H = H_0 + h$. So if one were to start with zero total helicity initially, in a system with large R_m, one will always have $H_0 + h \approx 0$, or $|H_0| \approx |h|$.

For a given amount of helicity, the energy associated with the field is inversely proportional to the scale over which the field varies. If for example, the small-scale field were maximally helical, and varied on a single scale, with associated wave number, k_f, we will have $k_f < \mathbf{a}.\mathbf{b} >=< \mathbf{b}^2 >$. Similarly in a periodic box, a maximally helical large scale field with wave number k_m, satisfies, $k_m \int \mathrm{d}^3 x \mathbf{A_0}.\mathbf{B_0} = \int \mathrm{d}^3 x \mathbf{B_0^2}$. We denote the volume average of mean field quantities X_0 over the scale of the system, $\int X_0\, \mathrm{d}^3 x / V$, by $\overline{X_0}$. So, helicity conservation, with $|H_0| \approx |h|$, implies $\overline{\mathbf{B_0^2}} \approx (k_m/k_f) < \mathbf{b}^2 >$. Now in general, $< \mathbf{b}^2 >^{1/2}$, will saturate near the equipartition field strength, say $B_{eq}^2 = 4\pi \rho \mathbf{v}_T^2$. (Here ρ is the fluid density). So one obtains $\overline{\mathbf{B_0^2}} \approx (k_m/k_f) B_{eq}^2 \ll B_{eq}^2$, for $k_m/k_f \ll 1$. The mean field is expected to attain at most sub-equipartition values for $R_M \gg 1$, if helicity is strictly conserved.

80 A. Mangalam

2.4 Quasi Linear Saturation of the Dynamo

As a crude model of how the dynamo saturates, when the dynamo is not too supercritical, one may use the quasi-linear theory applicable to weak mean fields (cf. [27, 11, 28, 20]).

This gives a re-normalized turbulent emf, with $\alpha = \alpha_0 + \alpha_m$, where $\alpha_m = (\tau/3)\langle \mathbf{b} \cdot \nabla \times \mathbf{b}\rangle/(4\pi\rho)$, is proportional to the small-scale current helicity. One can now look for a combined steady state solution to the helicity conservation equation (36), and the mean-field dynamo equation. Assume again that the small-scale field has a scale k_f^{-1}. We then have $< \mathbf{b}.\nabla \times \mathbf{b} >= k_f^2 < \mathbf{a}.\mathbf{b} >$ (although, for fields which are not maximally helical, these helicities can not be related to the energy). Using this, we can write dh/dt in Eq. (36), in terms of $d\alpha_M/dt$ and hence in terms of $d\alpha/dt$. Further, for R_m large enough that helicity is conserved, and for a large-scale field which varies on scale k_m^{-1}, we can approximately write, $\overline{(\nabla \times \mathbf{B}_0).\mathbf{B}_0} \approx k_m^2 \overline{\mathbf{A}_0.\mathbf{B}_0} = -k_m^2 < \mathbf{a}.\mathbf{b} >= -(k_m/k_f)^2 < \mathbf{b}.\nabla \times \mathbf{b} >$. This relation is of course strictly valid only if H_0 is well defined. This in turn requires that negligible mean magnetic flux escape.

The helicity conservation equation then gives a dynamical equation for α-quenching

$$\frac{1}{\eta_T k_f^2}\frac{d\alpha}{dt} = -2\alpha\frac{\overline{\mathbf{B}_0^2}}{B_{eq}^2} - 2\frac{k_m^2}{k_f^2}(\alpha - \alpha_0) - 2\frac{\eta}{\eta_T}(\alpha - \alpha_0) \qquad (37)$$

We see that non-linear effects leads to a decrease in α with time, till the RHS of Eq. (37) becomes zero. As α decreases, the effective dynamo number of the dynamo, D, will decrease from an initial value D_0 and lead to a saturation of the mean field growth when $D = D_c$. This happens when $\alpha = \alpha_s = \alpha_0(D_c/D_0)$. The stationary solution for *both* the dynamical α quenching equation and the mean field dynamo equations is then obtained by equating the RHS of Eq. (37) to zero, and substituting a value of $\alpha = \alpha_{sat}$ given above. For $\eta/\eta_T \ll (k_m/k_f)^2$, this gives an estimated mean field strength $B_m = |\mathbf{B}_0|$,

$$B_m \approx [(D_0/D_c) - 1]^{1/2}\frac{k_m}{k_f}B_{eq}. \qquad (38)$$

The estimate of the mean field can be made for a given dynamo model by supplying D_0, D_c, and by taking typical turbulence parameters k_f and B_{eq}. The mean field maximal helical scale length is calculated from the mean magnetic field configuration and this is shown in §2.5 for a general toroidal-poloidal split.

2.5 Calculation of the Helical Scale Length of the Mean Field

We define the mean field helical scale length to be

$$k_m \equiv \frac{\langle B_0^2\rangle}{\langle A_0 \cdot B_0\rangle}. \qquad (39)$$

In a general toroidal-poloidal split, where the vector potential is described by

$$\boldsymbol{A} \equiv \boldsymbol{A}_P + \boldsymbol{A}_T = \nabla \times (\psi \nabla \chi) + \Phi \nabla \chi, \tag{40}$$

(eg. where $\chi = \phi$ for cylindrical symmetry and $\chi = z$ for planar symmetry), we find that $\boldsymbol{B}_P = \nabla_{\parallel}^2 \psi \equiv (\nabla_\chi \times \nabla)^2 \psi$, so that $\boldsymbol{A}_P \cdot \boldsymbol{B}_P = \Phi \nabla_{\parallel}^2 \psi$. It can be shown *in general* that $\boldsymbol{A}_T \cdot \boldsymbol{B}_T = 0$, so that

$$H = \langle \boldsymbol{A}_0 \cdot \boldsymbol{B}_0 \rangle = \langle \boldsymbol{A}_P \cdot \boldsymbol{B}_P \rangle + \langle \boldsymbol{A}_T \cdot \boldsymbol{B}_T \rangle = \langle \Phi \nabla_{\parallel}^2 \psi \rangle, \tag{41}$$

and the cross terms are zero. Hence k_m is calculated to be

$$k_m = \frac{\left\langle (\nabla_{\parallel}^2 \Phi)^2 + (\nabla_{\parallel}^2 \psi)^2 \right\rangle}{\left\langle \Phi \nabla_{\parallel}^2 \psi \right\rangle} \tag{42}$$

Application to the galactic dynamo

We consider the analytic model of [9]:

$$D = \left| \frac{\alpha_0 G h^3}{\eta_T^2} \right|, \quad D_0 \sim 10 - 20, \quad D_c = 6$$

We find $k_m \sim 400 - 1000$ pc and hence $k_m/k_f \sim 1/2$. As a result the mean field strength $B_m \approx 0.5 B_{eq}$.

Application to the solar dynamo

For simplicity, we consider the simplest solar dynamo model ([4, 29]), in which the convection zone is approximated by a thin slab and the generation process sources are concentrated at different levels

$$\partial_z v = -v_0 \delta(z)$$
$$\alpha = \alpha_s \delta(z - 1)$$

Here, the z axis is perpendicular to the slab and $B^2 \simeq B_y^2$, the velocity shear ($v = v_y$) describes the differential rotation of the star. Here the dynamo number, $D = \alpha_0 v_0 h^2 \eta_T^{-2} \sim 10^3$, where h is the slab thickness and the field performs undamped oscillations and the problem admits an analytic solution.

While an application to a more detailed model is pending, we can get some preliminary estimates. We take $k_m \sim 1/h = 0.3 R_\odot$, $k_f = 1/\ell$, where the mixing length, $\ell \sim 0.01 R_\odot$. For a typical $D_0/D_c \sim 10$, this indicates sub-equipartition (about 0.1) volume averaged mean fields. This estimate has to be checked by fully non linear models.

2.6 Conclusions and caveats on constraints from helicity

- We established a framework for estimating mean field strength using relative helicity in a general context for a given dynamo model.
- It is shown that the choice of C gauge is useful, in that with appropriate boundary and gauge conditions, the helicity may calculated using $H_R = H = \int C_0 \cdot B_0 \, \mathrm{d}^3 x$.
- A formula for the helical scale length for a general geometry of the field configuration that was calculated is useful for deriving the scale length from dynamo models or simulations.
- The question of how much the surface liberation of small scale twist and large scale writhe of opposite sign via coronal mass ejections (CMEs) help in preventing the dynamo can be addressed if the surface terms in eqns (35, 36) are taken into account.
- It is expected that qualitative predictions of the sigmoid shapes (pre-CME) can be made. The predictions can be tested by numerical simulations and observations.

References

1. Cowling, T. G., Mon. Not. R. Astron. Soc., 94, 39 (1934).
2. Backus, G. E. & Chandarsekhar, S., Proc. Nat. Acad. Sci., 42, 105 (1956).
3. Zeldovich, Ya. B., JETP, 31, 154 (1956).
4. Moffatt, H. K., *Magnetic Field Generation in Electrically Conducting Fluids*, Cambridge, Cambridge University Press (1978).
5. Lortz, D., Phys. Fluids, 11, 913 (1968).
6. Zeldovich, Ya. B. & Ruzmaikin, A. A., JETP, 78, 980 (1980).
7. Bullard, E. C. & Gellman, H., Phil. Trans. R. Soc. Lond., A247, 213 (1954).
8. Backus, G. E., Ann. Phys., 4, 372 (1958)
9. Ruzmaikin, A. A. & Sokoloff, D. D., Geophys. and Astrophys. Fluid Dynamics, 16, 73, (1980).
10. Mangalam, A., arXiv:astro-ph/0501041
11. Zeldovich, Ya. B., Ruzmaikin, A. A. & Sokoloff, D. D., *Magnetic Fields in Astrophysics*, New York, Gordon & Breach (1983).
12. Chandrasekhar, S., *Hydrodynamic and Hydromagnetic Stability*, Dover, New York (1961).
13. Vekua, I. N., *Generalized Analytic Functions*, Oxford, Pergamon Press (1962).
14. Hide, R. & Palmer, T. N., Geophys. and Astrophys. Fluid Dynamics, 19, 301 (1982).
15. James, R. W., Roberts, P. H. and Winch, D. E., Geophys. and Astrophys. Fluid Dynamics, 15, 149 (1980).
16. Cowling, T. G., Q. J. Mech. Appl. Math., 10, 129 (1957).
17. Ivers, D. J. & James, R. W., 1995, GAFD, 79, 259
18. Donner, K. J. & Brandenburg, A., 1990, GAFD, 50, 121
19. Blackman, E.G. & Brandenburg, A., 2003, ApJ, 584, L99
20. Subramanian, K., BASI, 2002, 30, 715

21. Berger, M. & Field, G. B., 1984, JFM, 147, 133
22. Berger, M., A & A, 201, 355
23. Finn, J. H. and Antonsen, T. M., 1985, Comments on Plasma Phys and Control Fusion, 9, 111
24. Jensen, T. & Chu, M. S., 1984, J. Plasma Physics, 25, 489
25. Kulsrud, R. M. & Anderson, S. W., 1992, ApJ., 396, 606
26. Brandenburg, A., 2001, ApJ, 550, 824
27. Pouquet, A., Frish, U. & Leorat, J., 1976, JFM, 77, 321
28. Gruzinov, A. V. & Diamond, P. H., 1994, PRL, 72, 1651
29. Kleeorin, N. and Ruzmaikin, A., 1981, GAFD, 17, 281

Planetary Dynamos

Vinod K. Gaur

Indian Institute of Astrophysics, Bangalore 560034, India

Summary. The article begins with a reference to the first rational approaches to explaining the earth's magnetic field notably Elsasser's application of magneto-hydrodynamics, followed by brief outlines of the characteristics of planetary magnetic fields and of the potentially insightful homopolar dynamo in illuminating the basic issues: theoretical requirements of asymmetry and finite conductivity in sustaining the dynamo process. It concludes with sections on Dynamo modeling and, in particular, the Geo-dynamo, but not before some of the evocative physical processes mediated by the Lorentz force and the behaviour of a flux tube embedded in a perfectly conducting fluid, using Alfvén theorem, are explained, as well as the traditional intermediate approaches to investigating dynamo processes using the more tractable Kinematic models.

Key words: planetary magnetism, self excited dynamo, dynamo modeling

1 Introduction

Magnetic fields are observed almost everywhere in the universe on a wide variety of scales: planetary, stellar and galactic. Most planets are surrounded by extensive magnetic structures called *magnetospheres* which are 10-100 times larger than their sizes. Magnetospheres of all four giant planets and of Earth and Mercury have sources in their internal fields whereas those of Venus and the comets are induced by interactions between their ionospheres and the solar wind. Although our knowledge about these magnetic structures is largely furnished by *in situ* spacecraft measurements, charged particles and atoms have been observed via the emitted photons at radio and higher frequencies ranging from a few Hz to several GHz. Thus was detected the magnetic field of Jupiter in the 1950s, producing the first evidence that planets other than earth might have their intrinsic magnetic fields.

Speculations about the origin of the internal magnetic fields of planets soon homed on to two possibilities: fossil ferromagnetism of an earlier inducing field

86   Vinod K. Gaur

retained by magnetic rocks such as those bearing magnetite (Fe_3O_4), or by an as yet dimly understood dynamo action of conducing fluids circulating in the planetary interior, as first proposed by Larmor [1]. It was soon realized, however, that remnant magnetism was unlikely to produce the large observed dipolar fields, and would require the persistence of long sustained uniform fields whilst the bulk of the planetary magnetic material cooled below the curie point. Yet, Larmor's suggestion remained unheeded for over two decades until Walter Elsasser [2] in the 1940s applied magneto-hydrodynamics (MHD) – a fusion of Maxwell's electromagnetic equations with those of fluid dynamics – to the convective motion of the earth's liquid core stimulated by a small catalytic magnetic field, to derive a self sustained field. An important reason for the continued disregard of Larmor's proposal of circulating conducting fluids in the earth's interior being the likely cause of its magnetism, was the proof by Thomas Cowling [3] of an *anti-dynamo* theorem which showed that a perfectly axi-symmetric flow of a conducting fluid can not generate or sustain an axisymmetric magnetic field as exhibited by planetary magnetism. The problem of building in the necessary symmetry breaking mechanisms in the dynamos of axi-symmetric astrophysical bodies, whether theoretical or experimental, remains a fundamental issue in all dynamo models.

The MHD equations, however, are far too complex to yield analytic solutions although much insightful understanding of the possible dynamo processes have been gleaned by seeking constrained solutions. Meanwhile, considerable intellectual effort has been invested in studying this outstanding problem using supercomputers even as it poses a formidable challenge because of the great disparity between the typical scales of magnetic processes, of a few hundred kilometers in the earth's core, for example, compared with hydrodynamic turbulence of the order of ~ 10 meters, requiring 10^{15} grid points for a realistic simulation.

2 Characteristics of Planetary Magnetic Fields

The structure of extant magnetic fields of planets or fossil expressions of their earlier regimes enable one to explore educated speculations about the state of matter in the planetary interior as well as of its physical properties and dynamics via the magneto-hydrodynamic theory and solid state physics. Our knowledge of the magnetic fields of planets, except for the synchrotron source Jupiter, is largely derived from flyby and orbiter spacecrafts. They are, therefore, necessarily restricted in content, although the comparatively better knowledge of the Earth's present and historical field and fossil expressions of its past regimes are yet to distill some of the potentially implicit knowledge of its internal state and dynamics, notably the thermodynamic evolution of its inner and outer cores and its mantle.

The observed surface fields of planets are largely dipolar except for Uranus and Neptune, which have largely quadrupolar fields. However, their magnitudes show a wide variation from 10^{-4} T for Jupiter to 10^{-5} T (Tesla) for Earth, Saturn, Uranus and Neptune, 10^{-7} for Mercury and less than 10^{-8} for Venus. The dipole axes of Earth, Jupiter, Saturn are within 10* of their rotation axes but those of Uranus and Neptune diverge by 40-60*. The latter planets are also distinguished by their large quadrupole moments indicating different dynamo systems.

Hazarding a possible assumption that planet earth's magnetic field may be regarded as being proto- typical of planetary magnetism generally, we explore the Geo-dynamo hypothesis and test its implications against some of the prominent features of the observed terrestrial magnetic field. The latter include: its predominantly internal (>90%) long lived (>3.5 Ga) dipolar source, near congruence of its dipolar and rotation axes (~11*), episodal polarity reversals, and the westward drift. The last represents a coherent westward displacement of some specific features of the surface magnetic field, a veritable window into the dynamics of the core.

The inferred persistence of the earth's magnetic field from the remnant magnetization of rocks over 3 billion years old, requires a dynamo process whereby such long lived magnetic fields could be sustained by the circulation of conducting fluids (molten iron in the earth's outer core, metallic hydrogen in the cores of Jupiter and Saturn and ionized gases in stars). The dynamo process may thus be visualized to consist in sustaining a magnetic field through a positive feedback of currents initially induced in a moving conductor by some fortuitously present seed magnetic field that would eventually be dispensable for maintaining the dynamo process. A prototypical example of such a self sustaining dynamo is furnished by the Homo-polar disc dynamo (Figure 1) which despite its toy like simplicity captures some essential features of planetary and astrophysical dynamos. From the figure, one can write the equation of the current circuit as:

$$L\frac{dI}{dt} + R \times I = \text{Magnetic Flux} \times \text{frequency} = M \times I \times \Omega/2\pi, \qquad (1)$$

where Ω is the angular frequency of rotation, L the self inductance of the loop and R its resistance, I the induced current, and M the mutual inductance between the current loop and the rim of the rotating disc. From Eq. (1) the expression for $I(t)$ can be written as:

$$I(t) = I_o \exp[-(R - \alpha)t/L], \qquad (2)$$

where $\alpha = M \times \Omega/\pi$.

Thus, as long as $[(R-\alpha)/L]$ is negative, that is, $\alpha >$R or $\Omega > 2\pi R/M$, there would be an exponential growth of I(t) and thereby of the magnetic field. Therefore, for a self sustaining dynamo, R has to be low and the rotation adequate. Two important peculiarities of a self sustaining dynamo are immediately apparent from an analysis of this simple device:

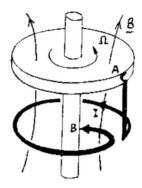

Fig. 1. Principle of a self excited dynamo [taken from [7]].

Firstly, the dynamo action can not be maintained if $\Omega < 2\pi R/M$ i.e., the sense of rotation is reversed. Therefore, axisymmetric flows can not sustain the dynamo process (Cowling's anti-dynamo theorem). Secondly, the dynamo action will cease if $R = 0$ (perfectly conducting fluid), because magnetic fields would be unable to diffuse [η in Eq. (5) would be large and so would be the diffusion time constant] in the region where magnetic induction is operating. Thus whilst the condition of low R is necessary to sustain the dynamo action, a perfect conductor would stall it.

3 Basic Equations

Interactions between circulating conducting fluids and the evolving magnetic field through a possible dynamo action can be mathematically described by incorporating the Lorentz magnetic force ($\mathbf{J} \times \mathbf{B}$) in the classical Navier-Stoke's equations. Accordingly,

$$\rho[(\partial \mathbf{V}/\partial t) + (\mathbf{V}.\nabla)\mathbf{V}] + 2\rho \mathbf{V} \times \Omega = [-\nabla p + (\mathbf{J} \times \mathbf{B}) + \rho \mathbf{g} + \rho \mathbf{f}] \quad (3)$$

subject to the condition of mass conservation

$$\partial \rho / \partial t + \nabla . (\rho V) = 0. \quad (4)$$

Convection driven motions of conducting materials in a planet's fluid core are assumed to be responsible for sustaining its dynamo. The convection, in turn, is generated by buoyancy forces ($\rho \mathbf{g}$) caused by thermal as well as compositional advection – diffusion;

The cooling and growth of the solid inner core in the case of the earth, not only supplies heat (thermal buoyancy) across the inner core-outer core boundary (ICB), but also expels the low density materials released during the solidification process of the iron alloy (compositional buoyancy). Other operating forces are the pressure gradient $(-\nabla p = \nabla \cdot T)$ force, the Coriolis force $(2\rho \mathbf{V} \times \boldsymbol{\Omega})$, the Lorentz force $(\mathbf{J} \times \mathbf{B})$ and the Inertial forces $(\rho D\mathbf{V}/Dt)$. The $(\rho \mathbf{f})$ term accounts for various other forcing terms that may be relevant to a specific situation being modeled, such as those due to compressibility.

Although, the above coupled non-linear partial differential equation is rather difficult to solve analytically, considerable headway has been made in simulating realistic low–Reynold number geodynamos. A particularly insightful approach is to study the behaviour of kinematic dynamos, i.e., of the evolution of the magnetic field for given velocity fields to explore the latter set that would sustain a self excited dynamo

action, without accounting for the counter influence on fluid motion through the Lorentz force. Next, one proceeds to address the question as to how a eligible velocity fields may be produced and are subsequently modified by the consequent $\mathbf{B}$ field via the Lorentz force.

Towards the first of these steps, we eliminate from Maxwell's quasi-stationary equations that ignore the displacement current, the electric field $\mathbf{E}$, and the current density field $\mathbf{J}$ through the generalized Ohm's law, $\mathbf{J} = \sigma[\mathbf{E} + \mathbf{V} \times \mathbf{B}]$, for moving conductors under the Galilean transformation, and thereby obtain the Induction equation for uniform conductivity $(\nabla \sigma = 0)$:

$$\frac{\partial \mathbf{B}}{\partial t} = \eta \nabla^2 \mathbf{B} + \nabla \times (\mathbf{V} \times \mathbf{B})], \tag{5}$$

where $\mu \sigma = (1/\eta)$.

An examination of the Induction Equation shows that the evolution of the $\mathbf{B}$ field is determined by the relative strengths of the two opposite terms on the r.h.s. of Eq. 5. Thus, whilst the $\nabla \times (\mathbf{V} \times \mathbf{B})$ term represents the tendency to advect and distort the magnetic field, enhancing its strength in the process, the $\eta \nabla^2 \mathbf{B}$ term describes the diffusion of the field relative to the fluid, tending to eliminate high field gradients in the advection mediated distortion of the B field. In the absence of sources to sustain the $\mathbf{B}$ field i.e., $\mathbf{V} = 0$, Eq. 5 reduces to a vector diffusion equation implying that the B field would decay in time of the order of the electromagnetic diffusion time constant $(\tau d = L2/\eta)$. The advection diffusion time constant, on the other hand, would be of the order of $(\tau a = L/V)$, where L and V represent typical length scale and velocity magnitudes. The ratio of these two time constants or, equivalently, of the advective and the diffusive tendencies $|\nabla \times (\mathbf{V} \times \mathbf{B})|/|\eta \nabla^2 \mathbf{B}|$ would determine whether a sustainable magnetic field would result. Their non-dimensional ratio is the magnetic Reynold's number:

$$R_m = |\nabla \times (\mathbf{V} \times \mathbf{B})|/|\eta \nabla^2 \mathbf{B}| = (\tau_d/\tau_a) = LV\eta, \tag{6}$$

which, clearly, must exceed 1, usually 10-100, in order to make the dynamo action possible. Rewriting Eq. 5 in terms of Rm for two limiting cases of scaling (eq. 5):

90   Vinod K. Gaur

[R* = LR; V* = ν V; t* = t dt] and dropping the asterisks, we obtain the non-dimensional form of the induction equation :

$$\partial \mathbf{B}/\partial t = \nabla^2 \mathbf{B} + R_m \nabla \times (\mathbf{V} \times \mathbf{B}) \qquad \text{for time scale, } t - t_d t \qquad (7)$$

$$\partial \mathbf{B}/\partial t = R_m^{-1} \nabla^2 \mathbf{B} + \nabla \times (\mathbf{V} \times \mathbf{B}) \qquad \text{for time scale, } t - t_a t \qquad (8)$$

Magnetic fields in astrophysical bodies unsupported by a continuously replenishing mechanism would, therefore, disappear in time of the order of the electromagnetic diffusion time constant which is proportional to the electrical conductivity. In the case of the earth ($L_{\text{thickness of core}} \sim 3.5 \times 10^6$ m, and conductivity $\sim 4 \times 10^5$ S/m), this would be of the order of $\sim 200,000$ years. On the other hand, a perfectly conducting fluid ($\eta = 0$), would stall diffusion of the magnetic field, thereby also stalling the dynamo action. Yet, the behaviour of the induction field in a highly conducting fluid exposes several insightful ideas. An especially important result that follows is that of the magnetic flux frozen to the conductor and moving with it so that any agency tending to distort it has to work against the magnetic stresses thereby creating additional magnetic energy and intensifying this field. However, as the dynamo process is essentially one of systematically amplifying the magnetic field *without an ultimate change in structure*, what is equally needed is the associated operation of a tempering process which would keep eliminating the complexities (kinks) that also grow with the field so that the initial structure is restored. This recognition underlines the crucial role of diffusion (the smoother of field gradients) in sustaining a dynamo operation even as its tendency must be necessarily much weaker than that of the advective processes needed for generating the field.

4 The Lorentz Force (J × B): A Digression

A moving conductor in the field of magnetic induction $\mathbf{B}$ experiences a force $\mathbf{F} = \mathbf{J} \times \mathbf{B}$. In a realistic dynamo, this appears as a back reaction of the evolving magnetic induction field on the velocity field. Eliminating J from the Lorentz force using the Ampere theorem equation of Maxwell i.e, $\nabla \times \mathbf{B} = \mu_o J$, we obtain:

$$\mathbf{J} X \mathbf{B} = [(\nabla \times \mathbf{B})/\mu_o] \times \mathbf{B} = -(1/\mu_o)\mathbf{B} \times (\nabla \times \mathbf{B})$$
$$= (1/\mu_o)[(\mathbf{B} \cdot \nabla)\mathbf{B} - \nabla(B^2/2)].$$

The second term is like the pressure gradient term signifying an isotropic pressure $p = B^2/2\mu_o$. The first term can be shown to be equal to:

$$(1/\mu_o)(\mathbf{B} \cdot \nabla)\mathbf{B} = [\chi\{\partial/\partial\chi(B^2/2\mu_o)\} + \mathbf{n}\{[B^2/2\mu_o R\}], \qquad (9)$$

where χ and $\mathbf{n}$ represent unit vectors respectively, tangential and normal to the B force line, at the point denoted by the radius vector R, and R measures the radius of curvature of the field line at that point. The terms on the right hand side of Eq. 10 are in fact the force components of the deviatoric magnetic tensor and together represent a tensional force of magnitude (B^2/μ_o) acting on the field lines. Therefore, if we visualize a flux tube as a topologically cylindrical volume with sides parallel to the magnetic field lines, the pressure gradient force would be seen to resist compressions or rarefactions of the magnetic flux tubes that may be forced by fluid motions, whilst the tensional force as that of a stretched rubber band, would resist their being bent or twisted.

5 Alfvén's Theorem

Diffusion plays a crucial role in smoothening out the growing topological complexity of an evolving dynamo field in order to steadily preserve its structure, and therefore requires the moving conductor to have nonvanishing resistivity. Yet, it proves instructive to examine the case of a perfectly conducting dynamo fluid ($\eta=0$), which reduces Eq. 5 to:

$$\partial\mathbf{B}/\partial t = \nabla \times (\mathbf{V} \times \mathbf{B}). \tag{10}$$

This is exemplified by the Alfvén theorem as follows: Consider the time behaviour of the magnetic flux $\phi = \int \mathbf{B}(R,t) \cdot d\mathbf{S}$, across an open surface S bounded by a closed curve C. As S(t), at time t spanned by C, evolves into S'(t +δt) spanned by C,' in response to fluid motions, it generates a volume $\mathbf{v}$ bounded by the closed surface formed by S, S' and the connecting curved surface S". An element of S", d S" = ($\mathbf{V}\,\delta t \times d\mathbf{L}$) is then the area swept by the length element dL of the closed curve C. And, the flux crossing this area is $\mathbf{B} \cdot (\mathbf{V} \times d\mathbf{L})\delta t$. Therefore,

$$\partial\phi/\partial t = \int_c (\mathbf{B} \cdot \mathbf{V} \times d\mathbf{L}) = \int_c (\mathbf{B} \times \mathbf{V}) \cdot d\mathbf{L} = \int_s [\nabla \times (\mathbf{B} \times \mathbf{V})] \cdot d\mathbf{S}$$

or, using Eq. 11,

$$\partial\phi/\partial t = -\partial/\partial t \int_s \mathbf{B} \cdot d\mathbf{S} = [-\partial/\partial t \int_v (\nabla \cdot \mathbf{B})dv] = 0 \tag{11}$$

because according to Faraday's law [$\nabla \times \mathbf{B} = -\partial\mathbf{B}/\partial t$], and, ($\nabla \cdot \mathbf{B}$) must vanish everywhere in space if at some initial time it had been zero. Eq. 12 asserts that the magnetic flux through any closed contour in an infinitely conducting fluid is conserved, i.e., a flux tube embedded in such a flowing fluid at any initial time, would maintain its integrity by remaining frozen to it. Since magnetic field lines are the limiting case of an infinitely thin flux tube, this is also true of the magnetic field lines which are accordingly restrained from breaking and reconnecting - a condition? which? severely inhibits the dynamo process.

92   Vinod K. Gaur

Indeed, this analogy of a magnetic flux tube behaving like an elastic string has phenomenological expressions such as the Alfvn waves that propagate the energy of a disturbance, such as that caused by fluid motion, with a velocity that actually equals $(T/m)^{1/2} = B(\mu_o\rho)^{-1/2}$. This can be easily checked by recalling that a tube of force of cross sectional area A would have a mass/length $= \rho A$ and tensional force $= (AB^2/\mu_o)$. Another instructive deduction from flux conservation evocatively shows how magnetic energy of a system is intensified by the motion of a fluid in which a flux tube is embedded. Thus, if the length L of this flux tube is stretched to L' by fluid motion, its cross section A is reduced to $A' = (AL/L')$ to conserve mass, thereby densifying the field lines within to intensify an original undisturbed B field to $B' = (BA/A') = (BL/L')$ and its energy E to $E' = E(L/L')^2$. This exemplifies the basic process whereby the advective energy of a highly conducting fluid is translated into magnetic energy of the system through $\nabla \times (\mathbf{V} \times \mathbf{B})$. However, to sustain the dynamo process the tube must be restored to its original form through a process of Ohmic diffusion and reconnection of magnetic field lines, in defiance of Alfvn's theorem. In dynamo theory the prime goal is to explore the nature of fields that would persist far longer than the diffusion time constant, in turn, requiring finite even though small resistivity . Despite this lack of internal consistency, the frozen flux assumption proves evocative in visualizing the evolution of the field, especially when R_m is large.

The toroidal fluid velocity just beneath the core mantle boundary, assuming infinite conductivity of the core and infinite resistivity of the mantle, by inverting the downward- continued observed B field through the insulating source free mantle to the core- mantle boundary in the induction Eq. 11.

6 Kinematic Dynamos

As stated earlier, the simultaneous solution for both the fluid velocity field and that of the evolving magnetic induction which reacts on the former, is a most challenging task. A simpler approach is to work with the linear kinematic dynamo equations assuming a small magnetic field (to justify neglect of the complex back reaction on to the V field), to identify velocity fields that would sustain a specific dynamo situation, and then to test their legitimacy as solutions of the MHD equations. The kinematic dynamo problem is therefore posed by the question: whether the flow of an electrically conducting, and initially unmagnetized fluid would lead to the exponential growth of a small seed magnetic field. If the answer is yes, then the flow can generate a self sustaining field Since a plausible velocity field must replenish magnetic energy faster than it dissipates ($\tau_a < \tau_d$, i.e., large R_m), we seek, for a prescribed steady-state velocity field, a self sustained, exponentially growing solution of the Induction equation 6, or as often called, the Kinematic Dynamo Equation (Eq. 7):

$$\partial\mathbf{B}/\partial t = \nabla^2\mathbf{B} + R_m\nabla \times (\mathbf{V} \times \mathbf{B}). \tag{12}$$

The conditions of self sustenance demand, i) that all fields and currents must solely arise from motions of the MHD fluid, requiring, in turn, that both V & B vanish at an infinite distance from the circulating fluids, and ii) that they must persist indefinitely as long as the power sources of the motion exists.

Accordingly, we seek a solution of the form:

$$B \propto \exp(\gamma t), \tag{13}$$

where γ is a positive quantity, in the limit when the conductivity vanishes, since MHD fluids are generally of very high conductivity. Assuming a non-linear dependence of conductivity, therefore, $\gamma \propto \sigma^{-\beta}$ in the limit as the conductivity becomes infinite, where $0 \leq \beta \leq 1$. When β is zero, the growth rate is independent of conductivity and the dynamo is called the *fast* dynamo.

With the substitution of Eq. 14, Eq. 13 is reduced to solving the eigenvalue equation:

$$\gamma \mathbf{B} = [\nabla^2 \mathbf{B} + R_m \nabla \times (\mathbf{V} \times \mathbf{B})]. \tag{14}$$

Roberts [4] considers the solution of Eq. 15 for a planetary system consisting of a conducting fluid occupying a bounded volume V_a, surrounded by an exterior volume V* which is an insulator with no sources of B or E. The spectrum of γ in this case is found to be discreet with limit point at $-\infty$, and $B = \sum_k B_k(R) \exp(\gamma_k t)$. Once the value of B in V_a has been obtained using the largest real eigenvalues, E in V_a can be calculated using Faraday's equation and E^*, V_a^*, with the help of Boundary conditions. Despite its fractured approach to elucidating self sustained dynamos, kinematic dynamo models have played an important role in establishing that spherical homogeneous dynamos, considered counter- intuitive on the strength of Cowlings anti-dynamo theorem, were indeed possible.

7 Dynamo Modeling

The decades after the advent of computers in the sixties, witnessed a progressive sophistication in the simulation of numerical dynamos, showing that three dimensional self-consistent simulations were possible in principle, but it was only in 1995 that Glatzmaier and Roberts [7] showed that such models could reproduce several features of the earth's magnetic field, including polarity reversals. A large number of dynamo models have been explored since to elucidate significant characteristics of this field: its secular variation, morphology and even spatial spectra.

Whilst all planetary models are spherical, and incorporate three dimensional fluid flows and magnetic field, to meet Cowling's prohibition of 2-dimensional solutions, they necessarily include:

i) a plausible energy source for maintaining the flow of an electrically conducting fluid furnished by thermal or compositional or both, and

ii) rotation which through the Coriolis force, structures the fluid flow in a manner conducive to dynamo action.

A simple Geo-dynamo model, now adopted for benchmarking various numerical codes [8], consists of a rotating fluid shell bounded between the core-mantle boundary at radius r_0 and inner core-outer boundary at radius r_i. The solid inner core $(r \leq r_i)$ has conductivity as that of fluid outer core, and the mantle outside the shell $(r \leq r_0)$ is insulating mantle. All material properties in the fluid shell are represented by constants and the Boussinesq approximation is applied accounting all temperature and composition induced density changes to buoyancy in the Navier-Stokes equations. Physical dynamo models so constructed are then driven by the standard MHD equations 3–5, complemented by a thermal/compositional transport equation and appropriate boundary conditions. The above equations are usually solved for their non-dimensional forms which being non-unique lead to a variety of control parameters and solution characteristics. The fundamental length scale in the benchmark model is represented by the thickness of the fluid shell $[D = (r_o - r_i)]$, time by the viscous diffusion time $(D^2/\nu_{\text{kinematic viscosity}})$, temperature by ΔT, the difference between that of the inner and outer boundary of the shell, and magnetic induction by $(\Omega\rho/\sigma)^{1/2}$. In the case of the geodynamo, the two most characteristic quantities of the solution are: the Magnetic Reynold's number $R_m = LV_{rms}\eta$, and the Elsasser number $\Lambda = (\sigma B^2/\rho\Omega)$ which measures the ratio of the Lorentz to Coriolis forces and, in the scaling stated above, equivalent to the square of the expected non-dimensional magnetic field magnitude inside the region of field generation.

Besides, exposing possible velocity and magnetic induction fields that would sustain a dynamo and reproduce the observed space-time characteristics, numerical dynamo models, by offering the possibility of designing controlled experiments, have the farther reaching potential of elucidating the physics of fluid flows interacting with their consequent magnetic induction fields, even though their range of validity may be severely restricted by practical limits to computational schemes. Thus, it has been possible to study models with moderate values of the magnetic Reynold's number, Rm up to ~ 100 which is not too unrealistic considering that this umber for the geodynamo is ~ 1000. On the other hand, all models assume a viscosity that is far larger than expected to occur in planetary interiors, but are constrained by the necessity to suppress small scale turbulence which cannot be adequately resolved within the extant computational resources. All these caveats notwithstanding, numerical dynamos have scored some remarkable successes in matching the significant geomagnetic field features not only of the right form but of the right strength. Especially exciting has been the new knowledge about the role of the earth's inner core in stabilizing its dipole polarity. Indeed, the pace of recent advances in numerical dynamo modeling has been so rapid as to warrant the belief that with more observational geomagnetic data and features added to the presently available observational data our understanding of the yet unresolved aspects of dynamo processes is expected to sharpen progressively.

8 The Geodynamo

The sources of the earth's magnetic field lie in the convecting fluids of the outer core. All solar system planets are known to have a central core of denser material as inferred from their moments of inertia, but some like Venus and Mars which are most similar to the earth do not have a dynamo operating in their cores although remnant magnetization of Martian rocks requires the existence of an extinguished dynamo that operated in the past. Our ability to study the earth's dynamo is considerably more robust because of our greater knowledge of the space-time characteristics of its field, but more specifically, because of our substantial understanding of the earth's surface geological processes fashioned by plate tectonics. Plate velocities and therefore their creation and subduction rates powered by mantle convection, are known to have varied over geological times, thereby varying the rate of heat abstraction from the core-mantle boundary (CMB) and thereby influencing the regime of core convection. Two contrasting situations are distinguished according to whether the core heat flow crossing this boundary is above or below a threshold value estimated to be $\sim 6TW$. If this exceeds the heat conducted at the top of the core along the adiabatic gradient, a thermal boundary layer of colder material forms at the top driving a top-down convection. If, on the other hand the mantle does not conduct away all the heat arriving from below, the excess heat accumulates or is convected into the interior. Thus, whilst thermal buoyancy is necessarily produced only on the inner core-outer core boundary, the style of convection would change as the earth transited from a high heat flow regime in the past, or even in later epochs owing to fluctuations in the mantle heat transport mediated by plate tectonic cycles that alter its material composition.

Palaeomagnetic investigations in the 1950's revealed that the earth's dipolar polarity has been reversed several hundred times during the past 160 million years [5]. The average duration of the stable geomagnetic polarity during the past 15 million years is $\sim 200,000$ years with large variations from some tens of thousand years to millions of years. This is a puzzling phenomenon unlike the magnetic polarity of the Sun which reverses at near constant intervals of ~ 11 years, about equal to its electromagnetic diffusion time τ_d. Speculations about the possible origin of the earth's aperiodic reversals lying in the coupled plate tectonics-core system through mantle convection, have been tested using numerical models designed to investigate the effects of a non-uniform pattern of CMB heat flow [5]. Their model did reproduce reversals that have analogues in the palaeomagnetic records and in particular, showed that the geodynamo is more stable and stronger when the CMB heat

96   Vinod K. Gaur

flow is axi-symmetric and equatorially symmetric, with maxima in the Polar regions.

Additionally, the electromagnetic coupling between the earth's core dynamo and the lower mantle which has a finite electrical conductivity typically of $\sim 10 S/m$, generates forces that result in exchange of angular momentum between the core and the mantle. Detailed estimates of the core top flow and the resulting electromagnetic torque in the mantle by Stix and Roberts [6] showed that this is of the right order to explain the small observed fluctuations of the length of the day. Other dynamical variations in the earth's spin notably the nutation of its axis have also been inverted to study the physical state of the earth's lower mantle

More detailed studies of the earth's dynamo in recent years, exposing the role of various interactive processes operating in its various domains: the inner and outer cores, and the mantle, have thus opened up the possibility of studying not only the different dynamo conditions operating in other planets and the basic physical principles deducible from the consideration of possible sets of boundary conditions, but also of studying the physical and chemical states of their internal structure complementarity with other lines of investigation, notably broad band seismology and high temperature-pressure mineralogy.

Acknowledgments: The matter presented above has drawn heavily from published work on the subject, particularly the book on Lectures on Solar and Planetary Dynamos by Proctor and Gilbert et al., and papers by Buffet, Christensen, Glatzmaier and their co-workers. Ajay Manglik painstakingly reviewed the manuscript and provided many helpful suggestions.

References

1. Larmor, J., Rept. Brit. Assoc. Adv. Sci. P 159, (1919)
2. Elsasser, Walter M., Phys. Rev., Vol. 72, P 821 (1947)
3. Cowling, T. G., MNRAS, 94, 39 (1934)
4. Roberts, P. H., Lectures on Solar and Planetary Dynamos, Ed: M. R. E. Proctor and A. D. Gilbert (1994)
5. Merrill, R. T. McElhinny, M. W. McFadden, P. L. The Magnetic Field of the Eath: Paleomagnetism, the Core, and the Deep Mantle, Acdemic Press, San Diego, California, P 531 (1996)
6. Stix M., Roberts, P. H. Phys. Earth Planet Int., Vol. 36, P 49 (1984)
7. Fitzpatrick, R., Lectures on Magneto-hydrodynamic Theory: http://farside.ph.utexas.edu
8. Christensen, U. R, Aubert, J., Cardin, P., et al. Phys. Earth Planet. Int., Vol 128, P 25 (2001)

Part III

Pulsar Radiation Mechanism

Pulsars as Fantastic Objects and Probes

Jin Lin Han

National Astronomical Observatories, Chinese Academy of Sciences
Jia-20 DaTun Road, Chaoyang District, Beijing 100012, CHINA
hjl@bao.ac.cn

Summary. Pulsars are fantastic objects, which show the extreme states of matters and plasma physics not understood yet. Pulsars can be used as probes for the detection of interstellar medium and even the gravitational waves. Here I review the basic facts of pulsars which should attract students to choose pulsar studies as their future projects.

Keywords: Pulsars, ISM, Gravitational waves

1 Pulsars: General Introduction

Pulsars are sending us pulses – we can receive these pulses in radio bands. A small number of pulsars also emit high energy radiation in optical and X-ray or even γ-ray bands. Here let me discuss the radio pulsars, and leave the high energy emission and its explanation to Prof. Qiao in this volume.

After the first pulsar discovered by Hewish et al. [1] in 1968, it was soon realized that they are rotating neutron stars [2, 3] with a diameter of only 20 km but extremely high density (10^{15}g cm^{-3}) and extremely strong magnetic fields (10^8 to 10^{14} G). Because of revealing of this new state of matter in the universe, the pulsar discovery was awarded the Nobel Prize in 1974 in physics.

Pulsars take birth in supernova explosion, which is evident from the young pulsars and supernova remnants associations [4]. For example, the Vela pulsar is located in the center of Vela nebula, the Crab pulsar is acting as the heart of Crab nebula. Pulsars get high velocity (a few 100 km s^{-1}) [5, 6] in the explosion so that pulsars are running away quickly from their birthplace, even about half pulsars have escaped from our Galaxy in last 100 Myr [7].

The broadband radio emission of pulsars leaves from the emission region almost simultaneously. However, after these signals pass through the interstellar medium (ISM), the radio waves at a lower frequency, ν_1, in GHz, come later than these at a higher frequency ν_2, with a time delay of

$$dt = 4.15 \left(\frac{1}{\nu_1^2} - \frac{1}{\nu_2^2} \right) DM \quad \text{ms}, \tag{1}$$

where DM is known as the dispersion effect from the ionized gas in the ISM along the path from a pulsar to us, given by

$$DM = \int_0^D n_e dl \quad \text{pc cm}^{-3}, \tag{2}$$

where n_e is the electron density in units of cm^{-3} and D the distance from observer to pulsar in pc. In Fig. 1 (left panel), we have plotted the signals received in each frequency channel as function of pulse phase. It is evident that the signals at low frequency are delayed in arrival time compared to those at high frequencies. After making the delay corrections and adding the channels we get the pulse profile (see Fig. 1 right panel).

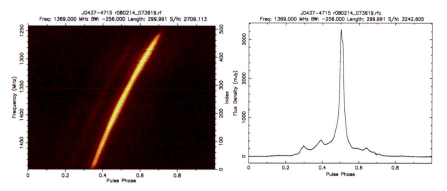

Fig. 1. The dispersion of pulsed signals – the example is observation of PSR J0437-4715 (left). The dedispersion and adding signals from all frequency channels gives a strong and stable profile (right). Data is obtained by the author with Parkes telescope.

How to find pulsars? Because a pulsar have a very accurate period – the period change due to slow-down is small enough in half an hour, one can make Fourier transform of the recorded power, and found the period in the power-spectrum. Nowadays, there are many terrestrial radio frequency interferences (RFI) which look like pulsar signals. However, these RFI normally show the maximum power at zero dispersion. Therefore, when radio power from many frequency channels is recorded, only after the power from all channels are properly de-dispersed the pulsar signal should show the maximum power. Therefore, here are several steps to find a new pulsar: **1)** record signals in time series at many frequency-channels; **2)** de-disperse the channel signals with many trial DM and add all channels together to get one time series for each trial DM; **3)** search for the periodicity from each dedispersed time series.

Because most pulsars have only narrow pulses, the power-spectrum should show not only the primary pulsar rotation frequency, but also its harmonics. Often the harmonics are added together to enhance the searching signal-to-noise ratio; **4)** verify the dedispersed pulse by searching for the best detection signal-to-noise ratio in the 2-D parameter space around the proposed P and DM and then folding data in the right period P and DM; **5)** finally re-observe again in this sky position and search the pulse again around the proposed P and DM. If the pulse can be found in the same DM but slightly evolved period, then a pulsar is definitely found! See [51, 16] for the pulsar searching strategy.

Up to now, about 1800 pulsars have been discovered [50], including some discovered using the GMRT [20, 37]. Most of pulsars are in our Milky way Galaxy, and only about 20 were discovered from extensive searches of nearby galaxies, the Large and Small Magellanic clouds. Pulsars have a period of 1.3 ms to $\sim$10 s. Some very fast rotating pulsars, so-called millisecond pulsars, have a period of only some milliseconds but very stable and very small period derivatives. From measurements of binary pulsars, it has been established that neutron stars normally have a mass [55] about 1.4 solar mass ($M_\odot$) though some are heavier (probably up to 2 $M_\odot$) and some are lighter (1.2 $M_\odot$).

Here are pulsar books I would like to recommend to readers: 1). "Handbook of Pulsar Astronomy", by D. Lorimer and M. Kramer [42], published by the Cambridge University Press (2005), contains many up-to-dated material of pulsar studies; 2) "Pulsar Astronomy" by A. Lyne & F. Graham-Smith [45], also published by the Cambridge University Press (2006), is an excellent text book for graduate students, covering all aspects of pulsar astrophysics.

2 Pulsar emission

Radio emission from pulsars is generated in pulsar magnetosphere. We define the boundary of this magnetosphere by the light-cylinder, e.g. at the radius where the rotation speed is equal to the light speed. The particles, i.e. positrons and electrons, are accelerated along the magnetic fields above polar cap or the outer gap. These particles radiate [22] so that we can see the emission in radio and high energy band. *However, it is not clear what physical processes are involved for the particles to radiate.* Pulsar wind or wind nebula [63] can be formed if particles flow out through the open magnetic field lines passing through the light-cylinder.

It is the rotation that provides the energy source for pulsar emission and particle outflowing. One can calculate the braking torque due to magnetic dipole radiation, which finally result in $\dot{\mu} = -K\mu^n$, here the μ is the rotation frequency of the neutron star, and the n is the braking index which theoretically should be equal to 3. However, the measurements show that it is usually significantly smaller than 3, which implies [81] that other reasons are also consuming the rotation energy, e.g. pulsar wind. Otherwise, magnetic fields

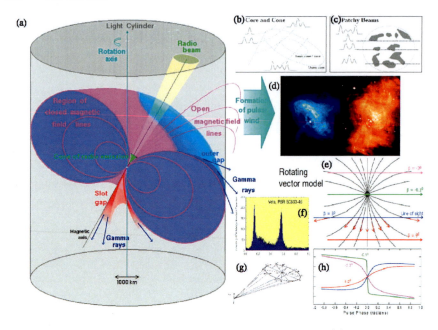

Fig. 2. The magnetosphere and radiation of neutron stars: (a) the particles are accelerated in the inner gap, outer gap or the slot gap, and radio emission beams out near the magnetic poles. The radio beam is not uniformly illuminated but maybe more bright in (b) conal regions or (c) randomly in patches. The outflowing particles can form the pulsar wind nebula. The polarization of the radio emission depends on (e) the magnetic field planes – when the (g) line of sight impacts the beam, the observed polarization angles show variations in the (h) "S-shaped" curves. This demonstration is composed by using the artwork produced by Drs. B. Link, D. Lorimer, A. Harding, G.J. Qiao, as well as (d) the Chandra and HST images of the Crab pulsar nebula, and (f) the γ-ray profile of the Vela pulsar.

may evolve. The smaller index means that the magnetic fields may be growing [41]. In Fig. 2, I provide an over view on pulsars, and models in understanding their action.

2.1 Pulse profiles and emission beam

Pulsars emit broadband radio waves. We receive these signals when the emission beam is sweeping towards us. The line of sight impacts the emission beam so that we see a pulse profile. Now it is clear that **1)** the average pulse profiles are very stable, except for a few pulsars with precession [65, 75]; **2)** profiles are the finger-prints of pulsars which differ from each other. Most pulsar profiles have one or two or three distinctive components or peaks, a small number of pulsars have 4 or 5, and occasionally more than 10 components [48]. Some components are difficult to be identified, but can be revealed by multi-Gaussian fitting [77] or "window-threshold technique" [24]; **3)** these profiles

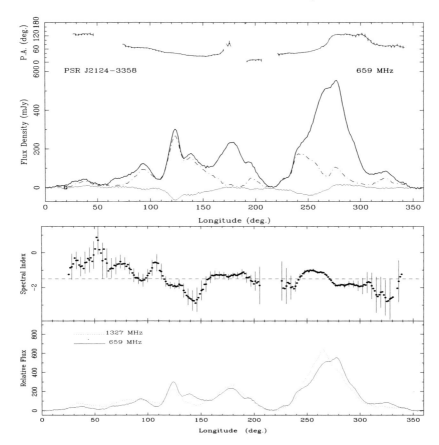

Fig. 3. The polarization profile of PSR J2124-3358 at 659 MHz (top panel) shows more than 12 distinguishable components, each with different fraction of linear (dash line) and circular polarization (lower solid line). Comparison of profiles at 659 MHz and 1327 MHz (lower panel) show that the spectrum varies against rotation phase. This is an extraordinary profile (data from [48]) and usually pulsars are not so complicated – see Fig.4.

vary with frequency, and each component has some-how independent spectral index [48], although in general pulsars have very steep spectrum [52]; **4)** some pulsars have a main pulse and an interpulse, separating about 180° in the rotation phase, which look like the emission from the two opposite magnetic poles; **5)** Some pulsars show profiles in two or three modes, which is so-called "mode-changing". Some pulsars even switch off their emission for sometime, which is called "nulling" [72].

It is naturally understandable that the profile peaks indicate the bright parts of the emission beam. That is to say, the pulsar emission beam is not uniformly illuminated, some parts brighter, some fainter. The line of sight only impacts one slice of beam for a given pulsar, so it is not possible to know the whole beam of any pulsar, except that one can fly in space (!) and observe a pulsar in different lines of sight with respect to its rotational axis! However, based on profiles of many pulsars, the emission beam has been suggested ideally to consist of one core and nested cones [59, 24]. This image gained some good support from the observational fact of the widening of the double profiles at lower frequencies [70]. This is so-called "radius-frequency mapping", which is explained as the emission comes from a pair cuts of the "outer cone" formed in the open dipole magnetic fields. However, if all peak emission comes from cones, how many cones are needed to explain more than 10 components of some pulsars? Can these emission cones be really formed above the magnetic polar caps? This problem can be eased in the so-called "patchy beam" model [44], where the emission comes from many bright parts of a beam. If the slices of emission beams of all pulsars are put together, one can not see distinct cones [31].

In fact, the averaged pulse profiles can only be used to reveal emission geometry or brightness distribution of the emission beam. A large amount of information on the emission process can be obtained from observations of individual pulses [18, 19]. It has been established that for pulsars with drifting subpulses, some emission zones is stably circulating around the magnetic axis of some pulsars with remarkably organized configuration. A detailed modeling hints that the imagined emission cones are patchy! To my understanding, *"the patchy cones"* seems to be the best description of the characteristics of pulsar emission beam: the cones only roughly define the emission region in the magnetosphere and "patches" are related to the non-random but preferred bunches of field lines for generation of emission spots which are probably physically related to sparking near the polar cap. This idea is similar but different from the idea of "window+sources" proposed by Manchester in 1995 [47].

2.2 Polarization

Pulsars are the strongest polarized radio sources in the universe. Almost 100% linear polarization is detected (see Fig. 4) for the whole or a part of profiles of some pulsars [78, 49, 74]. In fact, it is the polarization angle sweeping of the Vela pulsar that leads to the famous "rotating vector model" proposed by Radhakrishnan and Cooke in 1969 [57], which solidly established that the radio emission comes from region not far away from the magnetic poles of the neutron stars, and the observed polarization angle is related (either parallel or perpendicular) to the plane of magnetic field lines, which should have a "S"-shape (see PSR J2048−1616).

Assuming magnetic fields of pulsars are dominated by the dipole fields, then one can determine the emission geometry from the polarization observations [59]. The maximum polarization angle swept rate gives the information on the smallest impact angle of the line-of-sight from the magnetic axis [44]. After classification of pulsar profiles, Rankin has made extensive efforts to

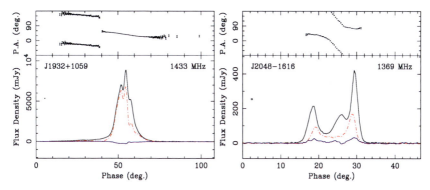

Fig. 4. The polarization profile of PSR J1932+1059(=B1929+10) main pulse and J2048−1616 at 1369 MHz, observed by the author using Parkes telescope. The dashed line is the linear polarization intensity, and the circular polarization is plotted by the lower solid line. The polarization angle curve is plotted on the top of a profile.

pin down the emission regions, by calculating the geometrical parameters for various types of pulsars [59, 60, 61, 62]. The geometrical parameters of a large number of pulsars were also calculated by Gould & Lyne [8]. From the phase shift between the pulse center and the largest sweep rate of polarization angle curve, the emission latitude can be estimated [12, 23].

The polarization angle curves often are not smooth but have some jumps (see PSR J1932+1059 in Fig. 4). When the polarization data samples of individual pulses are plotted against the phase bins, the orthogonal polarization modes can be solidly identified [46, 17, 66, 67]. Several possible origins of the orthogonal polarization modes have been suggested. These orthogonally polarized modes may reflect the eigenmodes of the magneto-active plasma in the open magnetic field lines above the pulsar polar cap, with different refraction indexes [10] to separate these modes in outwards emission, or with different conversion of different modes [56]. It is also possible that the pulsar emission of two modes is generated in one emission region by positrons and electrons respectively [9, 21], i.e. the intrinsic origin from the emission process rather than the propagation process. The non-orthogonal emission modes also have been observed from some pulsars [26], which could be the superposition of emission of two modes [54] or two regions [79].

Very special to the pulsars is the circular polarized emission, unique in the universe. Usually it is as strong as 10%, but in some pulsar components it could reach 70% [48]. The circular polarization measurements have been comprehensively reviewed in [32, 80]. In summary, we found that circular polarization is not restricted to core components and, in some cases, reversals of circular polarization sense are observed across the conal emission. For core components, there is no significant correlation between the sense change of circular polarization and the sense of linear position-angle variation. These results are contradictory to the conclusions given by Radhakrishnan & Rankin [58] based on early smaller sample of pulsar data. We found that in conal double profiles, the sense of circular polarization is correlated with the sense of position-angle variation. Pulsars with a high degree of linear polarization often have one hand of circular polarization across the whole profile. For most pulsars, the sign of circular polarization is same at 50 cm and 20 cm wavelengths, and the degree of polarization is similar, albeit with a wide scatter. Some pulsars are known to have frequency-dependent sign reversals. The diverse behavior of circular polarization may be generated in the emission process [25] or arise as a propagation effect [53].

3 Pulsars as probes for interstellar medium

The dispersion of the radio pulses is a very important tool, which can be used to not only identify radio emission from distant pulsars but also probe the ionized gas in the interstellar medium (ISM) which otherwise is difficult to detect and model. The scattering of radio signals measured from narrow radio pulses can be used to detect the randomness of electron distribution. The polarized radio signals of pulsars are Faraday rotated in the magnetized interstellar medium, so that pulsars act as key probes for the Galactic magnetic fields.

3.1 Electron density distribution in our Galaxy

The most knowledge of interstellar electron distribution comes from pulsar measurements. Using the observed dispersion measures of a sample of pulsars, $DM = \int_0^{\text{Dist}} n_e dl$, one can get the electron density distribution if pulsar distances can be independently measured; or vice versa. The models for electron density distribution of our Galaxy have been constructed and constrained from the DMs and independent distance estimates of a large number of pulsars [69, 27, 14, 15], which consists of the thin electron disk, thick disk and spiral arms as well as bulge component. It is such an electron density model that can be used to estimate distances of most of pulsars from observed DMs so that we can know approximately how far a pulsar is located.

In fact, the interstellar medium is not smoothly distributed, but more or less random or irregularly clumpy, with only some large-scale preferred distribution in spiral arms and concentrated towards the Galactic plane. Note also that pulsars are point radio sources. They move very fast [5]. The interstellar medium randomly refracts and diffracts the pulsar signal and makes the pulsar scintillating and scattering. When pulsars are observed in many frequency channels for some time, one can see that pulsar signals scintillate both in time and observation channel frequency dimension – see excellent examples of so-called dynamic spectrum shown by Gupta et al. [28]. The dynamic spectrum, which itself is the grey-plot of pulse intensity in the 2-D of observation frequency and time, shows the intensity correlation over both the time scale and frequency scale. Note that the randomness of the interstellar medium can be averaged over certain distance, so the distant pulsars do not scintillate much. The scintillation bandwidth is a function of frequency [71]. High sensitivity observations can produce parabolic arcs in the secondary spectrum of the dynamic spectrum [68], which depends on the velocity of scintillation pattern and geometry of the diffracting screen.

The more distant the pulsar is located, more the pulse signal is diffracted and scattered. This leads to the pulse broadening, especially at low frequencies. The most recent and largest data set for pulsar scattering have been obtained by Bhat et al [11], who observed 98 pulsars in several frequencies, from which, together with previous the pulse-broadening, τ, is found to scale with observation frequency ν and DM, in the form of

$$\log \tau(\text{ms}) \simeq -6.46 + 0.154 \log(DM) + 1.07 \log(DM)^2 - 4.4 \log(\nu/\text{GHz}). \quad (3)$$

We considered the scattering effected on the polarized emission, and found that scattering can indeed flatten the PA curves [40]. The simulations and the following observations by Camilo et al. [13] have confirmed such an effect.

3.2 Magnetic field structure of our Galaxy

For a pulsar at distance D (in pc), the rotation measure (RM, in radians m^{-2}) is given by

$$\text{RM} = 0.810 \int_0^D n_e \mathbf{B} \cdot d\mathbf{l}, \quad (4)$$

where n_e is the electron density in cm^{-3}, $\mathbf{B}$ is the vector interstellar magnetic field in μG and $d\mathbf{l}$ is an elemental vector along the line of sight from a pulsar toward us (positive RMs correspond to fields directed toward us) in pc. With the DM we obtain a direct estimate of the field strength weighted by the local free electron density

$$\langle B_{||} \rangle = \frac{\int_0^D n_e \mathbf{B} \cdot d\mathbf{l}}{\int_0^D n_e d\mathbf{l}} = 1.232 \frac{RM}{DM}, \quad (5)$$

where RM and DM are in their usual units of rad m^{-2} and cm^{-3} pc and $B_{||}$ is in μG. Previous analysis of pulsar RM data has often used the model-fitting method [34, 36], i.e., to model magnetic field structures in all of the paths from pulsars to us (observer), and fit the model to the observed RM data with the electron density model [14]. *Significant improvement* can be obtained when both RM and DM data are available for many pulsars in a given region with similar lines of sight. Measuring the gradient of RM with distance or DM is the most powerful method of determining both the direction and magnitude of the large-scale field local in that particular region of the Galaxy [33]. One can get

$$\langle B_{||}\rangle_{d1-d0} = 1.232\frac{\Delta\mathrm{RM}}{\Delta\mathrm{DM}}, \tag{6}$$

where $\langle B_{||}\rangle_{d1-d0}$ is the mean line-of-sight field component in μG for the region between distances $d0$ and $d1$, $\Delta\mathrm{RM} = \mathrm{RM}_{d1} - \mathrm{RM}_{d0}$ and $\Delta\mathrm{DM} = \mathrm{DM}_{d1} - \mathrm{DM}_{d0}$. From all available data, we found that [33] magnetic fields in all inner spiral arms are counterclockwise when viewed from the North Galactic pole. On the other hand, at least in the local region and in the inner Galaxy in the fourth quadrant, there is good evidence that the fields in inter-arm regions are similarly coherent, but clockwise in orientation. There are at least two or three reversals in the inner Galaxy, probably occurring near the boundary of the spiral arms. The magnetic field in the Perseus arm cannot be determined well. The negative RMs for distant pulsars and extragalactic sources in fact suggest the inter-arm fields both between the Sagittarius and Perseus arms and beyond the Perseus arm are predominantly clockwise. See my recent reviews [30, 29] for more details and references therein.

4 Pulsar timing and gravitational waves

Pulsars emit pulses with accurate period (P). After time-of-arrival (TOA) of every pulse is measured (timing observations), one has to correct all measurements to the barycenter of the Solar system. It is easy to find that pulsars are slowing down due to radiation. The rate of period change ($\dot{P}$) can be measured. From P and $\dot{P}$, one can estimate the magnetic field strength on the neutron star surface, $B = 3.2 \times 10^{19}\sqrt{P\dot{P}}$.

The timing data usually have some "noise", not from measurement uncertainty but from the stars them-self. More interesting is that from timing data of young pulsars, there are spectacular changes in the rotation period, known as "glitch". About 100 glitches have been observed from some 30 pulsars [73, 39, 82]. These glitches provide a penetrating means for investigation of pulsar interior structure.

Millisecond pulsars are very stable (small $\dot{P}$) and have very tiny timing-noise. By timing millisecond pulsars, especially pulsars in binary system provide tools to study gravitational theory. The orbital motion of pulsars and relativistic effects can be easily measured through the pulse time delay. The first extra Solar planet was discovered by using timing the millisecond pulsar PSR B1257+12 [76]. The gravitational redshift and time dilation as well as the Shapiro delay have been detected from a number of pulsar binary system [64], and the masses of pulsars as well as their companions can be measured very accurately.

The most exciting is the discovery of a relativistic binary pulsar, PSR B1913+16 [35] which have been used for examination of the general relativity. The orbital shrink, because of the gravitational wave radiation, has been detected from this pulsar – which reveals a new radiation form in the nature and was awarded the Nobel Prize in 1993. The newly discovered binary pulsar [43] can act as the testbed even better [38].

Acknowledgments

I am very grateful to Dr. R.T. Gangadhara for inviting me to participate the "First Kodai-Trieste Workshop on Plasma Astrophysics", and for their great local hospitality. I wish this lecture-notes can be useful for students to choose pulsar studies in future. India has many world famous pulsar astronomers and we wish to make tight connections on pulsar studies in future, given the fact that the radio telescopes, Indian GMRT and Chinese FAST, both make pulsar study as primary projects. My research in China at present is supported by the National Natural Science Foundation (NNSF) of China (10521001 and 10773016) and the National Key Basic Research Science Foundation of China (2007CB815403).

References

1. A. Hewish, S.J. Bell, J.D. Pilkington et al: Nature **217**, 709 (1968)
2. T. Gold: Nature **218**, 731 (1968)
3. F. Pacini: Nature **219**, 145 (1968)
4. V.M. Kaspi: Advances in Space Sciences **21**, 167 (1998)
5. G. Hobbs, D.R. Lorimer, A.G. Lyne, M. Kramer: MNRAS **360**, 974 (2005)
6. C. Wang, D. Lai, J.L. Han: ApJ **639**, 1007 (2006)
7. X.H. Sun, J.L. Han: MNRAS **350**, 232 (2004)
8. D.M. Gould, A.G. Lyne: MNRAS **301**, 235 (1998)
9. M.C. Allen, D.B. Melrose: Proc. Astron. Soc. Australia, **4**, 365 (1982)
10. J.J. Barnard, J. Arons: ApJ **302**, 138
11. N.D.R. Bhat, J.M. Cordes, F. Camilo, et al: ApJ **605**, 759
12. M. Blaskiewicz, J.M. Cordes, I. Wasserman: ApJ **370**, 643 (1991)
13. F. Camilo, J. Reynolds, S. Johnston et al: ApJ, in press=arXiv:0802.0494 (2008)
14. J.M. Cordes, T.J.W. Lazio: astro-ph/**0207156v3** (2002)
15. J.M. Cordes, T.J.W. Lazio: astro-ph/**0301598v1** (2003)
16. J.M. Cordes, P.C.C. Freire, D. Lorimer, et al: ApJ **637**, 446 (2006)
17. J.M. Cordes, J.M. Rankin, D.C. Backer: ApJ **223**, 961 (1978)
18. A.A. Deshpande, J.M. Rankin: ApJ **524**, 1008 (1999)
19. A.A. Deshpande, J.M. Rankin: MNRAS **322**, 438 (2001)
20. P.C. Freire, Y. Gupta, S.M. Ransom, C.H. Ishwara-Chandra: ApJ **606**, 53 (2004)
21. R.T. Gangadhara: A&A **327**, 155 (1997)
22. R.T. Gangadhara: ApJ **609**, 335 (2004)
23. R.T. Gangadhara: ApJ **628**, 923 (2005)
24. R.T. Gangadhara, Y. Gupta: ApJ **555**, 31 (2001)
25. M. Gedalin, P.E. Gruman, D.B. Melrose: MNRAS **325**, 715 (2001)
26. J.A. Gil, A.G. Lyne, J.M. Rankin et al: A&A **255**, 181 (1992)

27. G.C. Gómz, R.A. Benjamin, D.P. Cox: ApJ **122**, 908 (2001)
28. Y. Gupta, B.J. Rickett, A.G. Lyne: MNRAS **269**, 1035 (1994)
29. J.L. Han: IAU Symp. **242**, 55 (2007)
30. J.L. Han: Nuclear Physics B Proc. Supp. **175**, 62 (2008)
31. J.L. Han, R.N. Manchester: MNRAS **320**, L35 (2001)
32. J.L. Han, R.N. Manchester, R.X. Xu, G.J. Qiao: MNRAS **300**, 373 (1998)
33. J.L. Han, R.N. Manchester, A.G. Lyne et al: ApJ **642**, 868 (2006)
34. J.L. Han, G.J. Qiao: A&A, **288**, 759 (1994)
35. R.A. Hulse, J.H. Taylor: ApJ **201**, 55 (1975)
36. C. Indrani, A.A. Deshpande: New Astronomy, **4**, 33 (1999)
37. B.C. Joshi, M.A. Malaughlin, M. Kramer et al: arXiv **0710.2955** (2007)
38. M. Kramer, I.H. Stairs, R.N. Manchester et al: Science **314**, 97 (2006)
39. A. Krawczyk, A.G. Lyne, J.A. Gil et al: MNRAS **340**, 1087 (2003)
40. X.H. Li, J.L. Han: A&A **410**, 253 (2003)
41. J.R. Lin, S.N. Zhang: ApJ **615**, 133 (2004)
42. D. Lorimer, M. Kramer: *Handbook of Pulsar Astronomy*, Cambridge University Press (2005)
43. A.G. Lyne, M. Burgay, M. Kramer et al: Sciences **303**, 1153 (2004)
44. A.G. Lyne, R.N. Manchester: MNRAS **234**, 477 (1988)
45. A.G. Lyne, F. Graham-Smith: *Pulsar Astronomy*, Cambridge University Press (2006)
46. R.N. Manchester: PASA **2**, 334 (1975)
47. R.N. Manchester: JApA **16**, 107 (1995)
48. R.N. Manchester, J.L. Han: ApJ **609**, 354 (2004)
49. R.N. Manchester, J.L. Han, G.J. Qiao: MNRAS **295**, 280 (1998)
50. R.N. Manchester, G.B. Hobbs, A. Teoh, M. Hobbs: AJ **129**, 1993 (2005)
51. R.N. Manchester, A.G. Lyne, F. Camilo, et al: MNRAS **328**, 17 (2001)
52. O. Maron, J. Kijak, M. Kramer, R. Wielebinski: A&AS **147**, 195 (2000)
53. D. Melrose, Q. Lou: MNRAS **352**, 915 (2004)
54. D. Melrose, A. Miller, A. Karastergiou, Q. Lou: MNRAS **365**, 638 (2006)
55. D. Page, S. Reddy: ARNPS **56**, 327 (2006)
56. S.A. Petrova: A&A **378**, 883 (2001)
57. V. Radhakrishnan, D.J. Cooke: ApL **3**, 225 (1969)
58. V. Radhakrishnan, J.M. Rankin: ApJ **352**, 258 (1990)
59. J.M. Rankin: ApJ **274**, 333 (1983)
60. J.M. Rankin: ApJ **352**, 247 (1990)
61. J.M. Rankin: ApJS **85**, 145 (1993)
62. J.M. Rankin: ApJ **405**, 285 (1993)
63. P. Slane, AIP conf.proc. Vol. **968**, 143 (2008)
64. I.H. Stairs, Sciences **304**, 547 (2004)
65. I.H. Stairs, A.G. Lyne, S.L. Shemar: Nature **406**, 484 (2000)
66. D.R. Stinebring, J.M. Cordes, J.M. Rankin et al: ApJS **55**, 247 (1984a)
67. D.R. Stinebring, J.M. Cordes, J.M. Weisberg et al: ApJS **55**, 279 (1984b)
68. D.R. Stinebring, M.A. McLaughlin, J.M. Cordes, et al: ApJ **549**, 97 (2001)
69. J.H. Taylor, J.M. Cordes: ApJ **411**, 674 (1993)
70. S.E. Thorsett: ApL **377**, 263 (1991)
71. N. Wang, R.N. Manchester, S. Johnston et al: MNRAS **358**, 270 (2005)
72. N. Wang, R.N. Manchester, S. Johnston: MNRAS **377**, 1383 (2007)
73. N. Wang, R.N. Manchester, R.T. Pace et al: MNRAS **317**, 843 (2000)

74. J.M. Weisberg, J.M. Cordes, S.C. Lundgren et al: ApJS **121**, 171 (1999)
75. J.M. Weisberg, J.H. Taylor: ApJ **576**, 942 (2002)
76. A. Wolszczan, D.A. Frail: Nature **355**, 145 (1992)
77. X.J. Wu, X. Gao, J.M. Rankin, W. Xu, V.M. Malofeev: AJ **116**, 1984 (1998)
78. X.J. Wu, R.N. Manchester, A.G. Lyne, G.J. Qiao: MNRAS **261**, 630 (1993)
79. R.X. Xu, G.J. Qiao, J.L. Han: A&A **323**, 395 (1997)
80. X.P. You, J.L. Han: ChJAA **6**, 237 (2006)
81. Y.L. Yue, R.X. Xu, W.W. Zhu: Advance in Space Science **40**, 1491 (2007)
82. W.Z. Zou, N. Wang, R.N. Manchester, et al: MNRAS **384**, 1063 (2008)

Pulsar Radio Emission Geometry

R. T. Gangadhara

Indian Institute of Astrophysics, Bangalore 560034, India
E-mail: ganga@iiap.res.in

Summary. Pulsar radio emission is belived to come from relativistic plasma accelerated along the dipolar magnetic field lines in pulsar magnetosphere. The beamed emission by relativistic sources occur in the direction of tangents to the field lines in the corotating frame, but in an inertial (lab) frame it is aberrated toward the direction of rotation. To receive such a beamed emission line-of-sight must align with the source velocity within the beaming angle $1/\gamma$, where γ is the Lorentz factor of the source. By solving the viewing geometry, in an inclined and rotating dipole magnetic field, we find the coordinates of the emission region in corotating frame. Next, give a general expression for the phase shift in the intensity profile in lab frame by taking into account of aberration, retardation and polar cap currents.

By considering uniform and modulated emissions, we have simulated a few typical pulse profiles. The circular polarization of antisymmetric type is an intrinsic property of curvature radiation, and it survives only when there is modulation or discrete distribution in the emitting sources. Our model predicts a correlation between the polarization angle swing and antisymmetric circular polarization.

Key words: pulsars: general – radiation mechanisms: nonthermal — stars: magnetic fields – geometry

1 Introduction

The profile morphology and polarization of pulsars has been widely attempted to interpret in terms of emission in dipolar magnetic field lines (e.g., [1], [2],[3], [3] [4], [5], [6], [7], [8], [9] [10]). Most of the radio emission models assume (1) radiation is emitted by the relativistic secondary pair plasma, (2) beamed radio waves are emitted in the direction of field line tangents, and (3) emitted radiation is polarized in the plane of dipolar field lines or in the perpendicular directions.

We solve the viewing geometry to find the coordinates of emission spot of the radio waves in corotating frame in §2. Next, find the coordinates of emission spot in an inertial (laboratory) frame by finding a general expression for the phase shift in the intensity profile due to aberration, retardation and polar cap currents in §3, and the expression for emission altitude is give in §4. The radio emission from coherent sources and polarization is estimated in §6.

2 Coordinates of Emission Spot of Radio Waves in Corotating Frame

Consider a magnetosphere with dipole magnetic field inclined with respect to rotation axis. We assume that the magnetosphere is stationary or slowly rotating such that the rotation effects are negligible. Next, consider the relativistic plasma moving along the curved dipolar magnetic field lines. Let S be the source (bunch) moving along the curved trajectory (C) experiences acceleration ($\boldsymbol{a}$), and emits a relativistically beamed radiation in the direction of velocity ($\boldsymbol{v}$), see Figure 1. To receive the beamed emission, observer's line of

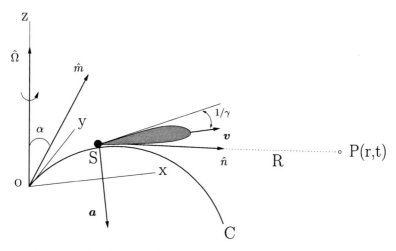

Fig. 1. Geometry for the calculation of radiation field at P. The unit vectors $\hat{m}$ and $\hat{\Omega}$ represent the magnetic and the rotation axes, respectively.

sight ($\hat{n}$) must align with the source velocity within the beaming angle $1/\gamma$, where γ is the Lorentz factor of the bunch. In other words a distant observer at P receives beamed emission only when $\hat{n} \cdot \hat{v} = \cos\tau \sim 1$ for $\tau \approx 1/\gamma$, where $\hat{v} = \boldsymbol{v}/|\boldsymbol{v}|$.

The position vector of a bunch moving along the dipolar magnetic field line is given by (see Eq. 2 in [12]):

$$\boldsymbol{r} = r_e \sin^2\theta \{\cos\theta\cos\phi'\sin\alpha + \sin\theta(\cos\alpha\cos\phi\cos\phi' - \sin\phi\sin\phi'),$$
$$\cos\phi'\sin\theta\sin\phi + \sin\phi'(\cos\theta\sin\alpha + \cos\alpha\cos\phi\sin\theta),$$

$$\cos\alpha\cos\theta - \cos\phi\sin\alpha\sin\theta\} \,, \tag{1}$$

where r_e is the field line constant, and the angles θ and ϕ are the magnetic colatitude and azimuth Q (see Fig. 2), respectively. Next, ϕ' is the rotation phase and α is the magnetic axis

$$\hat{m} = \{\sin\alpha\cos\phi', \ \sin\alpha\sin\phi', \ \cos\alpha\} \tag{2}$$

inclination angle relative to the rotation axis ($\hat{\Omega}$). The velocity of bunch is

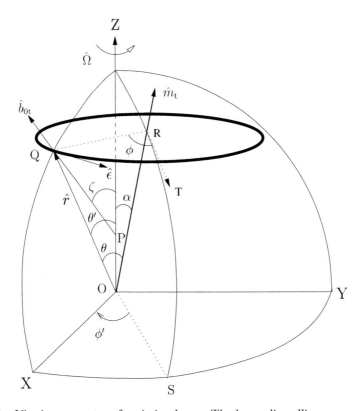

Fig. 2. Viewing geometry of emission beam. The heavy line ellipse represents the cone of emission centered on the magnetic axis $\hat{m}_t$. The arcs ZQX, ZRS, ZY and XSY represent the great circles centered at O (star center). The magnetic colatitude ϕ and the phase angle ϕ' of the emission spot are measured from the meridional (Ω, $\hat{m}_t$) plane. They have the signs in such a way that ϕ' is positive while ϕ is negative on the trailing side, and the vice versa on leading side.

given by

$$v = \frac{d\mathbf{r}}{dt} = \left(\frac{\partial \mathbf{r}}{\partial \theta}\right)\left(\frac{\partial \theta}{\partial t}\right) = \left(\frac{\partial \theta}{\partial t}\right)\mathbf{b} \,, \tag{3}$$

where $\boldsymbol{b} = \partial \boldsymbol{r}/\partial \theta$ is the magnetic field line tangent. Let s be the arc length of field line then $ds = |\boldsymbol{b}|d\theta$, and $|\boldsymbol{b}| = (r_e/\sqrt{2})\sin\theta\sqrt{5 + 3\cos(2\theta)}$. Therefore, the magnitude of velocity $v = ds/dt = \kappa c$. The parameter κ is assumed to be constant, which specifies the speed of bunch as a fraction of the speed of light c. Hence, we have

$$\boldsymbol{v} = \kappa c \hat{b} , \tag{4}$$

where

$$\begin{aligned}
\hat{b} = \boldsymbol{b}/|\boldsymbol{b}| = \{&\cos\tau\cos\phi'\sin\alpha + \sin\tau(\cos\alpha\cos\phi\cos\phi' - \sin\phi\sin\phi'), \\
&\cos\phi'\sin\tau\sin\phi + \sin\phi'(\cos\tau\sin\alpha + \cos\alpha\cos\phi\sin\tau), \\
&\cos\alpha\cos\tau - \cos\phi\sin\alpha\sin\tau\} .
\end{aligned} \tag{5}$$

Here τ is the angle between $\hat{m}$ and $\hat{b}$. In terms of polar angle θ, it is given by

$$\tan\tau = \frac{\sin\tau}{\cos\tau} = \frac{3\sin(2\theta)}{1 + 3\cos(2\theta)} , \tag{6}$$

where

$$\cos\tau = \hat{b}\cdot\hat{m} = \frac{1 + 3\cos(2\theta)}{\sqrt{10 + 6\cos(2\theta)}},$$

$$\sin\tau = (\hat{m}\times\hat{b})\cdot\hat{e}_\phi = \frac{3\sin(2\theta)}{\sqrt{10 + 6\cos(2\theta)}} ,$$

and

$$\begin{aligned}
\hat{e}_\phi = \{&-\cos\alpha\sin\phi\cos\phi' - \cos\phi\sin\phi', \quad \cos\phi\cos\phi' - \cos\alpha\sin\phi\sin\phi', \\
&\sin\alpha\sin\phi\}
\end{aligned} \tag{7}$$

is the binormal to the field line. We can solve equation (6) for θ and obtain

$$\cos(2\theta) = \frac{1}{3}\left(\cos\tau\sqrt{8 + \cos^2\tau} - \sin^2\tau\right) . \tag{8}$$

Next, the acceleration of bunch is given by

$$\boldsymbol{a} = \frac{\partial\boldsymbol{v}}{\partial t} = \frac{(\kappa c)^2}{|\boldsymbol{b}|}\frac{\partial\hat{b}}{\partial\theta} = (\kappa c)^2\boldsymbol{k}, \tag{9}$$

where $\boldsymbol{k} = (1/|\boldsymbol{b}|)\partial\hat{b}/\partial\theta$ is the curvature (normal) of the field line. Then the radius of curvature of field line is given by

$$\rho = \frac{1}{|\boldsymbol{k}|} = \left(2 - \frac{8}{3(3 + \cos(2\theta))}\right)|\boldsymbol{b}| . \tag{10}$$

Therefore, using $\boldsymbol{k} = \hat{k}/\rho$, we can write

$$a = \frac{(\kappa c)^2}{\rho}\hat{k},\qquad(11)$$

where

$$\hat{k} = \{(\cos\alpha\cos\phi\cos\phi' - \sin\phi\sin\phi')\cos\tau - \cos\phi'\sin\alpha\sin\tau,$$
$$(\cos\phi'\sin\phi + \cos\alpha\cos\phi\sin\phi')\cos\tau - \sin\alpha\sin\phi'\sin\tau,$$
$$- \cos\phi\sin\alpha\cos\tau - \cos\alpha\sin\tau\}\,.\qquad(12)$$

Let $\hat{n} = (\sin\zeta, 0, \cos\zeta)$ be the line of sight. where $\zeta = \alpha + \beta$ and β is the sight line impact angle relative to $\hat{m}$. The half–opening angle Γ of the emission beam is given by

$$\cos\Gamma = \hat{n}\cdot\hat{m} = \cos\alpha\cos\zeta + \sin\alpha\,\sin\zeta\,\cos\phi'\,.\qquad(13)$$

2.1 Magnetic Colatitude of Emission Spot

In the absence of rotation, the radiation is beamed in the direction of field line tangent. So, at any instant, observable radiation comes from a spot in the magnetosphere where the tangent vector $\hat{b}$ points in the direction of $\hat{n}$. That is $\Gamma \approx \tau$ at the emission spot. Therefore, using equation (8), we can find the relation for *magnetic colatitude* θ as a function of Γ :

$$\cos(2\,\theta) = \frac{1}{3}(\cos\Gamma\,\sqrt{8 + \cos^2\Gamma} - \sin^2\Gamma)\,,\qquad 0\le\Gamma\le\pi\,.\qquad(14)$$

For $\Gamma \ll 1$, equation (14) reduces to the well known approximate form $\theta \approx \frac{2}{3}\Gamma$.

2.2 Magnetic Azimuth of Emission Spot

Next, let μ be the angle between $\hat{n}$ and $\hat{b}$, then we have

$$\cos\mu = \hat{n}\cdot\hat{b}$$
$$= \cos^2\Gamma + (\cos\alpha\,\sin\zeta\,\cos\phi' - \sin\alpha\,\cos\zeta)\sin\Gamma\cos\phi -$$
$$\sin\zeta\,\sin\phi'\sin\Gamma\sin\phi.\qquad(15)$$

Again to receive the radiation $\mu \sim 0$, therefore, we find the *magnetic azimuth* ϕ of the emission spot by solving $\hat{n} \times \hat{b} \approx 0$:

$$\sin\phi = -\sin\zeta\,\sin\phi'\csc\Gamma,$$
$$\cos\phi = (\cos\alpha\,\sin\zeta\,\cos\phi' - \cos\zeta\,\sin\alpha)\csc\Gamma.$$

Therefore, we have

$$\phi = \arctan\left(\frac{\sin\zeta\,\sin\phi'}{\cos\zeta\,\sin\alpha - \cos\alpha\,\sin\zeta\,\cos\phi'}\right).\qquad(16)$$

Note that ϕ is measured with respect to the meridional plane defined by $\hat{\Omega}$ and $\hat{m}$.

2.3 Altitude of Emission Spot

Pulsar radio emission is generally believed to be coherent curvature radiation by secondary pair plasma streaming along the dipolar magnetic field lines. By having known the θ, we can find the *emission altitude r* by assuming the curvature radiation and some typical values for the Lorentz factor γ of the sources. The characteristic frequency of curvature radiation, at which the emission peaks, is given by [3]:

$$\nu = \frac{3c}{4\pi} \frac{\gamma^3}{\rho}, \tag{17}$$

Since particles closely follow the curved dipolar field lines, the curvature of particle trajectory can be approximated with the curvature of field lines.

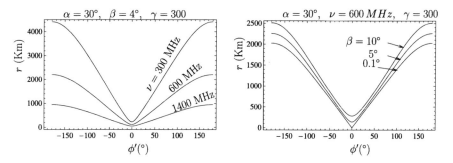

Fig. 3. Emission altitude r is plotted as a function of rotation phase ϕ' for different ν in left panel and different β in right panel.

We can solve equation (10) for the emission altitude

$$r = \frac{12(3 + \cos(2\theta))\sin\theta}{\sqrt{(10 + 6\cos(2\theta))^3}} \rho. \tag{18}$$

Next, substituting for ρ from equation (17), we obtain

$$r = \frac{9c\gamma^3}{\pi\nu} \frac{(3 + \cos(2\theta))\sin\theta}{\sqrt{(10 + 6\cos(2\theta))^3}}. \tag{19}$$

Emission altitude r is plotted as a function of rotation phase ϕ' for different ν in Fig. 3 using $\alpha = 30°$ and $\gamma = 300$. It shows the beviour of emission altitude for different ν and β. It is clear that low frequency emission comes from higher altitude than the higher. At lower β observer tends to receive radiation from lower altitude than at larger β.

3 Coordinates of Emission Spot in the Laboratory Frame

In the laboratory frame, the radiation emitted by the relativistic plasma experiences retardation and aberration phase shift due to the altitude differences and rotation of pulsar. The rotation velocity of the plasma particle (bunch) at Q is given by

$$\mathbf{v}_{\rm rot} = \mathbf{\Omega} \times \mathbf{r} = \Omega r \sin\theta'\,\hat{\epsilon}\,, \tag{20}$$

where $\mathbf{\Omega}$ is the pulsar angular velocity, and the unit vector $\hat{\epsilon}$ represents the direction of rotation. Consider a Cartesian coordinate system - XYZ (see Fig. 2), with Z-axis parallel to the rotation axis and X-axis lies in the fiducial plane defined by $\hat{n}$ and $\hat{\Omega}$. Let Θ be the angle between the field line tangent $\hat{b}_{0t}$ and $\hat{\epsilon}$, then we have

$$\hat{\epsilon} = \cos\Theta\,\hat{\epsilon}_\parallel + \sin\Theta\,\hat{\epsilon}_\perp\,, \tag{21}$$

where the unit vectors $\hat{\epsilon}_\parallel$ and $\hat{\epsilon}_\perp$ are parallel and perpendicular to $\hat{b}_{0t}$. Therefore, the angle Θ is given by

$$\cos\Theta = \hat{\epsilon}.\hat{b}_{0t} = \frac{a_1}{\sqrt{(1 + 4\,a_2^2)a_3}}\,, \tag{22}$$

where

$$a_1 = \sin\alpha\,\sin\phi, \quad a_2 = \cot\theta,$$

$$a_3 = \frac{\cos^2\alpha\cos^2\phi + a_2^2\sin^2\alpha + a_2\cos\phi\,\sin(2\,\alpha) + \sin^2\phi}{1 + a_2^2}.$$

We find $\Theta = 90°$ for the field lines which lie in the meridional plane ($\phi' = 0$) but for other field lines it is $< 90°$ on leading side and $> 90°$ in the trailing side. But for an aligned rotator ($\alpha = 0°$), it is $90°$ for all the field lines.

3.1 Phase shift of radiation emitted by a particle (bunch)

Since the radiation by a relativistic particle is beamed in the direction of velocity, to receive it sight line must align with the particle velocity within the angle $1/\gamma$, where γ is the Lorentz factor. The particle velocity is given by

$$\mathbf{v} = \kappa c\,\hat{b}_{0t} + \mathbf{v}_{\rm rot}, \tag{23}$$

where c is the speed of light. By substituting for $\mathbf{v}_{\rm rot}$ from equation (20) into equation (23) we obtain

$$\mathbf{v} = (\kappa c + \Omega r \sin\theta'\cos\Theta)\,\hat{b}_{0t} + \Omega r \sin\theta'\sin\Theta\,\hat{\epsilon}_\perp\,. \tag{24}$$

By assuming $|\mathbf{v}| \sim c$, from equation (24) we obtain the parameter

$$\kappa = \sqrt{1 - \left(\frac{\Omega r}{c}\right)^2 \sin^2\theta'\sin^2\Theta} - \frac{\Omega r}{c}\sin\theta'\cos\Theta\,. \tag{25}$$

We observe that $\kappa \sim 1$ for $r/r_{\mathrm{LC}} \ll 1$, but at large r it decreases from unity due to increase in rotation velocity, where r_{LC} is the light cylinder radius. Machabeli & Rogava ([11]), by considering the motion of a bead inside a rotating linear tube, have deduced a similar behavior in the velocity components of bead.

Using equation (20) we can solve equation (21) for $\hat{\epsilon}_\perp$, and obtain

$$\hat{\epsilon}_\perp = \frac{\hat{\Omega} \times \hat{r}}{\sin \theta' \sin \Theta} - \cot \Theta \, \hat{b}_{0t} \,. \tag{26}$$

Let ψ be the angle between the rotation axis and $\mathbf{v}$, then we have

$$\cos \psi = \hat{\Omega}.\hat{v} = \cos \zeta \left(\sqrt{1 - \left(\frac{\Omega r}{c}\right)^2 \sin^2 \theta' \sin^2 \Theta} - \frac{\Omega r}{c} \sin \theta' \cos \Theta \right)$$

$$= \kappa \, \cos \zeta \,, \tag{27}$$

where $\hat{v} = \mathbf{v}/|\mathbf{v}|$. For $r/r_{\mathrm{LC}} \ll 1$, it reduces to $\psi \sim \zeta$.

3.2 Aberration Angle

If η is the aberration angle, then we have

$$\cos \eta = \hat{b}_{0t} \cdot \hat{v} = \frac{\kappa c + \Omega r \sin \theta' \cos \Theta}{|\mathbf{v}|} \,, \tag{28}$$

$$\sin \eta = \hat{\epsilon}_\perp \cdot \hat{v} = \frac{\Omega \, r}{|\mathbf{v}|} \sin \theta' \sin \Theta \,. \tag{29}$$

Therefore, from equations (28) and (29), we obtain

$$\tan \eta = \frac{\Omega r}{c} \frac{\sin \theta' \sin \Theta}{\sqrt{1 - (\Omega r/c)^2 \sin^2 \theta' \sin^2 \Theta}}. \tag{30}$$

Hence the radiation beam, which is centered on the direction of $\mathbf{v}$, gets tilted (aberrated) with respect to $\hat{b}_{0t}$ due to rotation.

For $\Omega r/c \ll 1$, it can be approximated as

$$\tan \eta \approx \frac{\Omega r}{c} \sin \theta' \sin \Theta \,. \tag{31}$$

3.3 Aberration Phase Shift

Consider Figure 4 in which ZAD, ZBX, ZCY and DXY are the great circles centered on the neutron star. The small circle ABC is parallel to the equatorial great circle DXY. The unit vector $\hat{b}_{0t}$ represents a field line tangent, which makes the angle ζ with respect to the rotation axis ZO. The velocity unit vector $\hat{v}$ is inclined by the angles η and ψ with respect to $\hat{b}_{0t}$ and ZO, respectively. We resolve the vectors $\hat{b}_{0t}$ and $\hat{v}$ into the components parallel and perpendicular to the rotation axis:

$$\hat{b}_{0t} = \sin\zeta\, \hat{b}_{0t\perp} + \cos\zeta\, \hat{\Omega}\, , \tag{32}$$

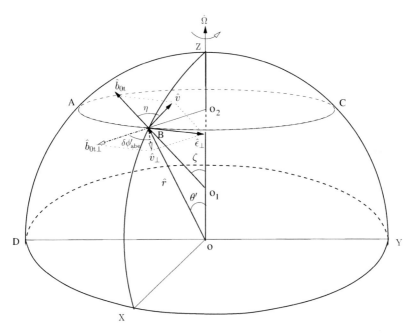

Fig. 4. Celestial sphere describing the aberration phase shift of pulsar radio emission, where η is the aberration angle, and $\delta\phi'_{\text{abe}}$ is the corresponding phase shift.

$$\hat{v} = \sin\psi\, \hat{v}_\perp + \cos\psi\, \hat{\Omega}\, , \tag{33}$$

where the unit vectors $\hat{b}_{0t\perp}$ and $\hat{v}_\perp$ lie in the plane of small circle ABC. Next, by solving for $\hat{b}_{0t\perp}$ and $\hat{v}_\perp$, we obtain

$$\hat{b}_{0t\perp} = \frac{1}{\sin\zeta}(\hat{b}_{0t} - \cos\zeta\, \hat{\Omega})\, , \tag{34}$$

$$\hat{v}_\perp = \frac{1}{\sin\psi}(\hat{v} - \cos\psi\, \hat{\Omega})\, . \tag{35}$$

By taking scalar product with $\hat{b}_{0t\perp}$ on both sides of equation (35), we obtain

$$\cos(\delta\phi'_{\text{abe}}) = \hat{v}_\perp \cdot \hat{b}_{0t\perp} = \frac{1}{\sin\psi}(\hat{v}\cdot\hat{b}_{0t\perp} - \cos\psi\, \hat{\Omega}\cdot\hat{b}_{0t\perp})\, . \tag{36}$$

122   R. T. Gangadhara

Since $\hat{\Omega}$ and $\hat{b}_{0t\perp}$ are orthogonal, we have

$$\cos(\delta\phi'_{\mathrm{abe}}) = \frac{1}{\sin\psi}(\hat{v}\cdot\hat{b}_{0t\perp}). \tag{37}$$

Using $\hat{b}_{0t\perp}$ from equation (34) we obtain

$$\cos(\delta\phi'_{\mathrm{abe}}) = \frac{1}{\sin\psi}\frac{(\hat{v}\cdot\hat{b}_{0t} - \cos\zeta\,\hat{v}\cdot\hat{\Omega})}{\sin\zeta}. \tag{38}$$

By substituting for $\hat{v}\cdot\hat{b}_{0t}$ and $\hat{v}\cdot\hat{\Omega}$ from equations (28) and (27), we obtain

$$\cos(\delta\phi'_{\mathrm{abe}}) = \frac{1}{\sin\psi}\frac{(\cos\eta - \cos\zeta\,\cos\psi)}{\sin\zeta}. \tag{39}$$

Substituting for η again from equation (28), we obtain

$$\cos(\delta\phi'_{\mathrm{abe}}) = \frac{1}{\sin\zeta\sin\psi}\left(\kappa + \frac{\Omega r}{c}\sin\theta'\cos\Theta - \cos\zeta\cos\psi\right). \tag{40}$$

By substituting for κ from equation (27), we obtain

$$\cos(\delta\phi'_{\mathrm{abe}}) = \frac{1}{\sin\zeta\sin\psi}\left(\frac{\cos\psi}{\cos\zeta} + \frac{\Omega r}{c}\sin\theta'\cos\Theta - \cos\zeta\cos\psi\right). \tag{41}$$

It can be further reduced to

$$\cos(\delta\phi'_{\mathrm{abe}}) = \tan\zeta\cot\psi + \frac{\Omega r}{c}\frac{\sin\theta'\cos\Theta}{\sin\zeta\sin\psi}. \tag{42}$$

The aberration phase shift $\delta\phi'_{\mathrm{abe}}$ is plotted as a function of β in Figure 5 for different α in the two cases: (1) $\phi' = 0°$ and $r/r_{\mathrm{LC}} = 0.1$, and (2) $\phi' = 60°$ and $r/r_{\mathrm{LC}} = 0.1$. The Figure 5a shows for $\alpha \sim 90°$, $\delta\phi'_{\mathrm{abe}} \approx r/r_{\mathrm{LC}}$, which is nearly independent of β. But, for other values of α, it does depend on β : larger for $\beta < 0$ and smaller for $\beta > 0$. Since θ' decreases with $|\phi'|$, and $\Theta < 90°$ on leading side and $> 90°$ on trailing side, $\delta\phi'_{\mathrm{abe}}$ is smaller in Figure 5b compared to its corresponding values in Figure 5a. Further, we note that $\delta\phi'_{\mathrm{abe}}$ has negative gradient with respect to $|\phi'|$ for both the signs of β. In Figure 6, $\delta\phi'_{\mathrm{abe}}$ is plotted as a function of ϕ' for different α and fixed $\beta = 4°$. It is highest at large α and small $|\phi'|$. For $r/r_{\mathrm{LC}} \ll 1$, we can series expand $\delta\phi'_{\mathrm{abe}}$ and obtain

$$\delta\phi'_{\mathrm{abe}} = b_1\,r_{\mathrm{n}} + b_2\,r_{\mathrm{n}}^2 + O(r_{\mathrm{n}})^3\,, \tag{43}$$

where $r_{\mathrm{n}} = r/r_{\mathrm{LC}}$,

$$b_1 = \csc^2\zeta\sin\theta'\sqrt{-\cos(\Theta - \zeta)\cos(\Theta + \zeta)},$$
$$b_2 = -\cot^2\zeta\cos\Theta\,\sin\theta'\,b_1\,.$$

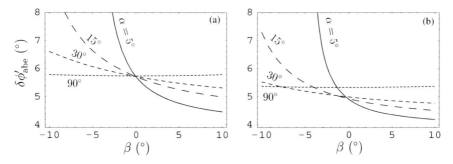

Fig. 5. Aberration phase shift $\delta\phi'_{abe}$ vs the sight line impact angle β for different α: panel (a) for $r_n = 0.1$ and $\phi' = 0°$, and (b) for $r_n = 0.1$ and $\phi' = 60°$.

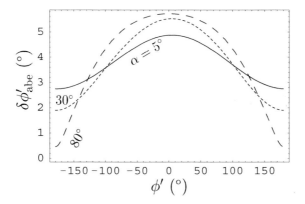

Fig. 6. Aberration phase shift vs the phase ϕ'. Chosen $r_n = 0.1$, $\beta = 4°$ and different α as marked on the figure.

3.4 Retardation phase shift

Let θ_e, ϕ_e and $\hat{r}_e$ be the magnetic colatitude, azimuth and the position vector of the emission spot at the emission time, respectively. In the expressions for θ and ϕ (see eqs. 9 and 11 in G04) we replace ϕ' by $\phi' + \delta\phi'_{abe}$ to obtain θ_e and ϕ_e. In Figure 7 using $r_n = 0.1$, $\alpha = 30°$ and $\beta = 4°$ we have plotted θ_e and ϕ_e as functions of ϕ'. It shows both the minimum of θ_e and zero crossing point of ϕ_e are at the phase $\phi' = -5.5°$. Using the values of θ_e and ϕ_e we find the unit vector $\hat{r}_e$ (see eq. 2 in G04). For brevity, we shall drop the suffix on $\hat{r}$. Note that $\phi' > 0$ on the trailing side (Fig. 1). Consider the emission radii r_1 and r_2 such that $r_1 < r_2$. The time taken by the signal emitted at the radius $\mathbf{r}_1$ is given by

$$t_1 = \frac{1}{c}(d - \mathbf{r}_1 \cdot \hat{n}), \quad (44)$$

where d is the distance to the pulsar, and $\hat{n} = (\sin\zeta, 0, \cos\zeta)$ is the unit vector pointing toward the observer. For another radius $\mathbf{r}_2$, the propagation

time is given by
$$t_2 = \frac{1}{c}(d - \mathbf{r}_2 \cdot \hat{n}), \tag{45}$$

The radiation emitted at the lower radius $\mathbf{r}_1$ takes more time to reach the

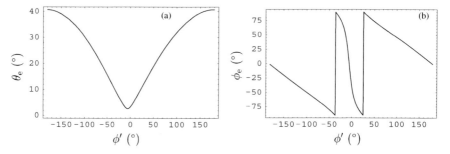

Fig. 7. Magnetic colatitude θ_e and azimuth ϕ'_e of emission spot vs the phase ϕ'. Chosen $r_n = 0.1$, $\alpha = 30°$ and $\beta = 4°$ for both panels.

observer than the one emitted at $\mathbf{r}_2$. The time delay between the two signals is given by
$$\delta t = t_1 - t_2 = \frac{1}{c}(\mathbf{r}_2 \cdot \hat{n} - \mathbf{r}_1 \cdot \hat{n}). \tag{46}$$
By considering the neutron star center ($\mathbf{r}_1 = 0$) as the reference and $\mathbf{r}_2 = \mathbf{r}$, we obtain
$$\delta t = \frac{r}{c}(\hat{r} \cdot \hat{n}). \tag{47}$$
Let σ be the angle between $\hat{r}$ and $\hat{n}$, then we have
$$\cos\sigma = \hat{r} \cdot \hat{n} = \cos\zeta(\cos\alpha\cos\theta_e - \cos\phi_e\sin\alpha\sin\theta_e) - \sin\zeta[\sin\phi'\sin\phi_e\sin\theta_e - \cos\phi'(\cos\theta_e\sin\alpha + \cos\alpha\cos\phi_e\sin\theta_e)]. \tag{48}$$

The time delay δt of components emitted at lower heights, shifts them to later phases of the profile by (e.g., [13])
$$\delta\phi'_{\text{ret}} = \Omega\delta t = \frac{\Omega r}{c}\cos\sigma . \tag{49}$$
In Figure 8, $\delta\phi'_{\text{ret}}$ is plotted as a function of ϕ' for different α and fixed $\beta = 4°$. At small α, it is nearly constant and larger. But at larger α it falls with respect to $|\phi'|$, as $\hat{r}$ inclines more from $\hat{n}$. For $r_n \ll 1$, we can find the series expansion:
$$\delta\phi'_{\text{ret}} = c_1 r_n + c_2 r_n^2 + O(r_n)^3 , \tag{50}$$
where
$$c_1 = \cos(\Gamma - \theta) ,$$
$$c_2 = \frac{\sin\alpha\sin\zeta\sin\phi'\sin(\Gamma - \theta)(4\cot\theta + \tan\theta)}{3\sqrt{8 + \cos^2\Gamma}} b_1 ,$$
and Γ is the half-opening angle of the emission beam (see eq. 7 in G04).

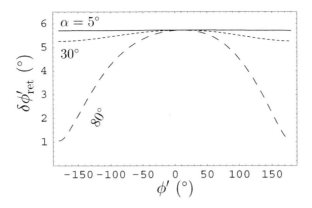

Fig. 8. Retardation phase shift vs the phase ϕ'. Chosen $r_n = 0.1$, $\beta = 4°$ and different α.

3.5 Relativistic Phase Shift

Since the retardation and aberration phase shifts are additive, they can collectively introduce an asymmetry into the pulse profile (e.g., GG01). Therefore, the relativistic phase shift is given by

$$\delta\phi'_{\rm rps} = \delta\phi'_{\rm ret} + \delta\phi'_{\rm abe}$$
$$= r_n \cos\sigma + \arccos\left[\tan\zeta \cot\psi + \frac{\Omega r}{c} \frac{\sin\theta' \cos\Theta}{\sin\zeta \sin\psi}\right]. \quad (51)$$

In Figure 9, we have plotted $\delta\phi'_{\rm rps}$ as a function of ϕ' for different α in the two cases of $\beta = \pm 4°$. It shows $\delta\phi'_{\rm rps}$ reaches maximum at $\phi' \sim 0$, and falls at large $|\phi'|$.

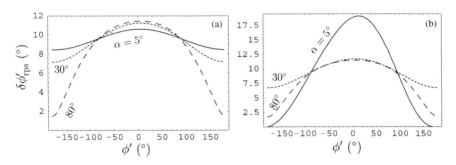

Fig. 9. Relativistic phase shift vs the phase ϕ' for different α at $r_n = 0.1$. Chosen $\beta = 4°$ for panel (a) and $-4°$ for (b).

In the limit of $r_n \ll 1$, we can series expand $\delta\phi'_{\rm rps}$ and obtain

126   R. T. Gangadhara

$$\delta\phi'_{\rm rps} = \mu_1\, r_{\rm n} + \mu_2\, r_{\rm n}^2 + O(r_{\rm n})^3\,, \tag{52}$$

where $\mu_1 = b_1 + c_1$ and $\mu_2 = b_2 + c_2$.

3.6 Phase Shift Due to Polar Cap Current

Goldreich and Julian [14] have elucidated that the charged particles relativistically stream out along the magnetic field lines of neutron star with aligned magnetic moment and rotation axis. Hibschman and Arons [10] have shown that the field-aligned currents can produce perturbation magnetic field $\mathbf{B}_1$ over the unperturbed dipole field $\mathbf{B}_0$, and it can cause a shift in the polarization angle sweep. Here we intend to estimate the phase shift in the intensity profile due to the perturbation field $\mathbf{B}_1$. We assume that the observed radiation is emitted in the direction of tangent to the field $\mathbf{B} = \mathbf{B}_0 + \mathbf{B}_1$. Using the equations D5 and D6 given by Hibschman and Arons ([10]), we find the Cartesian components of perturbation field:

$$\mathbf{B}_1 = \left[2\frac{\mu}{r_{\rm LC}}\frac{\cos\alpha\sin\theta\sin\phi}{r^2},\; -2\frac{\mu}{r_{\rm LC}}\frac{\cos\alpha\sin\theta\cos\phi}{r^2},\; 0\right], \tag{53}$$

where μ is the magnetic moment. The Cartesian components of unperturbed dipole is given by

$$\mathbf{B}_0 = \left[\frac{3}{2}\frac{\mu}{r^3}\sin(2\theta)\cos\phi,\; \frac{3}{2}\frac{\mu}{r^3}\sin(2\theta)\sin\phi,\; \frac{\mu}{r^3}(3\cos^2\theta - 1)\right]. \tag{54}$$

The magnetic field, which is tilted and rotated, is given by

$$\mathbf{B}_{\rm t} = \varLambda\cdot\mathbf{B}, \tag{55}$$

where $\varLambda$ is the transformation matrix given in G04.

The component of $\mathbf{B}_{\rm t}$ perpendicular to the rotation axis is given by

$$\mathbf{B}_{\rm t\perp} = \mathbf{B}_{\rm t} - (\mathbf{B}_{\rm t}.\hat{\varOmega})\hat{\varOmega}. \tag{56}$$

Similarly, we find the perpendicular components of $\mathbf{B}_{0\rm t} = \mathbf{\Lambda}\cdot\mathbf{B}_0$:

$$\mathbf{B}_{0\rm t\perp} = \mathbf{B}_{0\rm t} - (\mathbf{B}_{0\rm t}.\hat{\varOmega})\hat{\varOmega}. \tag{57}$$

If $\delta\phi'_{\rm pc}$ is the phase shift in $\mathbf{B}_{\rm t}$ due to the polar cap current then we have

$$\cos(\delta\phi'_{\rm pc}) = \hat{b}_{t\perp}\cdot\hat{b}_{0t\perp} = \frac{B_{\rm t\perp,x}}{|\mathbf{B}_{\rm t\perp}|}, \tag{58}$$

where $\hat{b}_{t\perp} = \mathbf{B}_{\rm t\perp}/|\mathbf{B}_{\rm t\perp}|$, the unit vector $\hat{b}_{0t\perp} = \mathbf{B}_{0\rm t\perp}/|\mathbf{B}_{0\rm t\perp}|$ is parallel to the unit vector $\hat{x}$ along the X-axis, and $B_{\rm t\perp,x}$ is the X-component of $\mathbf{B}_{\rm t\perp}$. If $\hat{y}$ is the unit vector along Y-axis, then we have

$$\sin(\delta\phi'_{\text{pc}}) = \hat{b}_{\text{t}\perp} \cdot \hat{y} = \frac{B_{\text{t}\perp,y}}{|\mathbf{B}_{\text{t}\perp}|}. \qquad (59)$$

Therefore, we have

$$\tan(\delta\phi'_{\text{pc}}) = \frac{B_{\text{t}\perp,y}}{B_{\text{t}\perp,x}} = \frac{d_1\,r_{\text{n}}}{d_2 + d_3\,r_{\text{n}}}, \qquad (60)$$

where

$$d_1 = \cos\phi'\sin(2\alpha) - 2\cos^2\alpha\tan\zeta,$$
$$d_2 = 3\cos\theta\tan\zeta,$$
$$d_3 = -\sin(2\alpha)\sin\phi'.$$

We have plotted $\delta\phi'_{\text{pc}}$ as a function of ϕ' in Figure 10 by choosing $\alpha = 10°$, $r_{\text{n}} = 0.01$ and 0.1. It decreases with the increasing $|\phi'|$ and it is mostly negative, except in the case of $\beta < 0$, where it is positive over a small range of ϕ' near $(\hat{\Omega}, \hat{m}_{\text{t}})$ plane. So, $\delta\phi'_{\text{pc}}$ tries to reduce the relativistic phase shift $\delta\phi'_{\text{rps}}$, except over a small range of ϕ' where it enhances the shift in the case of $\beta < 0$. In

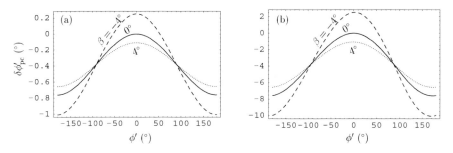

Fig. 10. The phase shift $\delta\phi'_{\text{pc}}$ due to polar cap current vs phase ϕ'. Chosen $\alpha = 10°$ and $r_{\text{n}} = 0.01$ for panel (a), and $r_{\text{n}} = 0.1$ for (b).

the limit of $r_{\text{n}} \ll 1$, we can series expand equation (60) and obtain

$$\delta\phi'_{\text{pc}} = \frac{d_1}{d_2}r_{\text{n}} - \frac{d_1 d_3}{d_2^2}r_{\text{n}}^2 + O(r_{\text{n}})^3. \qquad (61)$$

4 Emission Radius from Phase Shift

We can find the net phase shift due to aberration, retardation and polar cap current by adding equations (51) and (60):

$$\delta\phi' = \delta\phi'_{\text{rps}} + \delta\phi'_{\text{pc}}$$
$$= r_{\text{n}}\cos\sigma + \arccos\left[\tan\zeta\cot\psi + \frac{\Omega r}{c}\frac{\sin\theta'\cos\Theta}{\sin\zeta\sin\psi}\right] +$$
$$\arctan\left[\frac{d_1\,r_{\text{n}}}{d_2 + d_3\,r_{\text{n}}}\right], \qquad (62)$$

In Figure 11 we have plotted $\delta\phi'$ as a function of ϕ' in the four cases of r_n (0.01, 0.1, 0.2 and 0.3). It shows $\delta\phi'$ reaches maximum near the $(\hat{\Omega}, \hat{m}_t)$ plane and falls with respect to $|\phi'|$. Note that the magnitude of gradient of $\delta\phi'$ with respect to $|\phi'|$ is higher than that of $\delta\phi'_\text{rps}$. In the case of $\beta < 0$, we find $\delta\phi'$ becomes negative at large $|\phi'|$ as the magnitude of $\delta\phi'_\text{pc}$ exceeds the magnitude of $\delta\phi'_\text{rps}$. At higher r_n, we note that $\delta\phi'$ is slightly asymmetric about $\phi' = 0$. For $r_n \ll 1$, we obtain

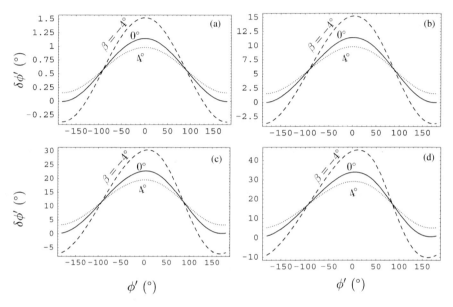

Fig. 11. The net phase shift due to aberration, retardation and polar cap current $\delta\phi'$ vs phase ϕ'. Chosen $\alpha = 10°$ and $r_n = 0.01, 0.1, 0.2$ and 0.3 for panels (a), (b), (c) and (d), respectively.

$$\delta\phi' = \nu_1 r_n + \nu_2 r_n^2 + O(r_n)^3, \tag{63}$$

where $\nu_1 = \mu_1 + (d_1/d_2)$ and $\nu_2 = \mu_2 - (d_1 d_3/d_2^2)$.

For $\delta\phi' \ll 1$, we can solve equation (63) for the emission radius and obtain

$$r = \frac{r_\text{LC}}{\nu_1} \delta\phi' - \frac{\nu_2 r_\text{LC}}{\nu_1^3} \delta\phi'^2 + O(\delta\phi')^3. \tag{64}$$

5 Electric field of curvature radiation

The relativistic bunch having charge q emits curvature radiation as it accelerates along the curved trajectory C (space curve) see Fig. 1 Then the electric field of the radiation at the observation point P is given by [15]:

$$\boldsymbol{E}(\boldsymbol{r},t) = \frac{q}{c}\left[\frac{\hat{n}\times[(\hat{n}-\boldsymbol{\beta})\times\dot{\boldsymbol{\beta}}]}{R\,\xi^3}\right]_{\text{ret}}, \tag{65}$$

where $\xi = 1 - \boldsymbol{\beta}\cdot\hat{n}$, R is the distance from radiating region to observer, $\hat{n}$ the observer's sight line, $\boldsymbol{\beta} = \boldsymbol{v}/c$ the velocity and $\dot{\boldsymbol{\beta}} = \boldsymbol{a}/c$ the acceleration of the bunch.

The radiation emitted by a relativistic bunch has a broad spectrum, and the range of frequency spectrum can be estimated by taking Fourier transformation of the electric field of radiation:

$$\boldsymbol{E}(\omega) = \frac{1}{\sqrt{2\pi}}\int\limits_{-\infty}^{+\infty}\boldsymbol{E}(t)\mathrm{e}^{i\,\omega t}dt. \tag{66}$$

In equation (65), ret means evaluated at the retarded time $t' + R(t')/c = t$. By changing the variable of integration from t to t', we obtain

$$\boldsymbol{E}(\omega) = \frac{1}{\sqrt{2\pi}}\frac{q}{c}\int\limits_{-\infty}^{+\infty}\frac{\hat{n}\times[(\hat{n}-\boldsymbol{\beta})\times\dot{\boldsymbol{\beta}}]}{R\,\xi^2}\mathrm{e}^{i\omega\{t'+R(t')/c\}}dt', \tag{67}$$

where we have used $dt = \xi\,dt'$. When the observation point is far away from the region of space where the acceleration occurs, the propagation vector or the sight line $\hat{n}$ can be taken to be constant in time. Furthermore, the distance $R(t')$ can be approximated as $R(t') \approx R_0 - \hat{n}\cdot\boldsymbol{r}(t')$, where R_0 is the distance between the origin O and the observation point P, and $\boldsymbol{r}(t')$ is the position of bunch relative to O.

Since bunches move with velocity κc along the dipolar field lines, over the incremental time dt, the distance (arclength) covered is $ds = \kappa c\,dt = |\boldsymbol{b}|d\theta$. Therefore, we have

$$t = \frac{1}{\kappa c}\int |\boldsymbol{b}|d\theta = \frac{r_{\mathrm{e}}}{\sqrt{2}\kappa c}\int \sin\theta\sqrt{5 + 3\cos(2\theta)}\,d\theta\ . \tag{68}$$

By choosing $t = 0$ at $\theta = 0$, we obtain

$$t = \frac{r_{\mathrm{e}}}{12\kappa c}\left[12 + \sqrt{3}\log\left(14 + 8\sqrt{3}\right) - 3\sqrt{10 + 6\cos(2\theta)}\cos(\theta) - \right.$$
$$\left. 2\sqrt{3}\log\left(\sqrt{6}\cos(\theta) + \sqrt{5 + 3\cos(2\theta)}\right)\right]. \tag{69}$$

Then equation (67) becomes

$$\boldsymbol{E}(\omega) \approx \frac{q\,\mathrm{e}^{i\omega R_0/c}}{\sqrt{2\pi}R_0\kappa c^2}\int\limits_{-\infty}^{+\infty}|\boldsymbol{b}|\frac{\hat{n}\times[(\hat{n}-\boldsymbol{\beta})\times\dot{\boldsymbol{\beta}}]}{\xi^2}\mathrm{e}^{i\omega\{t-\hat{n}.\boldsymbol{r}/c\}}d\theta, \tag{70}$$

130   R. T. Gangadhara

where the primes on the time variable t have been omitted for brevity. The integration limits have been extended to $\pm\infty$ for mathematical convenience, as the integrand vanishes for $|\theta - \theta_0| > 1/\gamma$. At any rotation phase ϕ', there exists a magnetic colatitude θ_0 and azimuth ϕ_0 at which the field line tangent $\hat{b}$ exactly align with $\hat{n}$, i.e., $\hat{b}_0 \cdot \hat{n} = 1$ and $\tau = \Gamma$, where Γ is the half-opening angle of the pulsar emission beam centered on $\hat{m}$. The expressions for θ_0 and ϕ_0 are given by Gangadhara ([12]; see Eqs. 9 and 11).

The polarization state of the emitted radiation can be determined using $\boldsymbol{E}(\omega)$ with the known $\boldsymbol{r}(t)$, $\boldsymbol{\beta}$ and $\dot{\boldsymbol{\beta}}$. Since the integral in equation (70) has to be computed over the path of particle, the line of sight $\hat{n}$ can be chosen, without loss of generality, to lie in the xz–plane:

$$\hat{n} = (\sin \zeta, \ 0, \ \cos \zeta) , \tag{71}$$

where $\zeta = \alpha + \sigma$ is the angle between $\hat{n}$ and $\hat{\Omega}$, and σ is the closest impact angle of $\hat{n}$ with respect to $\hat{m}$.

Let

$$\boldsymbol{A} = \{A_{\mathrm{x}}, A_{\mathrm{y}}, A_{\mathrm{z}}\} = \frac{1}{\kappa c} |\boldsymbol{b}| \frac{\hat{n} \times [(\hat{n} - \boldsymbol{\beta}) \times \dot{\boldsymbol{\beta}}]}{\xi^2} . \tag{72}$$

By substituting for $|\boldsymbol{b}|$, $\boldsymbol{\beta}$ and $\dot{\boldsymbol{\beta}}$, and series expanding in power of θ about θ_0 we obtain

$$A_{\mathrm{x}} = A_{\mathrm{x}0} + A_{\mathrm{x}1}\theta + A_{\mathrm{x}2}\theta^2 + A_{\mathrm{x}3}\theta^3 + O[(\theta - \theta_0)^4] ,$$
$$A_{\mathrm{y}} = A_{\mathrm{y}0} + A_{\mathrm{y}1}\theta + A_{\mathrm{y}2}\theta^2 + A_{\mathrm{y}3}\theta^3 + O[(\theta - \theta_0)^4] ,$$
$$A_{\mathrm{z}} = A_{\mathrm{z}0} + A_{\mathrm{z}1}\theta + A_{\mathrm{z}2}\theta^2 + A_{\mathrm{z}3}\theta^3 + O[(\theta - \theta_0)^4] . \tag{73}$$

where $A_{\mathrm{x}i}$, $A_{\mathrm{y}i}$ and $A_{\mathrm{z}i}$ with i = 0, 1, 2, 3 are the series expansion coefficients. Since their expressions are huge we have not included them to save space, however, one can reproduce them easily.

The scalar product between $\hat{n}$ and $\boldsymbol{r}$ is given by

$$\hat{n} \cdot \boldsymbol{r} = r_e \sin^2 \theta \Big[\cos \alpha \, (\cos \theta \cos \zeta + \cos \phi \cos \phi' \sin \theta \sin \zeta) -$$

$$\cos \zeta \cos \phi \sin \alpha \sin \theta + \sin \zeta \, (\cos \theta \cos \phi' \sin \alpha - \sin \theta \sin \phi \sin \phi') \Big] . \tag{74}$$

Next, substituting for t and $\hat{n} \cdot \boldsymbol{r}$ into the argument of exponential in equation (70), and series expanding it in powers of θ about θ_0 we obtain

$$\omega \left(t - \frac{\hat{n} \cdot \boldsymbol{r}}{c} \right) = c_0 + c_1\theta + c_2\theta^2 + c_3\theta^3 + O[(\theta - \theta_0)^4] , \tag{75}$$

where c_0, c_1, c_2, c_3 and c_4 are the series expansion coefficients. Again to save space we do not reproduce their expressions here, but one can reproduce them easily.

Now, by substituting the series expansions from equations (73) and (75) into equation (70), we obtain the components of $\boldsymbol{E}(\omega) = \{E_x(\omega), E_y(\omega), E_z(\omega)\}$:

$$E_x(\omega) = E_0 \int_{-\infty}^{+\infty} (A_{x0} + A_{x1}\theta + A_{x2}\theta^2 + A_{x3}\theta^3)e^{i(c_1\theta + c_2\theta^2 + c_3\theta^3)}d\theta \ ,$$

$$E_y(\omega) = E_0 \int_{-\infty}^{+\infty} (A_{y0} + A_{y1}\theta + A_{y2}\theta^2 + A_{y3}\theta^3)e^{i(c_1\theta + c_2\theta^2 + c_3\theta^3)}d\theta \ ,$$

$$E_z(\omega) = E_0 \int_{-\infty}^{+\infty} (A_{z0} + A_{z1}\theta + A_{z2}\theta^2 + A_{z3}\theta^3)e^{i(c_1\theta + c_2\theta^2 + c_3\theta^3)}d\theta \ , \quad (76)$$

where

$$E_0 = \frac{q}{\sqrt{2\pi R_0 c}} e^{i[(\omega R_0/c) + c_0]}.$$

Now by solving the integrals solutions, we obtain

$$E_x(\omega) = E_0(A_{x0}S_0 + A_{x1}S_1 + A_{x2}S_2 + A_{x3}S_3) \ ,$$
$$E_y(\omega) = E_0(A_{y0}S_0 + A_{y1}S_1 + A_{y2}S_2 + A_{y3}S_3) \ ,$$
$$E_z(\omega) = E_0(A_{z0}S_0 + A_{z1}S_1 + A_{z2}S_2 + A_{z3}S_3) \ . \quad (77)$$

To find the polarization angle of radiation field $\boldsymbol{E}$, we need to specify two reference directions perpendicular to the sight line $\hat{n}$. One could be the projected spin axis on the plane of the sky: $\hat{\epsilon}_\parallel = (-\cos\zeta, 0, \sin\zeta)$, then the other direction is specified by $\hat{\epsilon}_\perp = \hat{\epsilon}_\parallel \times \hat{n} = \hat{y}$, where $\hat{y}$ is a unit vector parallel to the y-axis. Then the components of $\boldsymbol{E}$ in the directions $\hat{\epsilon}_\parallel$ and $\hat{\epsilon}_\perp$ are given by

$$E_\parallel = \hat{\epsilon}_\parallel \cdot \boldsymbol{E} = -\cos\zeta \, E_x + \sin\zeta \, E_z \ ,$$
$$E_\perp = \hat{\epsilon}_\perp \cdot \boldsymbol{E} = E_y \ . \quad (78)$$

At any rotation phase ϕ', observer receives the radiation from all those field lines whose tangents lie within the angle $1/\gamma$ with respect to the sight line $\hat{n}$. Let η be the angle between the $\hat{b}$ and $\hat{n}$, then $\cos\eta = \hat{b}\cdot\hat{n}$, and the maximum value of η is $1/\gamma$. Therefore, at $\phi = \phi_0$ we solve $\cos(1/\gamma) = \hat{b}\cdot\hat{n}$ for τ, and find the allowed range $(\Gamma - 1/\gamma) \le \tau \le (\Gamma + 1/\gamma)$ of τ or $-1/\gamma \le \eta \le 1/\gamma$ of η, which in turn allows to find the range of θ with the help of equation (14). Next, for any given η within it's range, we find ϕ by solving $\cos\eta = \hat{b}\cdot\hat{n}$. It gives $(\phi_0 - \delta\phi) \le \phi \le (\phi_0 + \delta\phi)$, where

$$\cos(\delta\phi) = \frac{\sin\Gamma[\cos(1/\gamma)\csc(\Gamma + \eta) - \cos\Gamma\cot(\Gamma + \eta)]}{(\cos\zeta\sin\alpha - \cos\alpha\cos\phi'\sin^2\zeta)^2 + \sin^2\zeta\sin^2\phi'} \ .$$

Hence by knowing the ranges of θ and ϕ at any given ϕ', we can estimate the contributions to $\boldsymbol{E}$ from the emissions from all those field lines, whose tangents lie within the angle $1/\gamma$ with respect to $\hat{n}$. In Figure 12, we have plotted those regions at three phases; $\phi' = -30°$, $0°$ and $30°$ using $\alpha = 10°$, $\beta = 5°$ and

$\gamma = 300$. At the center of each region $\hat{b}$ exactly aligns with the sight line $\hat{n} = 1$. Further, in Figure 13, we have plotted them for $-180° \leq \phi' \leq 180°$ with step of 5° between the successive regions. We observe that the range of θ stays nearly constant (or decreasing negligibly) whereas that of ϕ gets narrower with respect to the increasing $|\phi'|$.

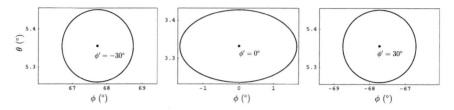

Fig. 12. Beaming regions specifying the range of magnetic colatitude θ and azimuth ϕ at the three selected phases $\phi' = -30°$, $0°$ and $30°$. The center of each region gives the values of ϕ_0 and θ_0. Chosen $\alpha = 10°$, $\beta = 5°$ and $\gamma = 400$.

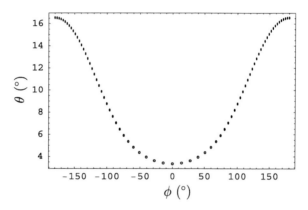

Fig. 13. The beaming regions specifying the range of magnetic colatitude θ and azimuth ϕ. They are plotted for the full range of phase: $-180° \leq \phi' \leq 180°$ with step of 5°. The center of each region gives the values of ϕ_0 and θ_0. Chosen $\alpha = 10°$, $\beta = 5°$ and $\gamma = 400$.

6 Polarization of radiation field

To understand the pulsar emission, we must model for all the Stokes parameters (I, Q, U & V —a set of parameters used to specify the phase and polarization of radiation) of the pulsar emission, and compare with the observations. They have been found to offer a very convenient method for establishing the association between the polarization state of observed radiation

Pulsar Radio Emission Geometry 133

and the geometry of emitting region. They are defined as follows:

$$I = E_\parallel E_\parallel^* + E_\perp E_\perp^* , \quad Q = E_\parallel E_\parallel^* - E_\perp E_\perp^* , \quad U = 2\,\mathrm{Re}[E_\parallel^* E_\perp] ,$$
$$V = 2\,\mathrm{Im}[E_\parallel^* E_\perp] . \tag{79}$$

The parameter I defines the total intensity, Q and U jointly define the linear polarization and it's position angle, and V describes the circular polarization.

It is well known that the components of a pulsar profile can be decomposed into individual Gaussians by fitting one to each of the sub-pulse component. For example the components in the pulse profile of PSR 1706-16 and PSR 2351+61 are fitted with appropriate Gaussians [16]. When the line-of-sight crosses the emission region, it encounters a pattern in intensity due to Gaussian modulation in the azimuthal direction. Because of the Gaussian modulation in the azimuthal direction, the intensity becomes nonuniform in the polar directions too. These arguments indicate that a Gaussian like intensity modulation exists in the polar directions too. So, we assume that the emission region of a pulse component has an intensity modulation both in the azimuthal and polar directions. Hence we define a modulation function f for a pulse component as

$$f(\theta, \phi) = f_0 \exp\left[-\left(\frac{\theta - \theta_\mathrm{p}}{\sigma_\theta} \right)^2 - \left(\frac{\phi - \phi_\mathrm{p}}{\sigma_\phi} \right)^2 \right] , \tag{80}$$

where θ_p and ϕ_p are the peak locations of Gaussian function and f_0 is the amplitude. If w_θ and w_ϕ are the full width at half maxima (FWHM) in the θ and ϕ directions, then $\sigma_\theta = w_\theta/(2\sqrt{\ln 2})$ and $\sigma_\phi = w_\phi/(2\sqrt{\ln 2})$.

Using a Gaussian with peak located at the meridional plane ($\phi' = 0°$), we computed the pulse profiles in the two cases of impact parameter ($\sigma = \pm 5°$), and plotted in Fig. 14. We observe the profile in the case of negative σ is broader than the positive case. This difference is due to the projection of emission region on to the equatorial plane of the pulsar. In the positive σ case, the χ_s swing is ccw and the V_s change is $-/+$ while in the case of negative σ the χ_s swing is cw and V_s change is $+/-$. Hence we find that *the polarization angle swing is correlated with the circular polarization sign reversal*. This correlation is invariant with respect the stellar spin directions.

7 Discussion

Observed pulsar radio luminosities, together with the small source size, imply extraordinarily high brightness temperatures, i.e., as high as 10^{31} K. To avoid implausibly high particle densities and energies, coherent radiation processes are invoked. Pacini & Rees [17], and Sturrock ([2]) among others were quick to point out that the observed coherence may be due to bunching of particles in the emission region of magnetosphere. The pair-production discharge

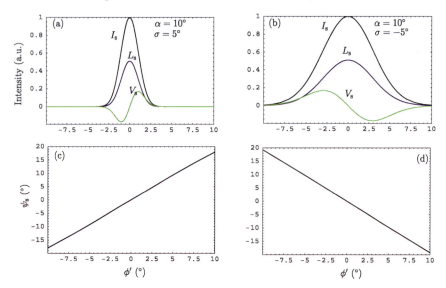

Fig. 14. Simulated pulse profiles. Chosen $P = 1$ s and $\gamma = 400$. For panels (a) and (c) used $\sigma_\theta = \sigma_\phi = 0.1$, $\theta_p = 3.33°$, $\phi_p = 0°$ and $f_0 = 1$ for the modulating Gaussian. Similarly, for panels (b) and (d) also used the same values, except $\phi_p = 180°$.

mechanism originally applied to pulsars by Sturrock automatically produces dense bunches that can produce coherence at radio frequencies with sufficient intensity to simulate pulsar action.

If the bunches of plasma particles with sizes much smaller than a wavelength of radiation exist then the net radiation field $\boldsymbol{E}(\omega) \approx N\boldsymbol{E}_o(\omega)$, where N is the number of charges present in the bunch. Hence the total radiation field due to a bunch of particles is equal to the vector sum of the fields radiated by each charge.

In an actual case, it is the combination of both incoherent and coherent superpositions determine the polarization state of the observed emission. Though the emissions from a single bunch is highly polarized, the radiation received by distant observer will be less polarized, as the radiation from many bunches is incoherently superposed. The degree of polarization is found depend on the time resolution chosen in the observations [18].

Circular polarization generally strongest in the central regions of a profile, but is by no means confined to these regions. It has been detected from conal components of many pulsars, for example, conal-double pulsars, and found to be highly correlated with the polarization angle swing ([20]). In most of the cases the sense reversal of circular polarization is nearly independent of frequency, suggesting that the circular polarization does not arise from propagation or plasma effects ([19], [21]). Radhakrishnan and Rankin [21] have argued that the circular polarization is intrinsically antisymmetric type and correlated with the polarization angle swing. Since radiation from many bunches is superposed at given pulse longitude, the circular polarization of different signs and magnitudes is added. The result of such an addition could be the reason for the diversities in the observed circular polarization.

8 Conclusion

We have derived the polarization state of the coherent curvature radiation due to relativistic plasma in pulsar magnetosphere. Though the emission from a single bunch is highly polarized, the net emission from many bunches within the beaming region is less polarized due to the incoherent superposition of radiation fields. If the emission within the beaming region is uniform then the net circular polarization goes to zero. Based on our model, we predict that curvature radiation has basically antisymmetric type of circular polarization, and it is correlated with the polarization angle swing.

References

1. Radhakrishnan, V., Cooke, D. J., 1969, Astrophys. Lett., 3, 225
2. Sturrock, P. A., 1971, ApJ, 164, 229
3. Ruderman, M. A., & Sutherland, P. G., 1975, ApJ, 196, 51
4. Lyne, A. G., Manchester, R. N., 1988, MNRAS, 234, 477
5. Blaskiewicz, M., Cordes, J. M., & Wasserman, I., 1991, ApJ, 370, 643
6. Rankin, J. M., 1983a, ApJ, 274, 333
7. Rankin, J. M., 1983b, ApJ, 274, 359
8. Rankin, J. M., 1990, ApJ, 352, 247
9. Rankin, J. M., 1993, ApJSupp., 85, 145
10. Hibschman, J. A., Arons, J., 2001, ApJ, 546, 382
11. Machabeli, G. Z., Rogava, A. D., 1994, Phy. Rev. A, 50, 98
12. Gangadhara, R. T., 2004, ApJ, 609, 335, (G04)
13. Phillips, J. A., 1992, ApJ, 385, 282
14. Goldreich, P., Julian, W. H., 1969, ApJ, 157, 869
15. Jackson, J. D., 1975, Classical Electrodynamics, (NY: Wiley)
16. Kramer M., Wielebinski, R., Jessner, A., Gil, J. A. & Seiradakis, J. H., 1994, A&AS, 107, 515
17. Pacini, F., Rees, M. J., 1970, Nat., 226, 622
18. Gangadhara, R. T., Xilouris, K. M., von Hoensbroech, A., et al. 1999, A&A, 342, 474
19. Michel, F. C., 1987, ApJ, 322, 822
20. You, X. P., Han, J. L., 2006, Chin. J. Astron. Astrophys., 6, 237
21. Radhakrishnan, V., Rankin, J. M., 1990, ApJ, 352, 258

Millisecond Pulsar Emission Altitude from Relativistic Phase Shift: PSR J0437-4715

R. T. Gangadhara and R. M. C. Thomas

Indian Institute of Astrophysics, Bangalore 560034, India
E-mail: ganga@iiap.res.in; mathew@iiap.res.in

Summary. We have analyzed the profile of the millisecond pulsar PSR J0437-4715 at 1440 MHz by fitting the Gaussians to pulse components, and identified its 11 emission components. We propose that they form a emission beam with 5 nested cones centered on the core. Using the phase location of component peaks, we have estimated the aberration–retardation (A/R) phase shift. Due to A/R phase shift, the centroid of intensity profile and the inflection point of polarization angle swing are symmetrically shifted in the opposite directions with respect to the meridional plane, which is defined by the rotation and magnetic axes. By recognizing this fact, we have been able to locate the phase location of meridional plane and estimate the absolute altitude of emission of core and conal components relative to the neutron star center. Using the expression for phase shift given recently by Gangadhara [1], we find that the radio emission comes from a range of altitude starting from the core at 7% of light cylinder radius to outer most cone at 30%.

Key words: Pulsars: general —pulsars: individual (PSR J0437-4715)

1 Introduction

The shape and polarization of integrated pulse of a pulsar are quite stable, and they are expected to the reflect the magnetic field geometry and the structure of pulsar emission beam. The common occurrence of odd number of components and their distribution across the pulse window has led to the nomenclature of core and cone structure for the pulsar emission beam [2]. However, Lyne & Manchester [3] and Manchester [4] have argued for a different interpretation, in which components on either side of core arise from a random distribution of emitting regions, which lead to the so-called patchy cone. Mitra & Deshpande [6] have attempted to test these models, and found the evidences in favor of conal structure. The frequency evolution of profile structure and polarization angle swing of pulsars has been interpreted by considering the radio emission by relativistic plasma moving in dipolar magnetic field lines (e.g.,[7],[8], [9], [2], [10], [11], [3], [14], [12], [13]).

For understanding the pulsar radio emission mechanism, the radio emission altitude bears an at most importance. There have been two kinds of methods proposed for estimating the radio emission altitudes: (1) *Purely geometric method,* which assumes the pulse edge is emitted from the last open field lines (e.g.,[15], [21], [27]), (2) *Relativistic phase shift method,* which assumes the asymmetry in the conal components phase location, relative to core, is due to the aberration-retardation phase shift, e.g.,[20]. Both the methods have merits and demerits: former method has an ambiguity in identifying the last open field lines while the latter rely on the clear identification of core-cone structure of the emission beam. Thomas & Gangadhara [33] have showed that the profiles becomes asymmetric due to the influence of rotation on the radio emitting sources moving along the dipolar magnetic field lines.

By assuming the pulsar radio emission comes from fixed altitude, Blaskiewicz et al. [14] have showed that polarization angle inflection point (PAIP) lags with respect to the centroid of the intensity profile by $\sim 4r/r_{\rm LC}$, where r is the emission altitude measured from the center of neutron star and $r_{\rm LC} = Pc/2\pi$ is the light cylinder radius. The parameters P and c are the pulsar spin period and speed of light, respectively. The results of purely geometric method are found to be in rough agreement with those given in [14]. However, the relativistic phase shift method clearly indicates that the emission altitude across the pulse window is not constant ([20], [23], [26]). Dyks, Rudak and Harding [16] by revising the relation for aberration phase shift given in [20], have re-estimated the emission heights. The average emission heights are found to be comparable with those of geometric method.

By considering the relativistically beamed radio emission in the direction of magnetic field line tangents, Gangadhara [19] has solved the viewing geometry in an inclined and rotating dipole magnetic field. He has showed that due to geometric restrictions in receiving the beamed emission, a distant observer can not receive the radio waves from a constant altitude. A more exact expression for relativistic phase shift (A/R) is given by Gangadhara [1], which includes the phase shift due to polar cap currents also. Under small angle approximations, the expression of [1] reduces to the one given in [16].

Recently, Johnston & Weisberg [26] have estimated the emission heights of the young pulsar PSR J1015-5719 using the expression for A/R phase shift given in [1]. In this paper, we intend to estimate the emission heights of the nearby, bright millisecond pulsar PSR J0437-4715. It was discovered in the Parkes southern survey ([25], [29]), has a pulse period of 5.75 ms. Navarro et al. [30] have presented the mean pulse observations of PSR J0437-4715 at multiple radio frequencies, and found a frequency dependent lag of the total-intensity profile with respect to the polarization profile. Jenet et al. [24] have presented the results based on the high time resolution single pulse observations of this pulsar. They find the single pulse properties of PSR J0437-4715 are similar to those of the common slow-rotating pulsars.

The outline of this paper is as follows. In §2, we discuss a method for identifying the phase location of meridional plane. Next, in §3 we consider the mean pulse data of millisecond PSR J0437-4715, and fit the Gaussian to pulse components. By assuming conal emission beam, we have estimated the A/R phase shifts and the emission heights.

2 Phase of Meridional Plane

In the nested cone structure of pulsar emission beam, the conal components are expected to be distributed symmetrically on either side of core when observed in the corotating frame. However, for an observer in the laboratory frame the profile becomes asymmetric due to rotation. A typical asymmetric profile is given in Figure 1, and such an asymmetry in the real profiles has been reported for many pulsars (e.g., [20], [23], [26]).

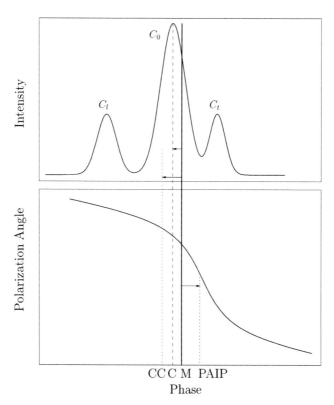

Fig. 1. A typical pulse profile. C_0 is the core. C_l and C_t are the components on leading and trailing sides, respectively. The dotted lines represent the phase of cone center (CC) and polarization angle inflection point (PAIP). The thick line mark the phase of meridional plane (M), and the broken line mark the phase of core peak (C). Arrows indicate direction of phase shift of CC, C and PAIP with respect to M.

140   R. T. Gangadhara and R. M. C. Thomas

Blaskiewicz et al. [14] have generalized the rotating vector model (RVM) by considering the first order relativistic effects, which arise due to the pulsar rotation. By assuming that the emission across the entire pulse comes from a constant height, they have showed that the centroid of the total intensity pulse advances to earlier phase by $\sim r/r_{\rm LC}$ while the polarization angle inflection point (PAIP) is delayed to later phase by $\sim 3\,r/r_{\rm LC}$, where r is the emission height measured from the center of neutron star. However, there are clear evidences to show that the emission do not come from a fixed altitude ([20], [23], [26]). Consider the intensity profile given in Figure 1, in which, C_l and C_t are the components of an arbitrary cone with center at CC. If the emission altitude is not constant across the pulse window then the retardation also contributes to the phase shift, and it is comparable to that of aberration [1]. So, due to A/R effects, the centroid intensity profile is advanced to earlier phase by $\sim 2\,r/r_{\rm LC}$ while the corresponding PAIP is delayed to later phase by $\sim 2\,r/r_{\rm LC}$ (e.g., [16], [18]). Hence, the cone center CC and PAIP lie symmetrically on either side of meridional plane M, which is at the phase $\phi' = 0°$. The meridional plane is defined by the plane containing the rotation axis and the magnetic axis. Note that the phase location of M is invariant with respect to r. In other words, both CC and PAIP come closer to M at smaller r, and move away from it at larger r. In the co-rotating frame, i.e., in the absence of rotation, the A/R effects vanish, and hence core peak and PAIP appear at the meridional plane M.

The field aligned polar cap current do not introduce any significant phase shift into the phase of PAIP but it does introduce a positive offset into polarization angle, however, it roughly cancels due to the negative offset by aberration [12]. The phase shift of pulse components due to polar cap current is estimated recently in [1], and found to be quite small compared to A/R phase shift.

3 Emission height in millisecond pulsar: PSR J0437-4715

Consider the nearest bright millisecond pulsar PSR J0457-4715, which has a period of 5.75 ms. It was discovered in Parkes southern survey by Johnston et al. [25], and is in a binary system with white dwarf companion. Manchester & Johnston [5] have presented the mean pulse polarization properties of PSR J0437-4715 at 1440 MHz. There is significant linear and circular polarization across the pulse with rapid changes near the pulse center. The position angle has a complex swing across the pulse, which is not well fitted with the rotating vector model, probably, due to the presence of orthogonal polarization modes. The mean intensity pulse has a strong peak near the center, where the circular polarization shows a clear sense reversal and the polarization angle has a rapid sweep. These two features strongly indicate that it is a core. The profile shows more than 8 identifiable components.

Table 1. Radio emission height in PSR J0437−4715 at 1440 MHz

Cone No.	ϕ_l'	ϕ_t'	$\delta\phi'$	Γ	r	s/s_{lof}
	(°)	(°)	(°)	(°)	(Km)	
(1)	(2)	(3)	(4)	(5)	(6)	(7)
0	—	—	−9.50±0.26	4.00±0.00	20.3±0.6	0.17±0.00
1	−29.50±0.24	8.50±0.46	−10.50±0.26	7.06±0.06	23.3±0.5	0.28±0.01
2	−50.00±0.50	21.50±0.54	−14.25±0.37	11.50±0.10	34.3±0.9	0.38±0.01
3	−81.50±0.15	37.94±0.08	−21.78±0.08	18.00±0.02	64.3±0.3	0.43±0.00
4	−102.01±0.36	56.14±0.29	−22.90±0.23	22.93±0.05	84.9±0.9	0.47±0.00
5	−129.50±1.01	89.54±4.15	−19.98±2.10	29.30±0.40	91.8±6.2	0.57±0.02

Consider the average pulse profile given in Figure 2 for PSR J0437-4715 at 1440 MHz. To identify pulse components and to estimate their peak locations we followed the procedure of Gaussian fitting to pulse components. Two approaches have been developed and used by different authors. Unlike Kramer et al. [28], who fitted the *sum* of Gaussians to the total pulse profile, we separately fitted a single Gaussian to each of the pulse components. We used the package Statistics'NonlinearFit' in Mathematica for fitting Gaussians to pulse components. It can fit the data to the model with the named variables and parameters, return the model evaluated at the parameter estimates achieving the leastsquares fit. The steps are as follows: (i) Fitted a Gaussian to the core component (VI), (ii) Subtracted the core fitted Gaussian from the data (raw), (iii) The residual data was then fitted for the next strongest peak, i.e., the component IV, (iv) Added the two Gaussians and subtracted from the raw data, (v) Next, fitted a Gaussian to the strongest peak (IX) in the residual data. The procedure was repeated for other peaks till the residual data has no prominent peak above the off pulse noise level. By this procedure we have been able to identify 11 of its emission components: I, II, III, IV, V, VI, VII, VIII, IX, X and XI, as indicated by the 11 Gaussians in Figure 2. The distribution of conal components about the core (VI) reflect the core-cone structure of the emission beam. We propose that they can be paired into 5 nested cones with core at the center.

Manchester & Johnston [5] have fitted the rotating vector model (RVM) of [7]to polarization angle data of PSR J0437-4715 and estimated the polarization parameters $\alpha = 145°$ and $\zeta = 140°$. The parameter α is the magnetic axis inclination angle relative to the pulsar spin axis, and $\zeta = \alpha + \beta$, where β is the sight line impact angle relative to the magnetic axis. However, for these values of α and ζ the colatitude s/s_{lof} (see eq. [15] in [20]) exceeds 1 for all the cones. The parameter s gives the foot location of field lines, which are associated with the conal emissions, relative to magnetic axis on the polar cap. It is normalized with the colatitude s_{lof} of last open field line. However, Gil and Krawczyk [22] have reported $\alpha = 20°$ and $\beta = -4°$ by fitting the average intensity pulse profile rather than by formally fitting the RVM. Further, they find the relativistic RVM by Blaskiewicz et al. [5] calculated with

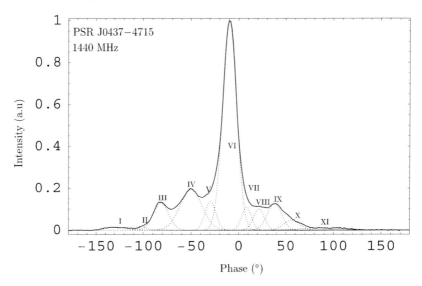

Fig. 2. Average pulse of PSR J0437-4715 at 1440 MHz and the model Gaussians (dotted line curves) fitted to emission components. Due to A/R effects, the core (VI) peak is shifted to the phase $\sim -9.5°$ from the meridional plane (M) which is at $0°$.

these α and β seems to fit the observed position angle quite well than the non-relativistic RVM. Further, Gil and Krawczyk [5] have reported that the centroid of intensity profile and PAIP are separated by $\sim 40°$. By taking the cone center (CC) of outermost cone as the centroid of intensity profile, we can locate the phase of meridional plane M. It is at the mid point, i.e., at $\delta\phi' \sim 20°$ between CC and PAIP. Using $\delta\phi' \approx 2r/r_{\rm LC}$, we get the emission height $r \approx 48$ km. We have adopted these values of α and β in our model as they also confirm $s/s_{\rm lof} \leq 1$ (see below). The phase $\phi' = 0°$, in Figure 2, corresponds to the phase of meridional plane (M), which is at the mid point between CC of outermost cone and PAIP of relativistic polarization angle swing.

In column 1 of Table 1 we have given the cone numbers. The cone number 0 stands for core component. The peak locations of conal components on leading and trailing sides are given in columns 2 and 3, respectively. The conal components are believed to arise from the nested hollow cones of emission ([2]-[31]), which along with the central core emission, make up the pulsar emission beam. Let ϕ'_l and ϕ'_t be the peak locations of conal components on leading and trailing sides of a pulse profile, respectively. Then, using the following equations, we estimate the phase shift ($\delta\phi'$) of cone center with respect to M and the phase location (ϕ') of component peaks in the absence of phase shift, i.e., in the corotating frame (see eq. [11] in [20]):

$$\delta\phi' = \frac{1}{2}(\phi'_t + \phi'_l), \qquad \phi' = \frac{1}{2}(\phi'_t - \phi'_l). \tag{1}$$

In column 4 of Table 1, we have given the values of $\delta\phi'$. It increases in magnitude from inner most cone to 4th cone, but for outermost cone i.e., for 5th cone it decreases slightly. We suspect this decrease in $\delta\phi'$ for the 5th cone is due to the fact that emission spot moves closer to the rotation axis, and hence aberration phase shift becomes smaller ([1]).

The half-opening angle Γ (see eq. [7] in [19]) of the emission beam is given in column 5, and it is about $30°$ for the last cone. In column 6, we have given the emission heights computed using the expression for A/R phase shift given in [1] (see eq. [45]). It shows the emission in PSR J0437-4715 occurs over a range of altitude starting from the core at 20 Km to the outermost cone at 90 Km. In column 7, we have given the colatitude s/s_{lof} of foot field lines, which are associated with the component emissions. It shows due to the relativistic beaming and geometric restrictions, observer tends to receive the emissions from the open field lines which are located in the colatitude range of 0.2 to 0.6 on the polar cap.

4 Discussion

A long standing question in pulsar astronomy has been the location of the radio emission region in magnetosphere. The emission heights of PSR B0329+54 given in [20] and six other pulsars in [23] are all relative to the emission height of core. They assumed that the core emission height is negligible compared to the light cylinder radius of classical pulsars, but it may not be so in the case of millisecond pulsars. Though Blaskiewicz et al. [14] have assumed a constant emission height in their model, they have indeed indicated the possibility of core emission at a lower altitude than conal emissions. We identify, due to A/R phase shift, that the meridional plane (M) is located at the mid point between the phase of core peak and polarization angle inflection point (PAIP). By recognizing this fact, we have been able to estimate the *absolute* emission heights of the core as well as the conal components. We find that the core emission in PSR J0437−4715 is indeed come from a lower altitude than the conal emissions.

We selected PSR J0437-4715 because of its strong core and clear conal emissions. Polarization angle (PA) is complicated near the core component but we can still locate the meridional plane by avoiding fitting that part of PA data. The idea is that without the knowledge of polarization angle data near the core it is possible to locate the meridional plane by just fitting the relativistic polarization angle formula of BCW91 to segments of PA angle data corresponding to the conal components. Since the emission altitude for each cone is different, one can not fit the entire PA data with a constant altitude. If we can best fit a segment of PA data corresponding to a outer cone or any other inner cone it is possible to locate the phase of the meridional plane. Since we used the formula of Blaskiewicz et al. [14], in which the aberration phase shift is independent of phase, our results are approximate.

144   R. T. Gangadhara and R. M. C. Thomas

The polarization angle curve obtained by fitting the segments of PA data under C_l and C_t gives a virtual inflection point (PAIP) corresponding certain height reported by fitting routine. Since the emission height vary for each cone, we shall have as many number of virtual inflection points as the cones. The point to be noted is that the cone center and PA virtual inflection point of each cone is approximately symmetrically located on either side of the meridional plane. Since the cone centers and the virtual PA inflection points are approximately symmetrically located on either side of meridional plane, it is possible to locate the phase of meridional plane even if we are unable to detect virtual PA inflection point for certain cones or core.

The geometrical method, which by comparing the measured pulse widths with geometrical predictions from dipolar models, can yield absolute emission heights. But such absolute values are likely to be underestimated due to an ambiguity in identifying the last open field lines. The profile widths generally measured at a 10%-level can give the lower limits for emission altitudes [34]. In PSR J0437−4715 the components I and XI are at the intensity level of 1 to 2%, and the field lines associated with them are not last open field lines, but they are found to be located at the colatitude $s/s_{\mathrm{lof}} \sim 0.6$ on polar cap.

It is possible that the A/R phase shift can be reduced by the rotational distortion of magnetic field sweep back of the vacuum dipole magnetic field lines. It was first considered in detail by Shitov [32]. Dyks and Harding [16] have investigated the rotational distortion of pulsar magnetic field by making the approximation of a vacuum magnetosphere. For $\phi' = 110°$, $\beta = -4°$ and $\alpha = 20°$ we computed the phase shift due to magnetic field sweep back $\delta\phi'_{\mathrm{mfsb}}$ (see eq. [49] in [1]). But it is found to be $< 0.6°$ for $r/r_{\mathrm{LC}} \leq 0.3$, which is much smaller than the aberration, retardation and polar cap current phase shifts in PSR J0437−4715.

5 Conclusion

We have analyzed the mean profile of the millisecond PSR J0437-4715 at 1440 MHz, and identified 11 of its emission components. We propose that they form a emission beam consist of 5 nested cones centered on the core. Using the phase location of component peaks, we have estimated the aberration–retardation (A/R) phase shift. Due to A/R phase shift, the center of each cone and the inflection point of corresponding polarization angle swing are symmetrically shifted in the opposite directions with respect to the meridional plane. By recognizing this fact, we have been able to locate the phase location of meridional plane and estimate the absolute altitudes of emission of core and conal components relative to neutron star center. We used the more exact expression for phase shift given recently in [1]. The radio emission at 1440 MHz in PSR J0437-4715 is found to come from an altitude range starting from the core at 7% of light cylinder radius to the outer most cone at 30%.

Acknowledgments

We used the data available on EPN archive maintained by MPIfR, Bonn. We are thankful to M. N. Manchester and S. Johnston for making their data available on web.

References

1. Gangadhara, R. T., 2005, ApJ, 628, 930
2. Rankin, J. M., 1983a, ApJ, 274, 333
3. Lyne, A. G., & Manchester, R. N., 1988, MNRAS, 234, 477
4. Manchester, M. N., 1995, A&A, 16, 107
5. Manchester, M. N. & Johnston, S., 1995, ApJ, 441, L6
6. Mitra, D., Deshpande, A. A., 1999 A&A, 346, 906
7. Radhakrishnan, V., & Cooke, D. J., 1969, Astrophys. Lett., 3, 225
8. Sturrock, P. A., 1971, ApJ, 164, 229
9. Ruderman, M. A., & Sutherland, P. G., 1975, ApJ, 196, 51
10. Rankin, J. M., 1983b, ApJ, 274, 359
11. Rankin, J. M., 1990, ApJ, 352, 247
12. Hibschman, J. A., & Arons, J., 2001, ApJ, 546, 382
13. Mitra, D., Li, X. H., 2004, A&A, 421, 215
14. Blaskiewicz, M., Cordes, J. M., & Wasserman, I., 1991, ApJ, 370, 643
15. Cordes, J. M., 1978, ApJ, 222, 1006
16. Dyks, J., Rudak, B., & Harding, A. K., 2004, ApJ, 607, 939
17. Dyks, J., & Harding, A. K., 2004, ApJ, 614, 869
18. Fung, F. K., 2005, Ph.D. thesis, Univ. of Utrecht, 82
19. Gangadhara, R. T. 2004, ApJ, 609, 335
20. Gangadhara, R. T., & Gupta, Y. 2001, ApJ, 555, 31
21. Gil, J. A., & Kijak, J., 1993, A&A, 273, 563
22. Gil, J. A., & Krawczyk, A., 1997, MNRAS, 285, 561
23. Gupta, Y., & Gangadhara, R. T. 2003, ApJ, 584, 418
24. Jenet, F. A., Anderson, S. B., Kaspi, V. M., Prince, T. A., Unwin, S. C., 1998, ApJ, 498, 365
25. Johnston, S., et al., 1993, Nature, 361, 613
26. Johnston, S., Weisberg, J. M., 2006, MNRAS, In press
27. Kijak, J., & Gil, J., 2003, A&A, 397, 969
28. Kramer M., Wielebinski, R., Jessner, A., Gil, J. A. & Seiradakis, J. H. 1994, A&As, 107, 515
29. Manchester, M.N., et al., 1996, MNRAS, 279, 1235
30. Navarro, J., Manchester, R. N., Sandhu, J. S., Kulkarni, S. R., Bailes, M., 1997, ApJ, 486, 1019
31. Rankin, J. M., 1993, ApJS, 85, 145
32. Shitov, Yu. P., 1983, Sov. Astron., 27, 314
33. Thomas, R. M. C., Gangadhara, R. T., 2005, A&A, 437, 537
34. Xilouris, K. M., Kramer, M., Jessner, A., et al., 1996, A&A, 309, 481

Magnetosphere Structure and the Annular Gap Model of Pulsars

G.J. Qiao[1], K.J. Lee[1], H.G. Wang[2], and R.X. Xu[1]

[1] Department of Astronomy, Peking University, Beijing 100871,
gjn@pku.edu.cn; k.j.lee@water.pku.edu.cn; r.x.xu@pku.edu.cn
[2] Center for astrophysics, Guangzhou University, Guangzhou 510006, China,
hgwang@gzhu.edu.cn

Summary. Pulsar radio and γ-ray emission has been studied for about forty years, yet no elaborated model is present to account for all the emission of pulsars from radio to γ-ray bands. A reasonable emission model should present the mechanism for wide band radiation. The magnetosphere structure and particle acceleration regions are the basis for solving problem, observations are important inputs to establish a reasonable model. In our work, both radio and gamma-ray observations have been considered. We will show how they can limit the radiation locations and radiation mechanisms. The advantages and disadvantages for different radiation locations (such as the polar gap, the outer gap and the annular gap) are discussed in comparison with observational facts. The annular gap model, an joint model for both radio and γ-ray emissions, was proposed recently and reproduce successfully some important observational phenomenon, e.g., bi-drifting. We will also discuss for radio and γ-ray observations of PSR B1055-52, and show that radiations come from the annular region and the core region. Success in reproducing many observational facts suggest that the annular gap is a promising model for pulsar multi-wavelength radiation.

Key words: pulsars: general — stars: neutron stars

1 Introduction

Since Hewish et al. [1] discovered the first radio pulsar, lots of magnetosphere theories and acceleration models for pulsars have been proposed, and large number of observation have been performed during the last 40 years. Goldreich & Julian (GJ) [2] have proposed that there is a magnetosphere outside the neutron star surface. The charge density is $\rho_{gj} = -\mathbf{\Omega} \cdot \mathbf{B}/(2\pi c)$ in the magnetosphere; the electric field along the magnetic field would be zero ($\mathbf{E} \cdot \mathbf{B} = 0$). This is just a "dead" or "static" magnetosphere. But the observed emission from pulsars hints that there exist some "accelerators" in pulsar magnetosphere, where electric field parallel to the magnetic field is not zero ($\mathbf{E} \cdot \mathbf{B} \neq 0$). These "accelerators" require the deviation of charge density ρ from the GJ charge density, i.e. $\rho - \rho_{gj} \neq 0$.

How can the charge density deviate from GJ charge density? There are two ways: if there is a charge flow, the velocity difference and the geometrical shape of the flow will make the local $\rho(r)$ different from the local GJ charge density; anther possibility is that charged particles can not flow easily at certain region in the magnetosphere, then a "charge starving" or "charge excess" region will appear.

For radio pulsars, one group of theories belongs to the "polar cap" models, which consider the formation of the parallel electric field in the region near magnetic poles. There are two kinds of models for the inner accelerators contingent upon whether free ejection of the charged particles is possible from the star surface. If electrons or ions can be pulled out from the star surface, the scheme called "space-charge-limited flow" model operates, in which the parallel electric field component is due to either the inertia of particles [3], the field line curvature and geometry [4, 6, 11, 17, 24, 71, 76], or the general relativistic frame-dragging effect [28, 53]). If the ion binding energy in the neutron star surface is large enough that charged particles can not be pulled out freely, a vacuum gap in the polar cap region will form, namely, the inner gap (Ruderman & Sutherland [70], hereafter RS75). The RS model can account for some features observed in radio bands, such as hollow conical emission, sub-pulse drifting, microstructure, etc., and is supported by observations for many years (see [40] for a review, [15, 79]). Furthermore, the "user-friendly" nature of this model is a virtue not shared by others [73]. Although there are still debates on the conical emission, it was agreed that there is a "core" emission beam [18, 19, 23, 41, 44, 52, 65, 66, 67], among others). The RS model, however, contradicts the observed "core" emission beam.

Even if the binding energy of the neuron star surface is not high enough, the inner core gap and annular gap can be formed [61] for situations both of $\mathbf{B} \cdot \mathbf{\Omega} < 0$ and $\mathbf{B} \cdot \mathbf{\Omega} > 0$. In some special conditions an inner vacuum gap (do not include the annular gap) can be formed also when $\mathbf{B} \cdot \mathbf{\Omega} < 0$ [20, 21]. If pulsars are the bare strange stars, the inner gap can be formed [59, 60, 83, 84, 85, 87] for situations both of $\mathbf{B} \cdot \mathbf{\Omega} < 0$ and $\mathbf{B} \cdot \mathbf{\Omega} > 0$.

Xia et al. [82] found that the Inverse Compton scattering (ICS) in strong magnetic fields dominates the production of high energy photons near the polar cap region. Resonant scattering could also be an important energy loss mechanism for the primary electrons, comparing with both electrostatic acceleration and curvature radiation losses. Inverse Compton scattering was found to influence the gap formation, since the pair production avalanche may be triggered by the up-scattered Compton photons rather than by curvature photons (e.g. [39, 64, 90, 91, 92]). These results have been confirmed by many authors (e.g. [14, 28, 29, 42, 75]). In fact, in this case the electrons will stop accelerating before they can emit significant curvature radiation. The standard models of polar cap acceleration thus need substantial revision [28, 91, 92, 93]. When the ICS process in strong magnetosphere is involved, the inner gap

Magnetosphere Structure and the Annular Gap Model of Pulsars   149

structure changes a lot, and the observed mode-changing phenomena can be understood [91, 92].

The ICS mechanism can also act as an important radiation mechanism for pulsar's radio radiation. In the RS model the inner gap breaks down and sparks periodically. The sparks produce low frequency waves. The low energy photons are inverse Compton scattered by the secondary particles produced in the sparking or pair cascades, and the up-scattered radio photons provide the observed radio emission from the pulsar. This is called the ICS model of pulsar radio emission. In the model, the "core" and conical emission beams can be obtained naturally [54, 55, 56, 57]. The observed polarization phenomena and pulse profiles changing with observing frequencies can also be calculated [58, 86]. The pulse files changing with observing frequencies can be checked observationally [94].

Some pulsars emit in both radio and γ-ray bands. Both the polar cap and the outer gap models of γ-ray pulsars have been suggested to explain the high-energy emission from pulsars [7, 10, 12, 13, 24, 28, 30, 38, 53, 68, 69, 95]. Qiao et al. [59, 60, 62, 63] propose a new scenario of the inner annular gap (IAG) beyond the conventional inner gap. The IAG or annular region is defined by magnetic field lines that intersect the null charge surface (NCS). As we know, the local magnetic field is perpendicular to the rotation axis at the NCS. So if the radiation is generated near the NCS (either inside or outside), a pulsar can easily produce a very wide pulse profile,which matches the observed γ-ray pulse profiles. Owing to the emission regions close to the star surface, it can fit the radio observations also at same time.

Besides these polar gap scenario, there are outer gap models. For a fully charge separated magnetosphere, i.e. only one sign of charge present at a given location (see [51]), several groups argue that an outer gap then will form around the the null charge surface [10, 12, 31, 35] instead of forming a polar gap. Whether the outer gap can form depends on whether the negative charged particles can not pass the positive GJ charge density (GJ69) in the annular region, which needs further consideration when the magnetosphere is not fully charge separated [61].

Growing evidence shows that the radio emission of pulsars is linked with the systematic drifting sub-pulses (grazing cuts of the line of sight) or periodic intensity modulations (central cuts of the line of sight) [15, 16, 79]. The drifting sub-pulse phenomena can only be understood based on the drifting sparks in the inner polar vacuum gap (RS75; [22, 60, 87]). Two questions rise: Firstly, in the free flow cases, can the "sparking-like" phenomena take place? Secondly, if the magnetosphere is not fully charge-separated, what will happen? Although, as it was mentioned above, the fully charge separated magnetosphere is considered conventionally, it may be questionable. Furthermore, even if the magnetosphere is strictly fully charge separated, the secondary pair plasma generated from the acceleration region will soon fill some regions in the magnetosphere, resulting in a quasi-neutral plasma. In this case, the so called annular region can play a very important role to produce gamma-ray emission in young pulsars.

This paper is arranged as follows. Magnetosphere structure and acceleration regions of pulsars are discussed in §2. The basic observational facts and comparison between radio & γ-ray emission properties are summarized in §3. In §4, radio observational properties and the inverse Compton scattering (ICS) model are discussed. In §5, several application of annular gap model are given such as γ-ray emission, bi-drifting sub-pulse of PSR J0815+09 and multi-band emission from PSR B1055-52. Discussion and conclusion are presented in §6.

2 Magnetospheric structure and acceleration region of pulsars

As it was pointed out that pulsar radiation needs high energy particles moving along the field lines. So the fundamental process is that the electric field parallel to the magnetic field should be non-zero in the acceleration regions, i.e. $\mathbf{E} \cdot \mathbf{B} \neq 0$. This "acceleration process" requires that local charge density ρ is not equal to the GJ charge density ρ_{gj} in certain region, where $\rho - \rho_{gj} \neq 0$.

There are two ways, as discussed above, to achieve the deviation. If the binding energy is low enough, the particles can flow out from the surface of the neutron star "freely", the flow process can make the $\rho(r)$ in some places different from the GJ charge density due to velocity contrast, field line curvature or general relativistic effect; another way is that there are some places where some kind of particles can not flow freely, then a "charge starving" or "charge excess" region will appear. We will discuss these in the following subsections.

2.1 Low binding energy of the neutron star surface: free flow situation

The models considering the free flow situation are (i) slot gap model (e.g. [5]), (ii) the polar cap model (e.g. [53]), and (iii) the annular gap model [61].

The slot gap model

The main assumptions for the electrodynamics in the co-rotating frame are: (1) The surface between the closed field line region and the open field lines region have equal potential $\Phi = 0$. (2) $\Phi = 0$ on the star surface. (3) Only one sign of charge is present at any point in the magnetosphere, that is magnetosphere fully charge separated. (4) $\rho_b(R) = \rho_{gj}(R)$, here $\rho_b(R)$ is the beam charge from the neutron star (NS) surface, and R is the radius of the NS.

In the co-rotating frame, Poisson's equation becomes

$$\nabla^2 \Phi = -4\pi(\rho_b - \rho_{gj}), \tag{1}$$

Magnetosphere Structure and the Annular Gap Model of Pulsars 151

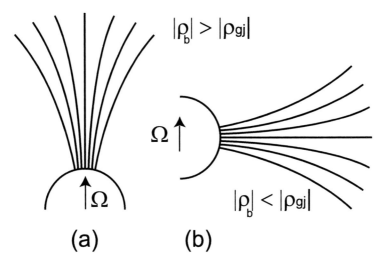

Fig. 1. An illustration for the basic picture of the slot gap model. Particles leave off the polar cap region with charge density ρ_b comparing with the locale GJ charge density ρ_{gj} for different geometry [4, 5, 6, 71]. With this figure and Eq.(1) and (2), Arons and Scharlemann [6] have emphasized that along the polar field lines, which curve toward the rotation axis of the star, are field lines with "favorable" curvature, Fig.(b), and particle acceleration occurs there. When the magnetic field lines curve away from the rotational axis, Fig.(a), negative particles can not be accelerated outward, it is so called "un-favorable" region.

The local co-rotating charge density at the point r is [5],

$$\rho_{gj}(r) = -\frac{\mathbf{\Omega} \cdot \mathbf{B}(r)}{2\pi c}\left[1 - k_g\left(\frac{R}{r}\right)^3\right], \qquad (2)$$

where $k_g = 2GI/R^3c^2 = 0.17(I_{45}/R_{10}^3)$. Relativistic space charge limited flow from the surface has a beam charge density

$$\rho_b(r) = -\frac{\mathbf{\Omega} \cdot \mathbf{B}(R)}{2\pi c}(1 - k_g)\frac{|\mathbf{B}(r)|}{|\mathbf{B}(R)|}. \qquad (3)$$

When the frame dragging effect is neglected, we have

$$\rho_b(r)/\rho_{gj}(r) = \frac{\cos \zeta(R)}{\cos \zeta(r)}, \qquad (4)$$

where ζ is the angle between the rotational axis $\mathbf{\Omega}$ and the locale magnetic field lines. The basic results are shown in Fig.1.

The "favorable" region

This is the region where $|\rho_b| < |\rho_{gj}|$ and $\rho_b - \rho_{gj} > 0$ are satisfied, and the magnetic field lines curve towards the rotational axis, as shown by the region

152   G.J. Qiao, K.J. Lee, H.G. Wang, and R.X. Xu

(a) in Fig.2. The electric field $\mathbf{E}_\parallel$ towards the inner part of the magnetosphere. So negative particles can be accelerated outward.

"Un-favorable" region

This is the region where $\mid \rho_b \mid > \mid \rho_{gj} \mid$, and $\rho_b - \rho_{gj} < 0$ are satisfied, and the magnetic field lines curve away from the rotational axis (region (b) in the Fig. 2). In this situation negative particles can not be accelerated. The electric field $\mathbf{E}_\parallel$ towards to the outer part of the magnetosphere, and hence negative particles can not be accelerated outward.

2.2 The polar cap model

The main assumptions are the same as that in the slot gap model [53]. But they included the frame-dragging effect, such that particles can be accelerated in both the "favorable" and the "un-favorable" regions. However the annular region is still not taken into account in this model.

2.3 The annular model

The open field line region of the pulsar magnetosphere is divided into two parts by the critical field lines (see Fig.2). The part that contains the magnetic axis is the core region, while the other part is the annular region. The pulsar polar cap is also correspondingly divided into the core and the annular polar cap regions (Fig. 2). For an aligned rotator, the radius from the magnetic pole to the edge of the core cap is $r_{cr} = (2/3)^{3/4} R (R\Omega/c)^{1/2}$, while the radius from the pole to the outer edge of the annular cap, i.e. the radius of the whole polar cap, is $r_p = R(\Omega R/c)^{1/2}$ (see RS75).

For easy discussion in the following, throughout the text we will focus on anti-parallel rotators, i.e. $\Omega \cdot \mu < \mathbf{0}$. The case of $\Omega \cdot \mu > \mathbf{0}$ could be derived by reversing the signs, and all the conclusions in this paper remain valid. μ is the magnetic axis.

For $\Omega \cdot \mu < 0$, the Goldreich-Julian charge density (ρ_{gj}) is positive in the region enclosed by the null charge surface (NCS). For the core-region field lines, positive charges flow out through the light cylinder. The supply of positive charges from the surface compensate for the deficit from the light cylinder. To maintain the invariance of the total charge of the pulsar, a current is needed to carry positive charges back to the star or a negative charges away from the star. For the fully charge separated magnetosphere, only one sign of charge presents at a given location [51]. In the annular region, the negative charged particles at lower altitude can not pass the region with positive GJ charge density at higher altitude. An outer gap then forms beyond the NCS [31, 35, 74].

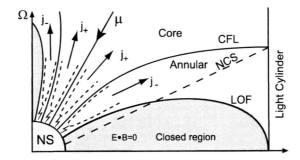

Fig. 2. A schematic diagram for the annular and the core regions for an inclined rotator. Ω and μ are the rotational and magnetic axis, respectively. The null charge surface (NCS) is the surface where the magnetic field is perpendicular to the rotation axis (i.e. $\Omega \cdot \mathbf{B} = 0$). Line 'CFL' and 'LOF' are the critical field line and the "last" open field line respectively. Charged particles with opposite signs leave off the annular and the core cap regions, respectively. The upper boundaries of the pair formation front are denoted by the dashed lines. The "flash" symbols indicate the locations with the maximal possibility to initiate the pair-production cascade [61].

2.4 High binding energy of star surface: polar gaps

If the binding energy of the star surface is high enough, then an inner vacuum gap will form (RS75). It is emphasized that both the core gap and the annular gap must be taken into account.

For neutron stars, the binding energy of positive particles could be high enough under some special conditions [20, 21] to form an inner vacuum gap. Only one of the inner vacuum gaps can form in this case: the conventional inner core gap (ICG) above the central part of the polar cap for a parallel rotator ($\Omega \cdot \mu < 0$). However, the physical conditions are different for bare strange stars. Beside the ICG above the core part of the polar cap another gap will form, no matter whether the star is a parallel ($\Omega \cdot \mu > 0$) or an antiparallel rotator ($\Omega \cdot \mu > 0$), because the binding energy is roughly infinite [87, 89]. The width of the IAG is a function of the inclination angle. If the width of the IAG is large enough, the potential drop in the IAG would be high enough so that sparking will take place there. The sparking leads to pair production and generates the secondary pairs, which will accelerate out of the IAG.

3 Basic observational results in radio and γ-rays bands

Forty years after the discovery of rotation-powered pulsars, many important discoveries shows that the pulsars are excellent laboratories for studying of particle acceleration as well as fundamental physics of strong gravity, strong magnetic fields and so on. But we still do not understand the fundamentals of their pulsed emission at any wavelength. Any theoretical model should take the basic observations into account. Here we list some basic observational facts in γ-ray and radio bands (see [25, 26, 32, 40, 41, 67, 77, 78]).

(1) The γ-ray light curves of most high-energy pulsars show two major peaks which are widely separated ($\sim 20\% - 40\%$ rotational period). The bridge emissions (the emission between the major peaks) are common.

In contrast to this, for most pulsars, the radio integrated pulse occurs over a small fraction of the period. The ratio of the pulse duration to pulse period are usually 2 percent to 6 percent, typical value is 3 percent, corresponding to between $5°$ and $20°$ of the pulse phase. The periods of pulsars scale over 3 orders, but the width of "pulse window" is narrow and independent of period. This very narrow radio emission beam together with the polarization observations show that the radiation location is near the magnetic pole, so that the beam is rigidly attached to the surface of the star, and the radio beam originates up to some tens of stellar radii above the surface.

It is clear that the high energy radiation (except crab pulsar) originates in a different region, where the emission beams are wide. To fit this requirement, the inclination angle can not be larger than $20°$ for polar cap model; or the emission region could be close to the light cylinder (outer gap model); or high energy emission region is in the annular region, where even if the radiation beam is close to the surface of the star, the radiation beam can still be very wide.

(2) In high energy band, the phase separated spectral hardness shows a systematic variation through profile, and the maximum is in the bridge. In radio bands, outside the main pulse profile the flux density usually drops rapidly to an undetectable level, and "bridge" radiation does not appear.

In γ-ray bands, the light curve together with the bridge emission having higher hardness are favorable to the polar cap model. The γ-ray light curves are consistent with single-pole, hollow-cone model (that is the polar cap and annular gap model) of the emission. Improved measurements and modeling of the phase-resolved spectra of pulsars are expected to be powerful tools in the study of the emission processes.

(3) The γ-ray spectra generally show a cutoff above several ten GeV. The lowest-field pulsars have no visible break in the EGRET energy range; the existence of a spectral change is deduced from the absence of TeV detections of pulsed emission. The highest-field pulsar among these, B1509-58, is seen in lower energy band [78]. These spectral observations are also consistent with single-pole, hollow-cone model (that is the polar cap and annular gap model) of the emission.

(4) Except the Crab pulsar, the other gamma-ray pulsars show significant phase offset between radio and gamma-ray pulses (not including the Geminga pulsar for its radio emission has not yet confirmed). Only the Crab pulsar has pulses which are (nearly) aligned across all the electromagnetic bands. This is a very serious observational fact for all emission theories. In normal case the

radio emission region is close the magnetic pole, if the high energy emission region is close to the light cylinder, how to produce the Crab-like radiation aligned across wide bands? Owing to many observational facts (such as pulse during a small fraction of the period, "S" shaped linear polarization and so no) shows that the radio emission region is close to the magnetic pole and the surface of the star. So only possible region for high energy will be close to the radio emission region such as scenario of polar cap and annular gap models.

4 Radio observational properties and the inverse Compton scattering (ICS) model

The radio beam of pulsars has the following properties: (1) radio beams can be divided into two kinds of components, i.e. the core and cone [41], or three kinds, i.e. the core, inner and outer cone [65, 66, 67]. (2) The core and conal components are emitted at different heights [18, 65]. (3) The shapes of pulse profiles change with observing frequencies [57, 58, 94]. (4) The radio emission is generally highly linearly polarized over all longitudes of profiles, but some pulsars may show depolarization and position angle jumps in the integrated profiles. For some pulsars, the position angle sweep is in an "S" shape. The sense of the circular polarization may reverse near the central phase of the integrated pulse [41, 66, 67, 86].

The debate between the "core-cone" beam structure and the "patchy" beam picture has persisted for a long time [19]. It is possible that the real pulsar emission beam is the convolution of a "patchy" source function and a "window" function [44]. The latter may itself be composed of a central core component plus one or more nested "conal" components.

Inverse Compton Scattering (ICS) model has been developed to reproduce the main observed properties. By now most theoretical models that explain pulsar radio emission are based on the picture of inner vacuum gap, e.g. the inverse Compton scattering model [56, 57, 58, 59], the spark model [21, 22] and plasma emission model [48]. To explain the γ-ray emission properties, the polar gap (PC) and the outer gap (OG) models are developed. But there are few efforts to establish a joint model for radio and γ-ray emission of pulsars. The IAG model is a model considering both radio and γ-ray at same time. The radio emission is produced above the polar cap at different heights for different components (core or cones), while the γ-ray emission originates at the places at the annular region near the null charge surface, but the location is different from the conventional outer gap model.

4.1 Radio emission in the ICS model

The basic picture of the ICS model [56, 57, 58, 86] is: low-frequency electromagnetic waves (with frequency $\sim 10^{5-6}$Hz) are assumed to be excited near the pulsar polar cap region by the periodic breakdown of the inner gap and propagate outwards in the open field line area. These low energy photons are inverse Compton scattered by the secondary particles produced in the pair cascades, and the up-scattered radio photons just provide the observed radio emission from the pulsar.

If such a RS-type gap does exist, its periodical breakdown due to pair formation avalanche can naturally provide the low-frequency electromagnetic waves required in the ICS model. However, the RS model confronts the basic difficulty known as the "binding energy problem". Recent investigations suggest that the inner vacuum gap can be formed in either neutron stars (or crusted strange stars) with extremely strong surface magnetic field [20] or bare strange stars [88]. Björnsson [9] pointed out that if ions are not bound, the outflow current is still not steady unless averaged over a sufficiently long time, since micro-instabilities should exist. Such micro-instabilities make the inner accelerator to show some features similar to what a RS-type vacuum gap shows, and the low frequency waves may also excited.

On a conventional view, the propagation of the low frequency waves may be prohibited by the dense plasma in pulsar magnetosphere. However, we note that the periodical breakdown of the gap causes the secondary plasma to eject in clumps so that a large space between adjacent clumps is analogous to a vacuum to allow low frequency waves to propagate, at least at low altitudes. Furthermore, owing to that the amplitude of the low frequency wave is very high, nonlinear effects could take a prominent role. In this case, the plasma frequency could be much lower [33], and the low frequency wave can propagate near the neutron star surface. We assume that low frequency wave can propagate to the reasonable heights of interest.

4.2 "Beam-frequency figure" in the ICS model

The fundamental output of the ICS model is the so-called "beam-frequency figure" or "beaming figure". Fig. 3 shows typical beam-frequency figures of the ICS model. It shows that along a certain field line of the dipole field, when the beaming angle (the angle between the radiation direction and the magnetic axis) increases (the emission altitude gets higher effectively), the emission frequency will decrease sharply from a high level at first, then rise again after reaching a minimum, and finally drop again after a maximum. Thus if an observer is observing at a certain frequency ν, and if the line-of-sight (hereafter LOS) of the observer is very close to the center of the beam, i.e. the impact angle is very small, all three emission regions, which correspond to the core, inner and outer conal emission respectively, can be observed. Hereafter, the three branches of the beam-frequency figure are referred to as the "core branch", the "inner conal branch" and the "outer conal branch", respectively. Even if observing at a same frequency ν, different observers with different impact angles will look at distinct emission branches so that various pulse profiles are observed.

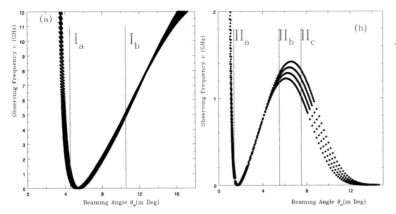

Fig. 3. Two typical "beam-frequency figures" in the ICS model. The observing frequency is plotted versus the so called beaming angle, the angle between the radiation direction and the magnetic axis. (a) For the pulsars with short period, we call it as Type I of "beam-frequency figures", the lines of Ia and Ib correspond to different observational result as different impact angle. b.) For the pulsars with longer period, we call it as Type II of "beam-frequency figures", the lines of IIa , IIb and IIc correspond to different observational result as different impact angle, see Qiao et al. [58] for details.

Physically, the shape of the beam-frequency figure is just due to the frequency formula of the ICS mechanism under the condition $B \ll B_q$, which reads

$$\nu \simeq 2\gamma^2 \nu_0 (1 - \beta \cos\theta_i), \qquad (5)$$

where $B_q = 4.414 \times 10^{13}$G is the critical magnetic field, ν is the scattered frequency, ν_0 is the frequency of the incident low frequency wave, $\beta = v/c$ is the velocity of the secondary particles in unit of light velocity c, $\gamma = 1/(1-\beta^2)^{1/2}$ is the Lorentz factor of the particles, and θ_i is the incident angle (the angle between the direction of a particle and an incoming photon). The Lorentz factor shows a decreasing behavior above the gap due to the energy loss of the particles. Commonly we assume $\gamma = \gamma_0 \exp[-\xi(r-R)]$, where γ_0 is the initial Lorentz factor of the particles, R is the radius of of the neutron star, $r - R$ is the emission height from the neutron star surface, and ξ reflects the scale factor of the deceleration.

The typical pulsar classification is established by [34, 41, 65, 67]. It is emphasized that pulse profiles of radio pulsars are changed with observed frequencies. So any classification should take this effect into account. A new classification (take the observing frequency evolution into account) can be see in [58]. There are very different evolution types. For example, the PSR B1933+16 shows the single-to-triple profile evolution with increasing frequency (Fig. 4),

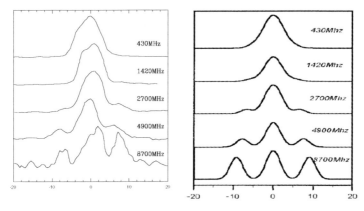

Fig. 4. A simulation for the type pulsar PSR B1933+16. P = 0.3587 s, the inclination angle $\alpha = 45°$, the impact angle $\beta = 0°$, $\gamma_0 = 3000$. The left plot is the observed pulse profiles at different frequencies. The observational data of pulsar PSR B1933+16 from Lyne & Manchester [41] for 1420 MHz, 2700 MHz and 8700 MHz, and from Sieber et al. [72] for 430 MHz and 4900 MHz. The right plot is the simulation result with ICS model(see Qiao et al. [58]). The horizontal axis is the longitude (in degrees).

and at higher frequencies the pulse profiles are much wider. In the normal curvature radiation picture: high observed frequencies are radiated from lower altitudes, so pulse profiles are narrow. But for lower observed frequencies, the radiation location should be higher and pulse profiles should be wider, the pictures are very different from observations. In the "beam-frequency figures" of ICS model, the simulation fit observations well.

From the "beam-frequency figures" one can see how many components in a pulse profile exist and what the positions of these components are. The detailed classification and simulation can be found in Qiao et al.[58].

4.3 The frequency evolution of the integrated pulse profiles

If one assumes that the shape of each emission component is Gaussian, as has been widely adopted in many other studies [34, 36, 81, 94], then we can get the emission component positions in the integrated pulse profiles. Some simulation examples for several different frequency evolution behavior (such as single pulse profiles at low frequencies but triple profiles at high frequencies, such as PSR B1933+16; or at low frequencies the pulsars show core-triple profiles and it evolves to conal-double profiles at higher frequencies, such as PSR B1845-01 and so on) are presented by Qiao et al. [58].

A three-dimensional (3D) method to calculate emission heights is developed in Zhang et al.[94]. As an example, radiation regions for different components at different frequencies have been calculated for PSR B2111+46. Emission components at seven frequencies are fitted with Gaussian components,

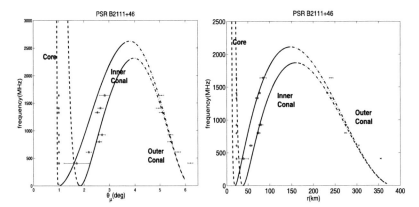

Fig. 5. The data points are obtained from the Gaussian fitting at different frequencies. In the fitting, $\alpha = 9°$ and $\beta = 1.4°$ are adopted [67]. Two curves were calculated for the two sparking points at 74% of polar cap radius in the $\Omega - \mu$ plane for the ICS model with a linear decay of the Lorentz factor as $\gamma = \gamma_0(1 - \kappa\frac{r-R}{R_e})$, where γ_0 is the initial Lorentz factor, k is a factor for the energy loss of particles, and R_e is the curvature radius of the magnetic field line, r is the height of emission, R the radius of the star. Chosen $\gamma_0 = 10000$ and $\kappa = 225$.

then the radiation heights for emission components are calculated. It is found that different emission components, even at the same radio frequency, are radiated from different heights. This is probably a common phenomena and challenges any emission mechanisms. The ICS model for pulsar radio emission fits the results well. In the Fig. 4 the dots are the positions of emission components obtained though the Gaussian fitting at different frequencies. The theoretical lines in the ICS model are presented in the Fig.5 also.

4.4 Bunching coherence of the ICS model

Extremely high brightness temperatures observed from radio pulsars require that any radiation mechanism should be coherent. Pulsar emission could be coherent through particle bunching (e.g. RS75, for a review[47]), maser behavior (e.g. [43]), or intrinsic growing of plasma waves (e.g. [8]). The earliest suggestion, i.e., coherent curvature radiation by bunches (RS75), was criticized both by its formation and its maintenance [46]. However, effective bunching could be possible for the ICS mechanism. If the a large-amplitude low frequency waves excited by gap breakdown exist and propagate at low altitudes, such waves can act as a fundamental oscillator to excite the particles to radiate in accord with each other. Particles with a same Lorentz factor will see an incident wave with the same frequency in their rest frames, and therefore re-emit the secondary waves with the same frequency and the same phase. An important difference of such an ICS bunching mechanism is fundamentally different from the curvature radiation (CR) bunching mechanism in that the requirement for physically localizing the particles in both the same momentum space and a coordinate space is much weaker. The fundamental oscillator can always "pick up" certain particles in a certain coordinate space region with an energy around a certain value to radiate coherently [86]. The disperse of the particles in both the momentum and the coordinate spaces can only reduce the number of particles to radiate coherently, i.e., reduce the degree of coherence, but never induce a sudden collapse of a bunch as is the case of CR bunching mechanism. Thus ICS bunching coherence should be free of the criticisms about the formation and the maintenance of the bunches [46]. Another advantage of ICS coherence over the CR coherence is that the characteristic frequency of CR, i.e., $\omega_{cr} = (3/2)(\gamma^3 c/\rho)$, only depends on the Lorentz factor γ of the particle at a certain point with the fixed curvature radius of the field line ρ while that of ICS depends both on γ and θ_i [eq.(1)]. Therefore a modest disperse of γ can lead to a large dispersion of ν for the CR mechanism so that coherence could be easily destroyed. For the ICS process, due to the adjustment of θ_i, emission with the same frequency can be emitted by the particles of slightly different Lorentz factors, and this is helpful to reach coherence. When bunching coherence is adopted in all the profile simulations, and it can account for steep radio emission spectra of pulsars.

5 The annular gap model confronted with observations of γ-ray emission, bi-drifting sub-pulse and the emission model for PSR B1055-52

As pointed out above, any γ-ray emission theory of pulsars should take the radio observational properties into consideration. In the polar cap model, it is easy to understand the hard bridge emission and the cutoff at the high energy region. The disadvantage of this model is that the inclination angle of the pulsars has to be small (smaller than $20°$) to generate wide radiation beam. In the outer gap model, it is easy to get the wide emission beams, but it is difficulte to produce the nearly alignment pulse profiles at all wavelength for the Crab-like pulsar. The annular gap model is easy to fit these observational facts. Because, for this model, the emission region is generated close to the null charge surface (NCS), where the radiation emission directions are perpendicular to the rotational axis, the γ-ray emission beam can be wide even the emission region are not far from the stellar surface while hardness bridge emission can also be understood, and the basic idea is shown in Fig.6.

5.1 For the case of lower binding energy of the star surface: free flow situation

Instead of fully charge separated magnetosphere (normally it is adopted in all models), the annular model assumes that the plasma in magnetosphere is not fully charge-separated. This is a key point for the annular model, and the basic idea is showing in Fig. 2.

For the case of not fully charge separated magnetosphere with the low binding energy of the star surface, the acceleration region should be divided into two parts [61], as it was suggested for high binding energy case [59]: central or core acceleration region and annular acceleration region. We emphasized that: (1) Two kinds of acceleration regions (annular and core) need to be taken into account. (2) If the potential drop in the annular region of a pulsar is high enough (normally the cases of young pulsars), charges in both the annular and core regions could be accelerated and will produce primaries pairs. Secondary pairs are generated in both regions and stream outwards to power the broadband radiations. (3) The potential drop in the annular gap region grows more rapidly than that in the core region. The annular gap acceleration process is thus a key point to produce wide emission beams as what we observed. (4) The advantages of both the polar cap and outer gap models are kept in this model. The geometric properties of the γ-ray emission can match the observations well. (5) Since different kinds of charges leave off pulsar through the annular and the core region respectively, the current closure problem can be partially solved.

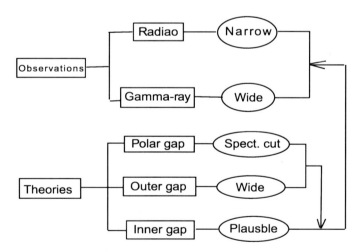

Fig. 6. A basic picture to show the advantages of the polar cap model, the outer gap model, and the basic request of the radio and γ-ray observations. The annular gap model may have the advantages of the both models and avoid the disadvantageous of them to fit the observation at radio as well as at the γ-ray bands at the same time.

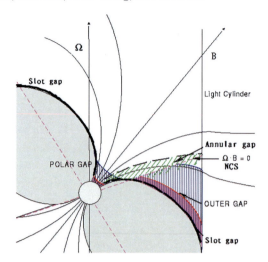

Fig. 7. The locations of the polar cap, outer gap, slot gap and the annular gap (see [27]). The annular gap is close to the null charge surface (NCS) where the direction of the magnetic field lines are perpendicular to the the rotational axis, which means that the emission beams can be very wide even if it is generated close the the star surface. The radio observations show that the emission region of it could be close to the magnetic axis and the star surface. One can see that the annular region is closer to the radio emission regions, so if in the annular region to produce γ-ray emission, then it well be benefit to fit Crab-like pulsar for nearly alignment emission at wide wavelength bands.

The locations of polar gap, annular gap, outer gap and slot gap are shown in Fig. 7.

5.2 For the case with high binding energy of the star surface: inner core gap and inner annular gap

Qiao et al. [59] proposed a inner gap model for the case with high binding energy of the star surface. Under this model, the observed of Crab-like, Vela-like, and Geminga-like light curves at both radio and γ-ray bands, can be understood theoretically (see Fig.8).

A new drifting pulsar (PSR J0815+09) was discovered in the Arecibo drift-scan searches [50]. An intriguing feature of this source is that within the four pulse components in the integrated pulse profile, the sub-pulse drifting direction in the two leading components is opposite from that in the two trailing components. In view that the leading theoretical model (RS75) for pulsar sub-pulse drifting can only interpret one-direction sub-pulse drifting, the observed "bi-drifting" phenomenon from PSR J0815+09 poses a great challenge to the pulsar theory. The inner annular gap (IAG), a new type of inner particle accelerator, was recently proposed to explain both γ-ray and radio emission from pulsars [59]. It was shown that the coexistence of the IAG and the conventional inner core gap (ICG) offers a natural interpretation to the bi-drifting phenomenon [60]. In particular, the peculiar drifting behavior in PSR J0815+09 can be reproduced within the inverse Compton scattering (ICS) model for pulsar radio emission.

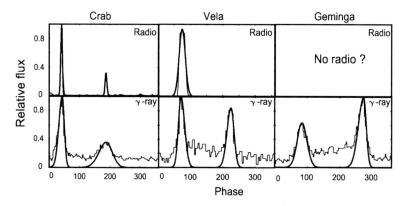

Fig. 8. The observed of Crab-like, Vela-like, and Geminga-like light curves at both radio and gamma-ray bands, as well as the theoretical fit in inner gap model. The gamma-ray and radio data are derived from Thompson [80] and the European Pulsar Network Data Archive, respectively. The parameters in the fitting are determined in Qiao et al.[59].

5.3 Observations of PSR B1055-52: radiation from the annular region

From observations of radio and γ-ray emission of PSR B1055-52, Wang at al. [80] show that the radio main pulse and the γ-ray pulse emissions come from the annular region while the inter pulse emission comes from the core region. Fig. 9 shows the result. At radio frequencies, PSR B1055-52 has a strong inter pulse (IP) follows the main pulse (MP) by a phase offset of 155° [41, 49]. The γ-ray light curve of PSR B1055-52 shows double peaks with a separation of $\sim 80°$. Multi-band observations reveal that the central phase of the radio MP lags the trailing γ-ray peak by $\sim 20°$ [79]; see Fig. 9).

5.4 The role of annular gap

The annular flow (see §2.3) and annular gap are very important for the γ-ray emission, as the simulation can match the observations well [59, 61]. If the binding energy of the surface is high enough, the consequences are: (1) Vacuum inner core gap (ICG) and inner annular gap (IAG) can form above

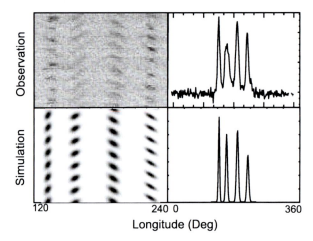

Fig. 9. The observed sub-pulse drifting patterns of PSR J0815+09 and pulse profiles and the simulation results given by Qiao et al. [60]. The left two panels are the patterns of drifting sub-pulses. The right two panels are the integrated pulse profiles. The observational data are derived from McLaughlin et al. [50].

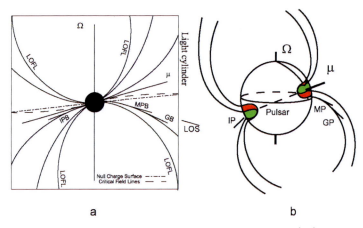

Fig. 10. Radio and gamma-ray pulse profiles of PSR B1055-52 [77] and an analysis result. It is shown that both main pulse (MP) emission and gamma-ray emission (GP) are come from the annular region, and the inter pulse emission (IP) comes from the core region (see Wang et al. [80]). In the figure red (dark grey) and green color (light grey) to show annular and core region respectively.

the polar cap region at the same time. (2) Sparking in the IAG plays a very important role for γ-ray emission. There is enough potential drop across the IAG for forming sparking, which is the prerequisite for radio and γ-ray emission. (3) The acceleration and radiation regions are mostly located between the NCS and the inner gap. (4) The γ-ray emission can be observed for both parallel rotator ($\mathbf{\Omega} \cdot \mu > 0$) and anti-parallel rotator ($\mathbf{\Omega} \cdot \mu < 0$). (5) Many observational characteristics observed at both radio and γ-ray bands can be reproduced in this model. The γ-ray emission beams of Crab-like, Vela-like and Geminga-like pulsars can be explained.

In the free flow case, similar results can be gotten and it can match observations. The consequences are: (1) Two kinds of acceleration regions (annular and core) need to be taken into account [61]. (2) Normally for young pulsars the potential drop in the annular region of a pulsar is high enough that charges in both the annular and the core regions could be accelerated and produce primary γ-rays. Secondary pairs are generated in both regions and stream outwards to power the broadband radiations. (3) The potential drop in the annular region grows more rapidly than that in the core region. The annular acceleration process will produce wide emission beams as it was observed in γ-ray bands. The geometric properties of the γ-ray emission from the annular flow is analogous to that presented in a previous work by Qiao et al. [59, 60, 62, 63], Wang et al.[80], Zhang et al. [94]. (4) The advantages of both the polar cap and outer gap models are retained in the annular model. (5) Since charges with different signs leave the pulsar through the annular and the core regions, respectively, the current closure problem can be partially solved.

6 Discussion and conclusion

1. The inverse Compton scattering (ICS) model of radio emission
   A straightforward conclusion of the ICS model is that the emission beams contain three different components, namely the core, the inner cone, and the outer cone. Different components are emitted at different altitudes. The core component is emitted in a region close to the neutron star surface, the location of the inner cone is higher, and that of the outer cone is even higher. All these are consistent with observed profile components and frequency evolution of the radio integrated pulse profiles [54, 55, 56, 57], as indicated by Zhang et al.'s results ([94], see Fig. 4 therein).

2. The annular gap model
   An annular gap model in the free flow situation is suggested by Qiao et al. [61], which can match the observations in a way that the vacuum annular gap model does [59].

3. Parallel ($\mathbf{\Omega} \cdot \mu > \mathbf{0}$) and anti-parallel ($\mathbf{\Omega} \cdot \mu < \mathbf{0}$).
   In the discussion above, we focus on the parallel rotator, owing to $\cos \zeta_{anti} = -\cos \zeta_{parll}$, and $\rho_{gj,anti} = -\rho_{gj,parall}$, all results are the same in anti-parallel case.

4. Does it have sparking in the free flow model?
   Observations show that there are sub-pulses, micro-structures and drifting sub-pulses, which can be understood when the sparking or bursting are taken into account. In normal free flow model, i.e. both polar cap and slot

gap model, sparking are not taken into account. However, sparking is a fundamental ingredient for vacuum gap model, e.g. the RS model or the inner annular gap model [59]. If the binding energy of the neutron star surface is not high enough, it can become a test for NSs and strange stars. Levinson et al. [37] pointed out that stability of models involving steady-state pair creation at a PFF is questionable. In fact early models did not make the steady-state assumption like what the later models did. In particular, Sturrock [76] argued that a steady state is not possible and the pairs escape in a sequence of sheets. RS75 invoked pair creation in bursts, called sparking. Instead of the steady-state pair creation, Levinson et al. [37] developed an oscillation or bursting model for the pair creation. Even if this treatment needs to be checked out, but it presents a very reasonable assumption and natural result.

Acknowledgments: The authors are grateful to Drs. Jin Lin Han and Bing Zhang for their long term contributions and thank the members in the pulsar group of Peking University for helpful discussions. This work is supported by NSFC (10778611, 10273001), the Special Funds for Major State Basic Research Projects of China (G2000077602), and the Key Grant Project of Chinese Ministry of Education (305001).

References

1. Hewish, A., Bell, S. J., Pilkington, J. D., et al., Nature, **217**, 709 (1969)
2. Goldreich, P., & Julian, W. H., ApJ, **157**, 869 (1969) (GJ)
3. Michel, F. C., ApJ, **192**, 713 (1974)
4. Arons, J., ApJ, **266**, 215 (1983)
5. Arons, J., ASP Conference Series, Vol. 202, eds. M. Kramer, N. Wex, and R. Wielebinski, P. 449 (1999)
6. Arons, J., Scharlemenn, E. T., ApJ, **231**, 854 (1979)
7. Ayasli, S., ApJ, **249**, 698 (1981)
8. Beskin, V. S., Gurevich, A. V., & Istomin, Y. N., Ap&SS, **146**, 205 (1998)
9. Björnsson, C.-I., ApJ, **471**, 321 (1996)
10. Cheng, K. S., Ho, C., Ruderman, M., 1986, ApJ, 300, 500
11. Cheng, A. F., & Ruderman, M. A., ApJ, **235**, 576 (1980)
12. Cheng, A., Ruderman, M., & Sutherland, P., ApJ, **203**, 209 (1976)
13. Cheng, K. S., Ruderman, M., & Zhang, L., ApJ, **537**, 946 (2000)
14. Daugherty, J. K., & Harding, A. K., ApJ, **336**, 861 (1989)
15. Deshpande, A. A., & Rankin, J. M., ApJ, **524**, 1008 (1999)
16. Deshpande, A. A., & Rankin, J. M., MNRAS, **322**, 438 (2001)
17. Fawley, W. M., Arons, J., & Scharlemann, E. T., ApJ, **217**, 227 (1977)
18. Gangadhara, R. T., & Gupta, Y. 2001, ApJ, 555, 31
19. Gil, J. A., & Krawczyk, A., MNRAS, **280**, 143 (1996)
20. Gil, J. A., & Melikidze, G. I., ApJ, **577**, 909 (2002)
21. Gil, J., & Mitra, D., A&A, **550**, 383 (2001)
22. Gil, J. A., & Sendyk, M. 2000, ApJ, 541, 351

23. Han, J. L., & Manchester, R. N., MNRAS, **320**, L35 (2001)
24. Harding, A. K., ApJ, **245**, 267 (1981)
25. Harding, A. K., ASP Conference Series Vol. 105, P. 315 (1996)
26. Harding, A. K., astro-ph/0503300 (2005)
27. Harding, A. K., astro-ph/0710.3517 (2007)
28. Harding, A. K., & Muslimov, A. G., ApJ, **508**, 328 (1998)
29. Hibschman, J. A., & Arons, J., ApJ., **554**, 624 (2001)
30. Hirotani, K., MNRAS, **317**, 225 (2000)
31. Holloway, N. J., Nature, **246**, 6 (1973)
32. Kanbach, G., astro-ph/0209021 (2002)
33. Kotsarenko, N. Ya., Stewart, G. A., & Vysloukh, V., Ap&SS, **243**, 427 (1996)
34. Kramer, M., Wielebinski, R., Jessner, A., Gil, J. A., & Seiradakis, J. H., A&AS, **107**, 515 (1994)
35. Krause-Polstorff, J., & Michel, F. C., MNRAS, **213**, 43 (1985)
36. Kuzmin, A. D., & Izvekova, V. A., Astronmy Letters, **22**, 394 (1996)
37. Levinson, A., Melrose, D., Judge, A., & Luo, Q. H., ApJ., **631**, 456 (2005)
38. Lu, T., & Shi, T., A&A, **231**, L7 (1990)
39. Luo, Q. H., ApJ, **468**, 338 (1996)
40. Lyne, A. G., & Graham-Smith, F., Pulsar Astronomy (2nd ed.: Cambridge: Cambridge Univ. Press) P. 233 (2005)
41. Lyne, A. G., & Manchester, R. N., MNRAS, **234**, 477 (LM88) (1988)
42. Lyubarskii, Y. E., & Petrova, S. A., A&A, **355**, 406 (2000)
43. Lyutikov, M., Blandford, R. D., & Machabeli, G., MNRAS, **305**, 338 (1999)
44. Manchester, R. N., J. Astrophys Astr. **16**, 107 (M95) (1995)
45. Manchester, R. N., & Taylor, J. H. 1977, *Pulsars* (Freeman: San Francisco)
46. Melrose, D. B., ApJ, **225**, 557 (1978)
47. Melrose, D. B., eds: T. H. Hankins, J. M. Rankin, & J. A. Gil IAU Colloq. 128, Pedagogical Univ. Press, P. 306 (1992)
48. Melrose, D. B., & Gedalin, M. E., ApJ, **521**, 351 (1999)
49. McCulloch, P. M., Hamilton, P. A., Ables, F. G., & Komesaroff, M. M., MNRAS, **175**, 71 (1976)
50. McLaughlin, M. A. et al., in proceedings of IAU Symp. 218, eds. F. Camilo & B. M. Gaensler (San Francisco: ASP), P. 127 (2004)
51. Michel, F. C., ApJ, **227**, 579 (1979)
52. Mitra, D. & Deshpande, A. A. 1999, A&A, 346, 906
53. Muslimov, A. G., & Harding, A. K., ApJ, **588**, 430 (2003)
54. Qiao, G. J., in: Vistas in Astronomy, **31**, 393 (1988a)
55. Qiao, G. J., in: High Energy Astrophysics, eds. G. Borner, New York, Springer-Verag, P. 88 (1988b)
56. Qiao, G. J., Proceedings of IAU Colloq. 128, Edited by T. H. Hankins, J. M. Rankin, and J. A. Gil. Pedagogical Univ. Press, P. 238 (1992)
57. Qiao, G. J., & Lin, W. P., A&A., **333**, 172 (1998)
58. Qiao, G. J., Liu, J. F., Zhang, B., & Han, J. L., A&A., **377**, 946 (2001)
59. Qiao, G. J., Lee, K. J., Wang, H. G., Xu, R. X., Han, J. L., ApJ, **606**, L49 (2004a)
60. Qiao, G. J., Lee, K. J., Zhang, B., Xu, R. X., & Wang, H. G., ApJ, **616**, L127 (2004b)
61. Qiao, G. J., Lee, K. J., Zhang, B., Wang, H. G., & Xu, R. X., ChjAA, **7**, 496 (2007)

168   G.J. Qiao, K.J. Lee, H.G. Wang, and R.X. Xu

62. Qiao, G. J., Lee, K. J., Wang, H. G.,& Xu, R. X., Proceedings of the 214th Symposium of the International Astronomical Union held at Suzhou, Eds: X. D. Li, V. Trimble, and Z. R. Wang. Published on behalf of the IAU by the Astronomical Society of the Pacific, P. 167 (astro-ph/0303231) (2003a)
63. Qiao, G. J., Lee, K. J., Wang, H. G., & Xu, R. X., 2003b, ASP Conference Proceedings, Vol. 302. Held 26-29 August 2002, Eds: Matthew Bailes, David J. Nice and Stephen E., P. 183 (2003b)
64. Qiao, G. J., & Zhang, B., A&A., **306**, L5 (1996)
65. Rankin, J. M., ApJ, **274**, 333 (R83) (1983)
66. Rankin, J. M., Proceedings of IAU Colloq. 128, Eds: by T. H. Hankins, J. M. Rankin, and J. A. Gil. Pedagogical Univ. Press, P. 133 (1992)
67. Rankin, J. M., ApJ, **405**, 285 (1993)
68. Ray, A., & Benford, G., Physical Review D., **23**, 2142 (1981)
69. Romani, R. W. 1996, ApJ, **470**, 469 (1996)
70. Ruderman, M. A., & Sutherland, P. G., ApJ, **196**, 51 (RS75) (1975)
71. Scharlemann, E. T., Arons, J., & Fawley, W. M., ApJ, **222**, 297 (1978)
72. Sieber, W., Reinecke, R. & Wielebiski, R., A&A, **38**, 169 (1975)
73. Shukre, C. S., in Magnetospheric Structure and Emission Mechanics of Radio Pulsars, Proceedings of IAU Colloq. 128, Lagow, Poland, 17-22 June 1990. Edited by T. H. Hankins, J. M. Rankin, and J. A. Gil. Pedagogical Univ. Press, p.412 (1992)
74. Smith, I. A., Michel, F. C., & Thacker, P. D., MNRAS, **322**, 209 (2001)
75. Sturner, S. J., ApJ, **446**, 292 (1995)
76. Sturrock, P. A., ApJ, **164**, 529 (1971)
77. Thompson, D. J. et al., ApJ, **516**, 297 (1999)
78. Thompson, D. J., in: Cosmic Gamma Ray Sources, eds K. S. Cheng & G. E. Romero, Kluwer ASSL Series (astro-ph/0312272) (2003)
79. Vivekanand, M., & Joshi, B. C., ApJ, **515**, 398 (1999)
80. Wang, H. G., Qiao, G. J., Xu, R. X., Liu, Y., MNRAS, **366**, 945 (2006)
81. Wu, X. J., Gao, X. Y., Rankin, J. M., Xu, W., & Malofeev, V. M., AJ, **116**, 1984 (1998)
82. Xia, X. Y., Qiao, G. J., Wu, X. J., & Hong, Y. Q., A&A, **152**, 93 (1985)
83. Xu, R. X., ChJAA, **3**, 33 (astro-ph/0211214) (2003b)
84. Xu, R. X. 2005, MNRAS, **356**, 359
85. Xu, R. X., in: Stellar astrophysics — a tribute to Helmut A. Abt, eds. K. S. Cheng, K. C. Leung & T. P. Li , Kluwer Academic Publishers, P. 73 (astro-ph/0211563) (2003a)
86. Xu, R. X., Liu, J. F., Han, J. L., & Qiao, G. J., ApJ, **535**, 354 (2000)
87. Xu, R. X., Qiao, G. J., & Zhang, B., ApJ, **522**, L109 (1999)
88. Xu, R. X., & Qiao, G. J., ApJ, **561**, L85 (2001)
89. Xu, R. X., Zhang, B., & Qiao, G. J., Astro Particle Physics, **15**, 101 (2001)
90. Zhang, B., & Qiao, G. J., A&A, **310**, 135 (1996)
91. Zhang, B., Qiao, G. J., Lin, W. P., & Han, J. L., ApJ, **478**, 313 (Zhang et al. 1997a)
92. Zhang, B., Qiao, G. J., & Han, J. L., ApJ, **491**, 891 (Zhang et al. 1997b)
93. Zhang, B., Harding, A. K., Muslimov, A. G., ApJ, **531**, L135 (2000)
94. Zhang, H., Qiao, G. J., Han, J. L., Lee, K. J., & Wang, H. G., A&A, **465**, 525 (2007)
95. Zhao, Y. H., Lu, T., Huang, K. L., Lu, J. L., & Peng, Q. H., A&A, **223**, 147 (1989)

Wave Modes in the Magnetospheres of Pulsars and Magnetars

C. Wang[1,2] and D. Lai[2,1]

[1] National Astronomical Observatories, Chinese Academy of Sciences. A20 Datun Road, Chaoyang District, Beijing 100012, China. `wangchen@bao.ac.cn`

[2] Center for Radiophysics and Space Research, Department of Astronomy, Cornell University. Ithaca, NY 14853, USA. `dong@astro.cornell.edu`

Summary. We study the wave propagation modes in the relativistic streaming pair plasma of the magnetospheres of pulsars and magnetars, focusing on the vacuum polarization effect. We show that the combined plasma and vacuum polarization effects give rise to a vacuum resonance, where "avoided mode crossing" occurs between the extraordinary mode and the (superluminous) ordinary mode. When a photon propagates from the vacuum-polarization-dominated region at small radii to the plasma-dominated region at large radii, its polarization state may undergo significant change across the vacuum resonance. Some possible applications of our results are discussed, such as polarization changes at high frequency for radio emission of pulsars and possibly magnetars.

Key words: plasmas – polarization – waves – star: magnetic fields – pulsars: general

1 Introduction

The magnetospheres of pulsars and magnetars consist of relativistic electron-positron pair plasmas, plus possibly a small amount of ions. These plasmas can affect the radiation produced in the inner region of the magnetosphere or the stellar surface. Understanding the property of wave propagation in pulsar/magnetar magnetospheres is important for the interpretation of various observations of these objects. Wave modes in pulsar magnetospheres have been studied in a number of papers under different assumptions about the plasma composition and the velocity distribution of electron-position pairs (e.g., [1], [2], [3], [4], [5]).

In this paper, we reinvestigate the property of wave propagation in the magnetospheres of pulsars and magnetars, focusing on the competition between the plasma effect and the effect of vacuum polarization. It is well known that in the strong magnetic field typically found on a neutron star, the electromagnetic dispersion relation is dominated by vacuum polarization (a prediction of quantum electrodynamics; e.g., [6], [7]; see [8] for extensive bibliography) at high photon frequencies (e.g., X-rays) and by the plasma effect at sufficiently low frequencies (e.g., radio waves). We show that the combined plasma and vacuum polarization effects give rise to a *vacuum resonance*: For a given plasma parameters and external magnetic field, there exists a special photon frequency at which the plasma effect and vacuum polarization effects "cancel" each other. The goal of our paper is to map out the parameter regimes under which the vacuum resonance may occur and to elucidate how wave propagation may be affected by the resonance. A more detailed work has been published in [9].

2 Dielectric tensor for an streaming electron-positron plasma

We consider a cold electron-positron plasma streaming along the magnetic field $\hat{\mathbf{B}}$ with the same velocity. In the coordinate system $x'y'z'$ with $\hat{\mathbf{B}}$ along $\hat{\mathbf{z}}'$, $\mathbf{k}$ in the $\mathbf{x}'$-$\mathbf{z}'$ plane and $\hat{\mathbf{k}} \times \hat{\mathbf{B}} = -\sin\theta_B \hat{\mathbf{y}}'$, we find the dielectric tensor caused by plasma effect is

$$\epsilon^{(\mathrm{p})} = \begin{bmatrix} 1 + f_{11} & if_{12} & \zeta f_{11} \\ -if_{12} & 1 + f_{11} & -i\zeta f_{12} \\ \zeta f_{11} & i\zeta f_{12} & 1 + f_{\eta} + \zeta^2 f_{11} \end{bmatrix}, \tag{1}$$

with

$$f_{11} = -\frac{v\gamma^{-1}}{1 - u\gamma^{-2}(1 - n\beta\cos\theta_B)^{-2}},$$

$$f_{12} = \frac{(1 - 2f)u^{1/2}v\gamma^{-2}(1 - n\beta\cos\theta_B)^{-1}}{1 - u\gamma^{-2}(1 - n\beta\cos\theta_B)^{-2}},$$

$$f_{\eta} = -\frac{v}{\gamma^3(1 - n\beta\cos\theta_B)^2},$$

$$\zeta = \frac{n\beta\sin\theta_B}{1 - n\beta\cos\theta_B}, \tag{2}$$

where β is the streaming velocity, γ the Lorentz factor, f the fraction of positions in the pair plasma, $n = ck/\omega$ the refractive index, and the other dimensionless quantities are

$$u = \frac{\omega_c^2}{\omega^2} = \left(\frac{eB}{mc\omega}\right)^2, \quad v = \frac{\omega_p^2}{\omega^2} = \frac{4\pi Ne^2}{m\omega^2}. \tag{3}$$

Vacuum polarization contributes a correction to the dielectric tensor:

$$\Delta\epsilon^{(\mathrm{v})} = (a - 1)\mathbf{I} + q\hat{\mathbf{B}}\hat{\mathbf{B}}, \tag{4}$$

where $\mathbf{I}$ is the unit tensor and $\hat{\mathbf{B}} = \mathbf{B}/B$ is the unit vector along $\mathbf{B}$. The magnetic permeability tensor $\boldsymbol{\mu}$ also deviates from unity because of vacuum polarization. The inverse permeability given by

$$\boldsymbol{\mu}^{-1} = a\mathbf{I} + m\hat{\mathbf{B}}\hat{\mathbf{B}}. \tag{5}$$

In the low frequency limit $\hbar\omega \ll m_e c^2$, general expressions for the vacuum polarization coefficients a, q, and m are given in [7] and [10]. For $B \ll B_Q = m_e^2 c^3/(e\hbar) = 4.414 \times 10^{13}$ G, they are given by

$$a = 1 - 2\delta_V, \qquad q = 7\delta_V, \qquad m = -4\delta_V, \tag{6}$$

where

$$\delta_V = \frac{\alpha_F}{45\pi}\left(\frac{B}{B_Q}\right)^2 = 2.650 \times 10^{-8} B_{12}^2 \tag{7}$$

and $\alpha_F = e^2/\hbar c = 1/137$ is the fine structure constant. For $B \gg B_Q$, simple expressions for a, q, m are given in [11] (see also [12] for general fitting formulae).

When $|\Delta\epsilon_{ij}^{(v)}| \ll 1$ or $B/B_Q \ll 3\pi/\alpha_F$ ($B \ll 5 \times 10^{16}$ G), the plasma and vacuum contributions to the dielectric tensor can be added linearly, i.e., $\epsilon = \epsilon^{(p)} + \Delta\epsilon^{(v)}$. In the frame with $\hat{\mathbf{B}}$ along $\hat{\mathbf{z}}'$,

$$[\epsilon]_{\hat{\mathbf{z}}'=\hat{\mathbf{B}}} = \begin{bmatrix} S' & iD & A \\ -iD & S' & -iC \\ A & iC & P' \end{bmatrix}, \tag{8}$$

with $S' = S + \hat{a}$, $P' = P + \hat{a} + q$ and $\hat{a} = a - 1$.

3 Wave Modes in a Cold Streaming Plasma

Using the electric displacement $\mathbf{D} = \epsilon \cdot \mathbf{E}$ and equation (5) in the Maxwell equations, we obtain the equation for plane waves with $\mathbf{E} \propto e^{i(\mathbf{k}\cdot\mathbf{r}-\omega t)}$ (henceforth we use $\mathbf{E}$ to denote $\delta\mathbf{E}$, and use $\mathbf{B}$ to denote $\mathbf{B}_0$)

$$\left\{\frac{1}{a}\epsilon_{ij} + n^2\left[\hat{k}_i\hat{k}_j - \delta_{ij} - \frac{m}{a}(\hat{k} \times \hat{B})_i(\hat{k} \times \hat{B})_j\right]\right\}E_j = 0, \tag{9}$$

where $n = ck/\omega$ is the refractive index and $\hat{\mathbf{k}} = \mathbf{k}/k$. We can solve this equation in coordinate system-$\mathbf{xyz}$ (with $\mathbf{k}$ along the z-axis and $\mathbf{B}$ in the x-z plane, such that $\hat{k} \times \hat{B} = -\sin\theta_B\hat{y}$), and find two wave modes labeled as "+" mode and "−" mode. We write the mode polarization vector as $\mathbf{E}_\pm = \mathbf{E}_{\pm T} + \mathbf{E}_{\pm z}\hat{z}$, with the transverse part

$$\mathbf{E}_{\pm T} = \frac{1}{(1 + K_\pm^2)^{1/2}}(iK_\pm, 1), \tag{10}$$

172   C. Wang and D. Lai

in which $iK_\pm = E_x/E_y$, and

$$K_\pm = \beta_{\rm p} \pm \sqrt{\beta_{\rm p}^2 + r} \simeq \beta_{\rm p} \pm \sqrt{\beta_{\rm p}^2 + 1}, \tag{11}$$

where $\beta_{\rm p}$ is the polarization parameter. Under the following two conditions

$$u\gamma^{-2} \left(1 - \beta \cos \theta_B\right)^{-2} \gg 1, \tag{12}$$

which implies that the Doppler-shifted frequency is lower than ω_c, and

$$v\gamma^{-1} \ll 1, \tag{13}$$

which is the weak dispersion condition, $\beta_{\rm p}$ is simplified to be

$$\beta_{\rm p} \simeq \frac{f_\eta}{2f_{12}} \frac{\sin^2 \theta_B}{(\cos \theta_B - \zeta \sin \theta_B)} \left(1 + \frac{q+m}{f_\eta}\right). \tag{14}$$

The last factor $1+(q+m)/f_\eta$ is the correction due to the vacuum polarization effect.

For $|\beta_{\rm p}| \gg 1$, the two modes are (almost) linearly polarized: the mode with $|K| \simeq 2|\beta_{\rm p}| \gg 1$ is polarized in the $\hat{\bf k}$-$\hat{\bf B}$ plane, and is usually called ordinary mode (O-mode); the mode with $|K| \simeq 1/(2|\beta_{\rm p}|) \ll 1$ is polarized perpendicular to the $\hat{\bf k}$-$\hat{\bf B}$ plane, and is called extraordinary mode (X-mode). But the linear polarization state is broken out at the point

$$f_\eta + q + m = 0, \tag{15}$$

where the two modes are exactly circular polarized ($\beta_{\rm p} = 0$), this defines the "vacuum resonance". For given ν, γ and B, the resonance occurs at the density

$$N_V = 9.905 \times 10^{11} B_{12}^2 \nu_1^2 \gamma_3^3 \left(1 - \beta \cos \theta_B\right)^2 F(b) \, {\rm cm}^{-3}, \tag{16}$$

where $F(b) \simeq 1$ for $B \ll B_{\rm Q}$. We can rewrite the resonance density in terms of the Goldreich-Julian density

$$\eta_V = \frac{N_V}{N_{\rm GJ}} = 14.15 P_1 B_{12} \nu_1^2 \gamma_3^3 \left(1 - \beta \cos \theta_B\right)^2 F(b), \tag{17}$$

where $N_{\rm GJ} = \Omega B/(2\pi ec)$. The physical meaning of the resonance is clear: For given ν, γ and B, the dielectric property of the medium is dominated by the plasma effect when $N \gg N_V$, while it is dominated by vacuum polarization when $N \ll N_V$; at $N = N_V$, the plasma effect and vacuum polarization compensate each other, and the wave modes become exactly circular polarized (see Fig.1).

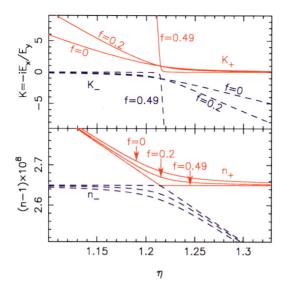

Fig. 1. The polarization ellipticity K (upper panel) and index of refraction n (lower panel) of the wave mode as a function of the plasma density parameter $\eta = N/N_{\rm GJ}$ near the vacuum resonance for $B = 10^{12}$ G, $\nu = 1$ GHz, $\gamma = 10^3$, and $\theta_B = 45°$. Three values of f are considered: $f = 0$, 0.2 and 0.49. The solid lines are for the "+" mode and dashed lines for the "−" mode.

4 Mode Evolution Across the Vacuum Resonance

Consider a photon (or electromagnetic wave) of a given frequency ν and polarization state propagating in the NS magnetosphere. The magnetosphere is inhomogeneous because of variations in **B**, N and possibly γ. How does the polarization of the photon evolve, particularly as the photon traverses the vacuum resonance region (e.g., from the plasma-dominated region to the vacuum-dominated region)? Clearly, if the variations of the magnetosphere parameters (**B**, N, etc.) are sufficiently gentle, the polarization state of the photon will evolve adiabatically, i.e. a photon in a definite wave mode will stay in that mode. Then Figure 1 shows that across the vacuum resonance, the photon polarization ellipse will rotate by $90°$, with the mode helicity unchanged.

To quantify the mode evolution, it is convenient to introduce the "mixing" angle θ_m via $\tan\theta_m = 1/K_+$, so that

$$\tan 2\theta_m = \beta_{\rm p}^{-1}, \qquad (18)$$

where we have used $|r - 1| \ll 1$. The transverse eigenvectors of the modes are $\mathbf{E}_{+T} = (i\cos\theta_m, \sin\theta_m)$ and $\mathbf{E}_{-T} = (-i\sin\theta_m, \cos\theta_m)$. Clearly, at the resonance, $\theta_m = 45°$, the X-mode and O-mode are maximally "mixed".

A general polarized electromagnetic wave with frequency ω traveling in the z-direction can be written as a superposition of the two modes:

174   C. Wang and D. Lai

$$\mathbf{E}(z) = A_+(z)\mathbf{E}_+(z) + A_-(z)\mathbf{E}_-(z). \tag{19}$$

Note that both $A_\pm$ and $\mathbf{E}_\pm$ depend on z. Substituting equation (19) into the wave equation

$$\nabla \times \left(\boldsymbol{\mu}^{-1} \cdot \nabla \times \mathbf{E}\right) = \frac{\omega^2}{c^2}\boldsymbol{\epsilon} \cdot \mathbf{E}, \tag{20}$$

we obtain the amplitude evolution equations (see [13])

$$i \begin{pmatrix} A'_+ \\ A'_- \end{pmatrix} \simeq \begin{pmatrix} -\Delta k/2 & i\theta'_m \\ -i\theta'_m & \Delta k/2 \end{pmatrix} \begin{pmatrix} A_+ \\ A_- \end{pmatrix}, \tag{21}$$

where $'$ stands for d/dz, $\Delta k = k_+ - k_-$. In deriving equation (21), we have assumed that $\mathbf{E}_\pm(z)$ and $A_\pm(z)\exp\left(-i\int^z k_\pm dz\right)$ vary on a length scale much larger than the photon wavelength, and we have used $k_+ \simeq k_-$ and $|k'_\pm/k_\pm| \ll |k_\pm|$. Clearly, when $|\theta'_m| \ll |\Delta k/2|$, or

$$\Gamma \equiv \left| \frac{(n_+ - n_-)\omega}{2\theta'_m c} \right| \gg 1, \tag{22}$$

the polarization vector will evolve adiabatically (e.g., a photon in the plus-mode will remain in the plus-mode).

5 Discussion

We now discuss possible implications of our results for various radiation processes in pulsars and magnetars. We assume a dipole magnetic field, with

$$B \approx B_* \left(\frac{R_*}{r} \right)^3, \tag{23}$$

where B_* is the surface magnetic field and R_* the radius of the NS star. Substituting equation (23) into equation (17), we obtain the location of vacuum resonance (assuming constant η and γ)

$$\frac{r_V}{R_*} \simeq 0.5 \left(\frac{\nu}{1\,\text{GHz}} \right)^{2/3} \left(\frac{B_*}{10^{12}\,\text{G}} \right)^{1/3} \left(\frac{\gamma}{10^3} \right) \left(\frac{\eta}{10^2} \right)^{-1/3} \left(\frac{P}{1\,\text{s}} \right)^{1/3} F^{1/3} \times$$
$$(1 - \beta\cos\theta_B)^{2/3}. \tag{24}$$

Since the dispersion due to vacuum polarization is of order $q + m \propto F(b)B^2 \propto r^{-6}$, while the plasma effect is measured by $\sim v\gamma^{-3} \propto N\gamma^{-3} \propto \eta B\gamma^{-3} \propto \eta\gamma^{-3}r^{-3}$, if $\eta\gamma^{-3}$ does not vary rapidly, we find that for a given photon frequency ν, the wave dispersion is dominated by the vacuum effect for $r \lesssim r_V$ and by the plasma effect for $r \gtrsim r_V$.

First consider the radio emission from the open field line region of a pulsar. The emission angle relative the local magnetic field line is $\theta_B \sim 1/\gamma$, so that

Wave Modes in the Magnetospheres of Pulsars and Magnetars   175

$1 - \beta \cos\theta_B \simeq \gamma^{-2}$. This would imply $r_V/R_* \ll 1$, even for $B_* \sim 10^{15}$ G and for high frequencies (e.g., $\nu = 20$ GHz). Along the ray trajectory, the angle θ_B increases. In the small angle approximation ($\theta_B \ll 1$), we have

$$\theta_B \approx \frac{3}{4}\sqrt{\frac{r_{\rm em}}{R_*}}\theta_0\left(1 - \frac{r_{\rm em}}{r}\right),\tag{25}$$

where $r_{\rm em}$ is the radius of the emission point, and θ_0 is the polar angle at the stellar surface of the emission field line. Thus θ_B increases from 0 (at $r = r_{\rm em}$) to $(3/4)(r_{\rm em}/R_*)\theta_0$. As an example, for $r_{\rm em} = 2R_*$ and $\theta_0 \sim \sqrt{R_*/R_{\rm LC}} \simeq 0.0145(P/1s)^{-1/2}$, equation (25) implies $\theta_B \lesssim 0.015(P/1s)^{-1/2}$. From equation (24), we find

$$\frac{r_V}{R_*} \lesssim 0.1\left(\frac{\nu}{20\,{\rm GHz}}\right)^{2/3}\left(\frac{B_*}{10^{15}\,{\rm G}}\right)^{1/3}\left(\frac{\gamma}{10^3}\right)^{-1/3}\left(\frac{\eta}{10^2}\right)^{-1/3}\left(\frac{P}{1\,{\rm s}}\right)^{-1/3}F^{1/3}.\tag{26}$$

This means that for radio emission along open, dipole field lines, plasma effects always dominate the property of wave propagation and vacuum resonance will not occur.

Radio emission may also come from the large-curvature magnetic field structure (e.g., field lines with curvature radius $\sim R_*$). In this case, even if $\theta_B \ll 1$ at emission, it will become significantly large ($\sim 45°$) after the wave propagates a short distance of order R_*. Thus, according to equation (24), vacuum resonance can occur for sufficiently high frequencies and strong surface magnetic fields. This could be the case with the high-frequency radio emission from the transient AXP XTE J1810-197 ([14]).

Finally, optical/IR radiation emitted from the neutron star surface or near vicinity may experience the vacuum resonance while propagating through the magnetosphere. According to equation (26), for $B_* = 10^{12}$ GHz, $\gamma = 10^3$, $\eta = 10^2$, $P=1$ s, $r_V \gtrsim r_{\rm em}$, vacuum resonance will occur when the photon frequency $\nu \gtrsim 10^5$ GHz, which is optical/IR radiation. The polarization of such radiation may probe the physical conditions of the magnetosphere.

References

1. Melrose, D. B. & Stoneham, R. J. 1977, PASAu, **3**, 120
2. Arons, J. & Barnard, J. J. 1986, ApJ, **302**, 120
3. Lyutikov, M. 1998, MNRAS, **293**, 447
4. Melrose, D. B., Gedalin, M. E., Kennett, M. P., & Fletcher, C. S. 1999, Journal of Plasma Physics **62**, 233
5. Asseo, E. & Riazuelo, A. 2000, MNRAS, **318**, 983
6. Heisenberg, W. & Euler, H. 1936, Z. Phys., **98**, 714
7. Adler, S.L. 1971, Ann. Phys., **67**, 599
8. Schubert, C. 2001, in Cantatore G., ed., Quantum Electrodynamics and Physics of the Vacuum, QED 2000, Second Workshop. AIP Conference Proceedings, V **564**, p. 28

9. Wang, C. & Lai, D. 2007, MNRAS, **377**, 1095
10. Heyl, J. S. & Hernquist, L. 1997, Phys. Rev. D, **55**, 2449
11. Ho, W. C. G. & Lai, D. 2003, MNRAS, **338**, 233
12. Potekhin, A. Y., Lai, D., Chabrier, G. & Ho, W. C. G. 2004, ApJ, **612**, 1034
13. Lai, D. & Ho, W. C. G. 2002, ApJ, **566**, 373
14. Camilo, F., Ransom, S. M., Halpern, J. P., Reynolds, J., Helfand, D. J., Zimmerman, N., & Sarkissian, J. 2006, Nature, **442**, 892

Polarization of Coherent Curvature Radiation in Pulsars

R. M. C. Thomas and R.T. Gangadhara

Indian Institute of Astrophysics, Bangalore – 560034, India
`mathew@iiap.res.in`

Summary. We consider the coherent curvature radiation by relativistic plasma in pulsar magnetosphere. The radiation emitted by relativistic particle bunches is beamed in the direction of field line tangents. For the first time, we have derived analytical expressions for the Stokes parameters of the coherent curvature emission, taking the detailed geometry into account. The emissions from individual bunches are incoherently superposed to find the net emission within the solid angle of radius $1/\gamma$ (beaming region) around the line of sight. By taking into account of the viewing geometry, we have estimated the polarization state of the emitted radiation.

Key words: pulsars: coherent radio emission - polarization

1 Introduction

A proper explanation for the observed features of polarization of pulsar radio profiles remains elusive in spite of decades of effort. Radhakrishnan and Rankin [1] have identified two types of circular polarization: 'antisymmetric' and 'symmetric'. In many pulsars the circular polarization is more or less concentrated around the center (core) region of the profile. Where it shows a sense reversal in most of the cases, while in a few cases the same sense is retained. Very few pulsars exhibit more complicated polarization behavior, which could be a mixer of the aforementioned features. Recently, You and Han [2] have found a strong correlation between the sign reversal of circular polarization and PPA swing in the case of the double conal pulsars. Circular polarization has been observed in individual pulses with mean value of typically around 30%, however, the average profiles show much lesser values. Frequency dependence of circular polarization has also been investigated ([2],[3]), however, the general unified trend in the frequency evolution is yet to be brought out.

Within the existing models, it is quite ambiguous to explain the circular polarization [4]. The circular polarization properties are often attributed to the pulsar emission mechanism or to the propagation effects in the pulsar magnetosphere. Cheng and Ruderman [5] have suggested that the expected asymmetry between the positively and negatively charged components of the magneto-active plasma in the far magnetosphere will convert the linear polarization into circular polarization. Kazbegi et al. ([6], [7]) have argued that it is the cyclotron instability rather than the propagation effect is responsible for the circular polarization in pulsars. Melrose and Luo [8] have explored the possibility of circular polarization because of the intrinsic relativistic effects, however, they could not provide a completely realistic model that can be directly related to observations. Gogoberidze and Machabeli [9] have invoked the propagation effects to explain the main circular polarization features. There are also strong arguments favoring intrinsic origin of circular polarization ([1], [10],[11]).

If we look back, we find in most of these models the inherent geometry of the emission region is not properly taken into account. It has become a draw back for the efforts on interpreting observations. Nevertheless, the geometry of emission region is expected to play an important role in the polarization of observed radio emission. Most of the pulsar observations emphatically support a nested conal structure for the pulsar emission beam. We adopt the model of a nested conal structure, as proposed by Rankin [12]. Gil and Krawczyk [13] have considered a nested conal structure for the pulsar emission region for simulating the pulse profile of PSR J0473-4075.

In this work, we consider the detailed geometry of the emission region in non-rotating or very slowly rotating dipole magnetic field, and study the polarization features of coherent curvature radiation. We adopt the model of coherent curvature radiation developed by Buschauer and Benford [14], as it can be easily generalized to include the detailed geometry of the emission region.

2 Radiation Electric Fields

The coherent emission from a bunch of particles moving along the curved trajectory with a local radius of curvature ρ has been discussed in vast detail, for example in [14]. The expression for the radiation electric field that is generated due to the accelerated motion of the particle bunch is given in [14]. By considering the electric fields and viewing geometry, we solve for the polarization state of the received radiation. Consider a Cartesian coordinate system$-xyz$ such that it's origin O is located at the center of the neutron star and the spin-axis $\hat{\Omega}$ is directed along the $z-$axis, as shown in Fig. 1. The $xy-$plane lies in the equatorial plane of the pulsar and the curve OP mark the trajectory of particle which is same as the shape of the field line. The unit vector $\hat{m}_t$ indicates the magnetic axis, and the line of sight

$$\hat{n} = \sin\zeta\,\hat{x} + 0\,\hat{y} + \cos\zeta\hat{z} \tag{1}$$

$$\zeta = \alpha + \beta \tag{2}$$

is chosen to lie in the xz–plane, where ζ is the angle between $\hat{n}$ and $\hat{\Omega}$. For the sake of simplicity we take that the trajectory of the particle bunch is identical to that of the flux tube along which it moves (or approximately same as that of the field line co-axial to it).

We chose a another Cartesian coordinate system–$x'y'z'$ such that z'–axis is parallel to the magnetic axis $\hat{m}_t$, and x'–axis lies in the meridional plane, i.e., plane defined by $\hat{m}_t$ and $\hat{\Omega}$. The velocity $\hat{v}$ is directed along the field line tangent $\hat{\tau}_t$ at the center of momentum (CM) of the bunch, and the acceleration $\hat{a}$ directed along the curvature vector of the field line. Let μ be the angle between the sight line $\hat{n}$ and the plane of motion of the particle bunch or osculating plane at the point CM, then

$$\sin \mu = \hat{n} \cdot (\hat{v} \times \hat{a}). \tag{3}$$

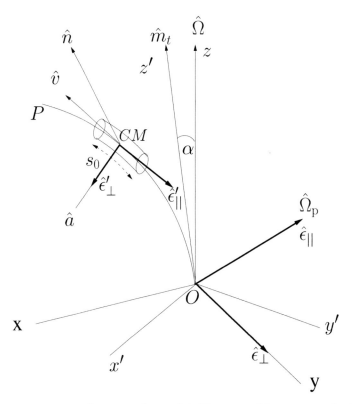

Fig. 1. The geometry relevant to the model. The curve OP represent the field line chosen to be the axis of the plasma column.

2.1 Radiation electric fields in frame–$x'y'z'$

We employ the slightly rearranged expression of [14], as given by Peyman and Gangadhara [15] , for the radiation electric field $\mathbf{E}' = \mathbf{E}'_{||}\hat{\epsilon}'_{||} + \mathbf{E}'_{\perp}\hat{\epsilon}'_{\perp}$ in the frame $x'y'z'$, where $E'_{||}$ and $E'_{\perp}$ are the components in the directions of $\hat{\epsilon}'_{||}$ and $\hat{\epsilon}'_{\perp}$, respectively. The unit vector $\hat{\epsilon}'_{\perp} = \hat{a}_0$, therefore, $\hat{\epsilon}'_{||} = \hat{n} \times \hat{\epsilon}'_{\perp}$. The expressions of $E'_{||}$ and $E'_{\perp}$ are given by

$$E'_{\perp} = -\mu\, C_{\perp}\, L_1(\mu)$$
$$E'_{||} = C_{||}\, L_2(\mu) \,, \tag{4}$$

where

$$C_{||} = \frac{\omega e^{i\omega R/c}}{2\sqrt{\pi}\, R\, c^2}(J_0\, s_0\, \zeta_0\, \eta_0)\frac{\sin[(k - k_{\mathrm{p}})s_0/2]}{(k - k_{\mathrm{p}})s_0/2}\frac{\sin(k\eta_0\mu/2)}{k\eta_0\mu/2}\left(\frac{36\rho}{\omega^2 c}\right)^{1/3}$$

$$C_{\perp} = C_{||}\sqrt{\frac{c_1}{3}}$$

$$c_1 = \frac{1}{2}\left(\frac{6\,\omega^2\rho^2}{c^2}\right)^{1/3} = 6^{1/3}\left(\frac{\rho\omega}{c}\right)^{2/3} \,, \tag{5}$$

ω and k are the radiation frequency and wavenumber, k_{p} is the wavenumber of the plasma perturbation, and s_0 is the coherence length. The parameters η_0 and ζ_0 specify the major and minor axes of elliptical cross section of the plasma column, respectively. J_0 is the current density. Since we are considering strong coherence (discussed in [14]), where $k\, s_0 \gg 10$ and the radial width of the coherently emitting plasma bunch is $\ll \lambda$, the wavelength of the radiation emitted, the radiation will be sharply peaked at $k \simeq k_{\mathrm{p}}$. Due to relativistic beaming, for the range of $\mu \ll 1$ an appreciable quantity of radiation is received from a single beam. By taking taking these factors into consideration, the sine terms in $C_{\perp}$ can be approximated to be unity. Following the definitions of [14], we write

$$\begin{pmatrix} L_1(\mu) \\ L_2(\mu) \end{pmatrix} = \int_{-\infty}^{+\infty} dt \begin{pmatrix} 1 \\ t \end{pmatrix} \exp[i\{(c_0 + c_1\,\mu^2)t + t^3\}], \tag{6}$$

where c_0 is defined as

$$|c_0| = \frac{c_1}{\gamma_0^2}\left|1 - 2\,h(\beta)\frac{k_{\mathrm{p}}}{k}\right| \tag{7}$$

and

$$h(\beta) = \frac{\beta}{1 + v_c/c\,\beta} \approx \frac{1}{2} \,.$$

Here $\beta = v/c$ and $v \approx c$ is the phase velocity of the plasma wave in the plasma frame, and v_c is the velocity of the center of momentum of the plasma column.

The coefficient of 't' in the exponential on the r.h.s. of Eq. (6) can be written as

$$z = c_0 + c_1 \mu^2 .$$
(8)

It is shown in [14] that z can be considered positive for a large range of practical parameters in a real situation. Hence the expressions of $L_1(\mu)$ and $L_2(\mu)$ can expressed as

$$L_1(z) = \frac{2}{3} \sqrt{z} K_{1/3} \left(\frac{2}{3^{3/2}} z^{3/2} \right)$$
(9)

and

$$L_2(z) = i \frac{2}{3^{3/2}} z \, K_{2/3} \left(\frac{2}{3^{3/2}} z^{3/2} \right) .$$
(10)

Let $\mathbf{E} = E_{||} \hat{\epsilon}_{||} + E_{\perp} \hat{\epsilon}_{\perp}$ be the corresponding radiation field observed in the laboratory frame–xyz, where $E_{||}$ and $E_{\perp}$ are the components of radiation field in the directions of $\hat{\epsilon}_{||}$ and $\hat{\epsilon}_{\perp}$, respectively (see Fig. 1). The components of radiation electric field in the frames $x'y'z'$ aneted using the following matrix relation:

$$\mathbf{E} = \mathbf{M} \cdot \mathbf{E}' ,$$
(11)

where the matrix

$$M = \begin{bmatrix} \cos \chi & \sin \chi \\ -\sin \chi & \cos \chi \end{bmatrix}$$
(12)

and $\chi = \cos^{-1} \left(\hat{\epsilon}_{||} \cdot \hat{e}_{\phi t} \right)$. The unit vector $\hat{e}_{\phi t} = \mathbf{T} \cdot \hat{\mathbf{e}}_{\phi}$, and the projected spin axis on the plane of sky:

$$\hat{\epsilon}_{||} = -\cos \zeta \, \hat{x} + 0 \, \hat{y} + \sin \zeta \, \hat{z}.$$

3 Stokes parameters of radiation field

Stokes parameters have been found to offer a very convenient method for establishing the association between the polarization state of observed radiation and the geometry of emitting region. Polarization and geometry are expected to play a vital (key) role in understanding the pulsar emission. For the mathematical convenience we represent the Stokes parameters as a column matrix:

$$ES = \begin{bmatrix} I \\ Q \\ U \\ V \end{bmatrix} .$$
(13)

182   R. M. C. Thomas and R.T. Gangadhara

The parameter I defines the total intensity, Q and U jointly define the linear polarization and it's position angle, and V describes the circular polarization. They can be determined using the following construction [16]:

$$ES = ST \cdot [(M \bigotimes M) \cdot (\mathbf{E}' \bigotimes \mathbf{E}'^*)], \qquad (14)$$

where the symbol '$\cdot$' represents the matrix inner product, $\bigotimes$ the outer product, and $*$ represents the complex conjugate. The matrix

$$ST = \begin{bmatrix} 1 & 0 & 0 & 1 \\ 1 & 0 & 0 & -1 \\ 0 & 1 & 1 & 0 \\ 0 & -i & i & 0 \end{bmatrix}. \qquad (15)$$

3.1 Polarization angle

The polarization angle of the radiation field $\mathbf{E}$ in the laboratory frame–xyz is given by

$$\tan(2\Psi) = \frac{U}{Q} = \tan[2(\chi + \Psi')] , \qquad (16)$$

where $\Psi' = (1/2)\tan^{-1}(U'/Q')$ is the polarization angle of $\mathbf{E}'$. Peyman and Gangadhara [15] have shown that the parameter $U' = 0$, for any given field line in the frame $x'y'z'$, therefore, the polarization angle Ψ' is either zero or $\pi/2$. In laboratory frame-xyz, the polarization angle is given by

$$\Psi = \chi = \tan^{-1} \left(\frac{\sin \chi}{\cos \chi} \right). \qquad (17)$$

3.2 The modulating gaussian

It is well known that the components of a pulsar profile can be decomposed into individual gaussians by fitting one to each of the sub-pulse component. For example the components in the pulse profile of PSR 1706-16 and PSR 2351+61 are fitted with appropriate by Kramer et al. [17]. When the line-of-sight crosses the emission region, it encounters a pattern in intensity due to gaussian modulation in the azimuthal direction. Because of the gaussian modulation in the azimuthal direction, the intensity becomes nonuniform in the polar directions to. These arguments indicate that a gaussian like intensity modulation exists in the polar directions too. So, we assume that the emission region of a pulse component has an intensity modulation both in the azimuthal and polar directions. we define a modulation function S_M for a pulse component as

$$S_\mathrm{M} = \exp[-(\phi_\mathrm{b}' + \phi_\mathrm{c}')^2/\sigma_{\phi'}^2 - (\theta_\mathrm{b}' + \theta_\mathrm{c}')^2/\sigma_{\theta'}^2] , \qquad (18)$$

where $\sigma_{\phi'}$ and $\sigma_{\theta'}$ are the gaussian widths in the azimuthal and polar directions, respectively. Here $\theta'_c = \zeta - \theta'_0$, $\phi'_c = -\phi'_0$, and θ'_0 and ϕ'_0 are the corresponding angles for the gaussian peak. The $\sigma_{\phi'}$ and $\sigma_{\theta'}$ need not be constant across the pulse profile, though they are assumed to be constant for the the small pulse widths of the order of a few times $1/\gamma$. The azimuthal width $\sigma_{\phi'}$ of the modulating gaussian may be directly inferred from the pulse profile whereas $\sigma_{\theta'}$ cannot be immediately deduced.

4 Resultant Stokes parameters

The resultant expressions are

$$I_{\rm sm} = C\, G_{\rm M} \tag{19}$$

$$L_{\rm sm} = \sqrt{Q_{\rm sm}^2 + U_{\rm sm}^2} = \frac{C}{2}\, G_{\rm M} \tag{20}$$

$$\chi_{\rm sm} = \frac{1}{2} \arctan\left(\frac{U_{\rm sm}}{Q_{\rm sm}}\right) \tag{21}$$

$$V_{\rm sm} = 3\, \Gamma\left(\frac{1}{3}\right)\, 2^{-4/3}\, c_1^{-1/2}\, C\, G_{\rm M} \left[\frac{\phi'_c}{\sigma_{\phi'}^2 |C_\phi|} + \frac{\theta'_c}{\sigma_{\theta'}^2 |C_\theta|}\right], \tag{22}$$

where the normalization constant

$$C = \frac{\gamma}{2} |C_{||}|^2 \frac{\pi}{\sqrt{c_1}} \left[\frac{2^{1/3}}{3}\right] \Gamma\left(\frac{2}{3}\right) \tag{23}$$

and

$$G_{\rm M} = \exp\left[-\left(\frac{\theta'_c}{\sigma_{\theta'}}\right)^2 - \left(\frac{\phi'_c}{\sigma_{\phi'}}\right)^2\right]. \tag{24}$$

5 Computation of polarization parameters

Consider a relativistic bunched plasma with Lorentz factor $\gamma = 1000$ accelerated along the dipolar magnetic field lines of pulsar with period $P = 1$ sec. We computed the intensity $I_{\rm sm}$, linear polarization $L_{\rm sm}$, circular polarization $V_{\rm sm}$ and polarization angle $\chi_{\rm sm}$ for the coherent curvature emission at 1.4 GHz. Chosen a modulating function $(G_{\rm M})$ with width $\sigma_{\theta'} = 4\,\sigma_{\phi'} = 4°$ and $\theta'_c \simeq (2/3)\,\beta$, and plotted in Figs. 2 and 3. These pulses resemble the pulses of conal singles or core component of multicomponent pulses. In all the cases, the polarization angle $\chi_{\rm sm}$ swing is found to be matching with the RC model[18].

The quantities $I_{\rm sm}$, $L_{\rm sm}$ and $V_{\rm sm}$ are different from the polarization parameters I', L' and V', and they satisfy the relation

$$I_{\text{sm}}^2 > L_{\text{sm}}^2 + V_{\text{sm}}^2.$$

In Figs. 2 & 3 we observe that the percentage of circular polarization relative to total intensity varies with respect to the geometric angles α and β. It shows a slight increase with respect to increasing β (see Fig. 2), and a steady decrease for increasing α (see Figs. 2 & 3)

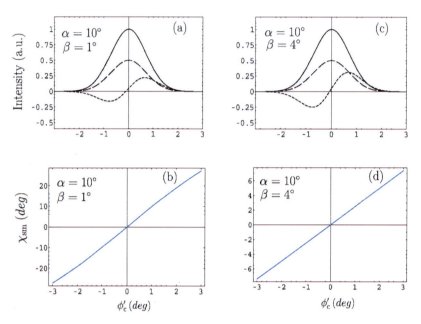

Fig. 2. The polarization parameters I_{sm} (thick line), L_{sm} (large dashed line) and V_{sm} (small dashed line) vs pulse phase ϕ'_c, are plotted in panels (a) and (c). They are normalized with the peak value $I_0 = I_{\text{sm}}(\phi'_c = 0)$. The angle χ_{sm} vs pulse phase ϕ'_c, is plotted on the panels (b) and (d). Chosen $\gamma = 1000$, $\sigma_{\theta'} = 4\,\sigma_{\phi'} = 4°$, $\theta'_c \simeq (2/3)\,\beta$ and $\rho_i/r_{\text{L}} = 20$.

6 Results and Discussions

Our model takes into account of (i) the detailed geometry of emission region, (ii) the incoherent addition within the beaming region having angular width $2/\gamma$, and (iii) possible modulations (gaussian) in the emission region. We transferred the coherent radiation fields from magnetic coordinates to laboratory frame. To relate the geometry of emission region with the polarization state of the radiation field, Stokes parameters have become convenient tools. If the size of the bunch is smaller than the wavelength of emitted radiation,

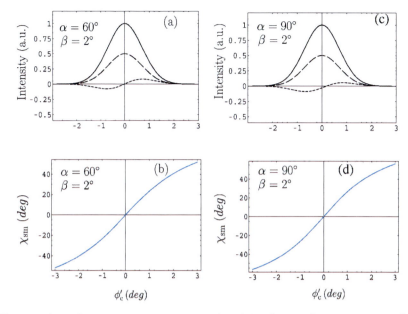

Fig. 3. The polarization parameters $I_{\rm sm}$ (thick line), $L_{\rm sm}$ (large dashed line) and $V_{\rm sm}$ (small dashed line) vs pulse phase ϕ'_c, are plotted in panels (a) and (c). They are normalized with the peak value $I_0 = I_{\rm sm}(\phi'_c = 0)$. The angle $\chi_{\rm sm}$ vs pulse phase ϕ'_c, is plotted on the panels (b) and (d). Chosen $\gamma = 1000$, $\sigma_{\theta'} = 4\,\sigma_{\phi'} = 4°$, $\theta'_c \simeq (2/3)\,\beta$ and $\rho_{\rm i}/r_{\rm L} = 20$.

then the radiation is expected to be coherent [19]. Since the arc-length corresponding to $2/\gamma$ is much larger than the wavelength, we have incoherently added the radiation fields from different bunches.

We considered a gaussian type of modulation in the intensity across the pulse phase. The mean profile of a pulsar can be split into components and they can be identified with some appropriate gaussians (e.g., [17], [20]). This hints that the individual components of a profile might have a basic unit of gaussian structure. The component shapes suggest that the emissions corresponding to each component is gaussian modulated.

On a closer look at the Eq. (22), we find that the expression for $V_{\rm sm}$ consists of a two dimensional gaussian $G_{\rm M}$ multiplied by the sum of two terms inside the square bracket: one has azimuthal variation and other the polar. The quantities $\sigma_{\phi'}$ and $\sigma_{\theta'}$ can vary significantly depending on the geometry of emission region, and for demonstration, we consider below a few cases. It can be immediately noted that when the $\sigma_{\phi'} \to \infty$ and $\sigma_{\theta'} \to \infty$ the circular polarization $V_{\rm sm} \to 0$. That means if there is no modulation, i.e., when the emission is uniform across the emission region, the resultant circular polarization goes to zero. The net circular polarization arises because of the non-uniformity in the emissions across the region above the polar cap.

Depending up on the values of $\sigma_{\phi'}$ and $\sigma_{\theta'}$ the two terms inside the square bracket (Eq. 22) describe the various behaviors of the circular polarization with respect to the pulse phase. If $\sigma_{\phi'} < \sigma_{\theta'}$ then the azimuthal term dominates over the polar term, and the circular polarization behaves like a sense reversing (antisymmetric) type with respect to pulse phase ϕ'_c. It is natural to guess that a relatively narrow pulse component should exhibit a sense reversing circular polarization, since $\sigma_{\phi'}$ is small. A typical example is the famous pulsar PSR B0329+54 which has a narrow core component with clearly sense reversing circular polarization. A probable region causing the sense-reversal can be traced from Fig. 4 as the line-of-sight cuts through the emission region (LC). The second term appearing within the square bracket in the expression

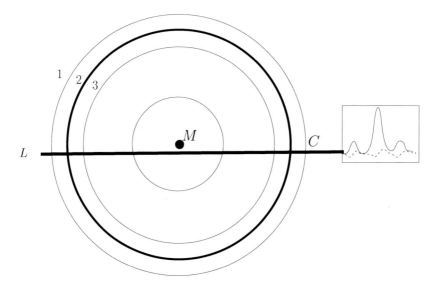

Fig. 4. The schematic diagram showing the possible origin of anti-symmetric circular polarization profile. The line LC approximately gives the sweep of the line of sight across the emission region.

of $V_{\rm sm}$ in Eq. (22) varies with respect to the polar angle θ'_c as well. For a given β, θ'_c will have a single sign, as the line of sight never cuts across the emission region in the polar directions. If $\sigma_{\theta'} \ll \sigma_{\phi'}$ then the polar term will dominate over the azimuthal term, and hence $V_{\rm sm}$ will remain with a single sign through out the pulse phase ϕ'_c.

7 Conclusion

We have explored the basic features of polarization of the coherent curvature radiation by relativistic bunched plasma accelerated along the dipolar magnetic field lines. The effects of geometry of emission region on the profile is taken into account in a detailed manner. We considered the gaussian type modulation for the emissions due to coherent curvature radiation by plasma columns. Our model predicts that the non-uniformities in the emissions within the beaming region is responsible for the sense reversing circular polarization. Our model predicts sense reversal of circular polarization under each component of a pulse profile. We find the magnitude of circular polarization dependence on the geometric angles α and β.

References

1. V. Radhakrishnan, J. M. Rankin: Astrophysical Journal, **352**, 258 (1990)
2. X. P. You, J. L. Han: Chinese Journal of Astronomy and Astrophysics, **6**, 56 (2006)
3. von A. Hoensbroech, H. Lesch: Astronomy and Astrophysics, **342**, 57 (1999)
4. J. L. Han, R. N. Manchester, R. X. Xu, G. J. Qiao: Mon. Not. Royal Astro. Soc., **300**, 373 (1998)
5. A. F. Cheng, M. A. Ruderman: Astrophysical Journal, **212**, 800, (1977)
6. A. Z. Kazbegi, G. Z. Machabeli, G. I. Melikidze: Mon. Not. of Royal Astro. Soc., **253**, 37 (1991)
7. A. Z. Kazbegi, G. Z. Machabeli, G. I. Melikidze: 1992, In: *Proceedings of IAU Colloq. 128*, held in Lagow, Poland, 17-22 June 1990. Edited by T.H. Hankins, J.M. Rankin, and J.A. Gil (Pedagogical Univ. Press) p.373
8. D. B. Melrose, Q. Luo: Mon. Not. of Royal Astro. Soc., **352**, 915 (2004)
9. G. Gogoberidze, G. Z. Machabeli: Mon. Not. of Royal Astro. Soc., **364**, 1363 (2005)
10. J. A. Gil, J. K. Snakowski: Astronomy and Astrophysics, **234**, 269 (1990)
11. F. C. Michel: Astrophysical Journal, **322**, 822 (1987)
12. J. M. Rankin: Astrophysical Journal, **405**, 285 (1993)
13. J. A. Gil, A. Krawczyk: Mon. Not. of Royal Astro. Soc., **285**, 561 (1997)
14. R. Buschauer, G. Benford: Mon. Not. of Royal Astro. Soc., **177**, 109 (1976)
15. A. Peyman, R. T. Gangadhara: Astrophysical Journal, **566**, 365 (2002)
16. S. Johnston: Publ. Astron. Soc. Aust. **19**, 277 2002
17. M. Kramer, R. Wielebinski, A. Jessner, J. A. Gil, J. H. Seiradakis: Astronomy and Astrophysics Supplement, **107**, 515 (1994)
18. V. Radhakrishnan, D. J. Cooke: Astrophysical Letters, **3**, 225 (1969)
19. G. B. Rybicki, A. P. Lightman: *Radiative Processes in Astrophysics*, (New York: Wiley, 1976)
20. X. Wu, X. Gao, J. M. Rankin, W. Xu, V. M. Malofeev: Astronomical Journal, **116**, 1984 (1998)

Part IV

Quantum Plasmas

Nonlinear Quantum Plasma Physics

Padma K. Shukla[1], Bengt Eliasson[2], and Dastgeer Shaikh[3]

[1] Theoretische Physik IV, Ruhr-Universität Bochum, D-44780 Bochum, Germany
ps@tp4.rub.de
[2] Theoretische Physik IV, Ruhr-Universität Bochum, D-44780 Bochum, Germany
bengt@tp4.rub.de
[3] Institute of Geophysics and Planetary Physics, University of Californina,
Riverside, CA 92521, USA shaikh@ucr.edu

Summary. We present simulation studies of the formation and dynamics of dark
solitons and vortices, and of nonlinear interactions between intense circularly polarized electromagnetic (CPEM) waves and electron plasma oscillations (EPOs) dense
in quantum electron plasmas. The electron dynamics in the latter is governed by
a pair of equations comprising the nonlinear Schrödinger and Poisson system of
equations, which conserves electrons and their momentum and energy. Nonlinear
fluid simulations are carried out to investigate the properties of fully developed
two-dimensional (2D) electron fluid turbulence in a dense Fermi (quantum) plasma.
We report several distinguished features that have resulted from our 2D computer
simulations of the nonlinear equations which govern the dynamics of nonlinearly
interacting electron plasma oscillations (EPOs) in the Fermi plasma. We find that a
2D quantum electron plasma exhibits dual cascades, in which the electron number
density cascades towards smaller turbulent scales, while the electrostatic potential
forms larger scale eddies. The characteristic turbulent spectrum associated with the
nonlinear electron plasma oscillations determined critically by quantum tunneling
effect. The turbulent transport corresponding to the large-scale potential distribution is predominant in comparison with the small-scale electron number density
variation, a result that is consistent with the classical diffusion theory. The dynamics of the CPEM waves is also governed by a nonlinear schrödinger equation, which
is nonlinearly coupled with the nonlinear Schrödinger equation of the EPOs via the
relativistic ponderomotive force, the relativistic electron mass increase in the CPEM
field, and the electron density fluctuations. The present governing equations in one
spatial dimension admit stationary solutions in the form a dark envelope soliton. The
dynamics of the latter reveals its robustness. Furthermore, we numerically demonstrate the existence of cylindrically symmetric two-dimensional quantum electron
vortices, which survive during collisions. The nonlinear equations admit the modulational instability of an intense CPEM pump wave against EPOs, leading to the
formation and trapping of localized CPEM wave pipes in the electron density hole
that is associated with a positive potential distribution in our dense plasma.

Key words: Quantum plasma - Vortices and Soitary waves: Electromagnetic waves

1 Introduction

About forty five years ago, Pines [1] had laid down foundations for quantum plasma physics through his studies of the properties of electron plasma oscillations (EPOs) in a dense Fermi plasma. The high-density, low-temperature quantum Fermi plasma is significantly different from the low-density, high-temperature "classical plasma" obeying the Maxwell-Boltzmann distribution. In a very dense quantum plasma, there are new equations of state [2, 3, 4] associated with the Fermi-Dirac plasma particle distribution function and there are new quantum forces involving the quantum Bohm potential [5] and the electron-1/2 spin effect [6] due to magnetization. It should be noted that very dense quantum plasmas exist in intense laser-solid density plasma interaction experiments [7, 8, 9, 10], in laser-based inertial fusion [11], in astrophysical and cosmological environments [12, 13, 14, 15], and in quantum diodes [16, 17, 18].

During the last decade, there has been a growing interest in investigating new aspects of dense quantum plasmas by developing the quantum hydrodynamic (QHD) equations [5] by incorporating the quantum force associated with the Bohm potential [5]. The the Wigner-Poisson (WP) model [19, 20] has been used to derive a set of quantum hydrodynamic (QHD) equations [2, 3] for a dense electron plasma. The QHD equations include the continuity, momentum and Poisson equations. The quantum nature [2] appears in the electron momentum equation through the pressure term, which requires the knowledge of the Wigner distribution for a quantum mixture of electron wave functions, each characterized by an occupation probability. The quantum part of the electron pressure is represented as a quantum force [5, 2] $-\nabla \phi_B$, where $\phi_B = -(\hbar^2/2m_e\sqrt{n_e})\nabla^2\sqrt{n_e}$, $\hbar$ is the Planck constant divided by 2π, m_e is the electron mass, and n_e is the electron number density. Defining the effective wave function $\psi = \sqrt{n_e(\mathbf{r}, t)}\exp[iS(\mathbf{r}, t)/\hbar]$, where $\nabla S(\mathbf{r}, t) = m_e\mathbf{u}_e(\mathbf{r}, t)$ and $\mathbf{u}_e(\mathbf{r}, t)$ is the electron velocity, the electron momentum equation can be represented as an effective nonlinear Schrödinger (NLS) equation [2, 3, 4], in which there appears a coupling between the wave function and the electrostatic potential associated with the EPOs. The electrostatic potential is determined from the Poisson equation. We thus have the coupled NLS and Poisson equations, which govern the dynamics of nonlinearly interacting EPOs is a dense quantum plasmas. This mean-field model of Ref. [2, 3] is valid to the lowest order in the correlation parameter, and it neglects correlations between electrons. The QHD equations is useful for deriving the Child-Langmuir law in the quantum regime [17, 18] and for studying numerous collective effects [2, 3, 4, 21, 22, 23, 24] involving different quantum forces (e.g. due to the Bohm potential [5] and the pressure law [2, 3] for the Fermi plasma, as well as the potential energy of the electron$-1/2$ spin magnetic moment in a magnetic field [44]). In dense plasmas, quantum mechanical effects (e.g. tunnelling) are important since the de Broglie length of the charge carriers (e.g. electrons and holes/positrons) is comparable to the dimensions of the system. Studies of collective interactions in dense quantum plasmas are relevant for the next generation intense laser-solid density plasma experiments [8, 10, 25], for superdense astrophysical bodies [12, 14, 15, 26] (e.g. the interior of white dwarfs and neutron stars), as well as for micro and nano-scale objects (e.g. quantum diodes [17, 18], quantum dots and nanowires [27], nano-photonics [28, 29], ultra-small electronic devices [30]) and micro-plasmas [31]. Quantum transport models similar to the QHD plasma model has also been used in superfluidity [32] and superconductivity [33], as well as the study of metal clusters and nanoparticles, where they are referred to as nonstationary Thomas-Fermi models [34]. The density functional theory [35, 36, 37] incorporates electron-electron correlations, which are neglected in the present paper.

It has been recently recognized [25, 38, 39] that quantum mechanical effects play an important role in intense laser-solid density plasma interaction experiments. In the latter, there are nonlinearities [40] associated with the electron mass increase in the electromagnetic (EM) fields and the modification of the electron number density by the relativistic ponderomotive force. Relativistic nonlinear effects in a classical plasma is very important, because they provide the possibility of the compression and localization of intense electromagnetic waves. In this Letter, we consider nonlinear interactions between intense CPEM waves and EPOs in dense quantum plasmas, which are relevant for a variety of applications in laboratories [9, 10].

In this paper, we investigate, by means of computer simulations, the formation and dynamics of dark/gray envelope solitons and vortices in quantum electron plasmas with fixed ion background. The results are relevant for the transport of information at quantum scales in micro-plasmas as well as in micro-mechanical systems and microelectronics. For our purposes, we shall use an effective Schrödinger-Poisson model [2, 21, 22, 23, 24], which was developed by employing the Wigner-Poisson phase space formalism on the Vlasov equation coupled with the Poisson equation for the electric potential. Such a model was originally derived by Hartree in the context of atomic physics for studying the self-consistent effect of atomic electrons on the Coulomb potential of the nucleus. The properties of 2D electron fluid turbulence and associated electron transport in quantum plasmas are investigated numerically by simulations. We find that the nonlinear coupling between the EPOs of different scale sizes gives rise to small-scale electron density structures, while the electrostatic potential cascades towards large-scales. Finally, we present theoretical and simulation studies of the CPEM wave modulational instability against EPOs, as well as the trapping of localized CPEM waves into a quantum electron hole in very dense quantum plasmas, which may be relevant for the next generation intense laser-plasma interaction experiments.

2 Dark solitons and vortices in a dense quantum plasma

In this section, we discuss the nonlinear properties and dynamics of dark solitons and vortices in a quantum plasma [4]. Generalizing the one-dimensional

194 Padma K. Shukla, Bengt Eliasson, and Dastgeer Shaikh

Schrödinger-Poisson system of equations [2] to multi-space dimensions, we have

$$i\frac{\partial \Psi}{\partial t} + A\nabla^2\Psi + \varphi\Psi - |\Psi|^4\Psi = 0,$$ (1)

and

$$\nabla^2\varphi = |\Psi|^2 - 1,$$ (2)

where the wave function Ψ is normalized by $\sqrt{n_0}$, the electrostatic potential φ by T_F/e, the time t by $\hbar T_F$ and the space $\mathbf{r}$ by λ_D. We have introduced the notations $\lambda_D = (T_F/4\pi n_0 e^2)^{1/2}$ and $A = \Gamma_Q/2$, where the quantum coupling parameter $\Gamma_Q = 4\pi e^2 m/\hbar n_0^{1/3}$ can be both smaller and larger than unity for typical metallic electrons [2]. Here n_0 is the equilibrium electron particle density, $T_F \simeq \hbar n_0^{2/3}/m_e$ is the Fermi temperature, m_e is the electron mass, e is the magnitude of the electron charge, and $\hbar$ is the Planck constant divided by 2π. Strictly speaking, the nonlinearity $|\Psi|^4$ in the last term in the left-hand side of Eq. (1) was derived for the one-dimensional model [2] and takes the form $|\Psi|^{4/D}$ in D dimensions. However, our numerical investigations of the profiles of dark solitons and vortices have shown very small differences if we use $D = 2$ (for two dimensions) instead of $D = 1$; henceforth, we will keep Eqs. (1) and (2) in the present form. The system (1) and (2) is supplemented by the Maxwell equation

$$\partial\mathbf{E}/\partial t = iA(\Psi\nabla\Psi^* - \Psi^*\nabla\Psi),$$ (3)

where the electric field $\mathbf{E} = -\nabla\varphi$. The system of equations (1)–(3) conserves the number of electrons $N = \int |\Psi| d^3x$, the electron momentum $\mathbf{P} = -i\int \Psi^*\nabla\Psi d^3x$, the electron angular momentum $\mathbf{L} = -i\int \Psi^*\mathbf{r}\times\nabla\Psi d^3x$, and the total energy $\mathcal{E} = \int(-\Psi^* A\nabla^2\Psi + |\nabla\varphi|^2/2 + |\Psi|^6/3) d^3x$. We note that one-dimensional version of Eq. (1) without the φ-term has also been used to describe the behaviour of a Bose-Einstein condensate [41].

Let us first consider a quasi-stationary, one-dimensional structure moving with a constant speed v_0, and make the ansatz $\Psi = W(\xi)\exp(iKx - i\Omega t)$, where W is a complex-valued function of the argument $\xi = x - v_0 t$, and K and Ω are a constant wavenumber and frequency shift, respectively. By the choice $K = v_0/2A$, we can then write the coupled system of equations as

$$\frac{d^2W}{d\xi^2} + \lambda W + \frac{\varphi W}{A} - \frac{|W|^4 W}{A} = 0,$$ (4)

and

$$\frac{d^2\varphi}{d\xi^2} = |W|^2 - 1,$$ (5)

where $\lambda = \Omega/A - v_0^2/4A^2$ is an eigenvalue of the system. From the boundary conditions $|W| = 1$ and $\varphi = 0$ at $|\xi| = \infty$, we determine $\lambda = 1/A$ and $\Omega = 1 + v_0^2/4A$. The system of Eqs. (4) and (5) supports a first integral in the form

$$H = A\left|\frac{dW}{d\xi}\right|^2 - \frac{1}{2}\left(\frac{d\varphi}{d\xi}\right)^2 + |W|^2 - \frac{|W|^6}{3} + \varphi|W|^2 - \varphi - \frac{2}{3} = 0, \quad (6)$$

where we have used the boundary conditions $|W| = 1$ and $\varphi = 0$ at $|\xi| = \infty$.

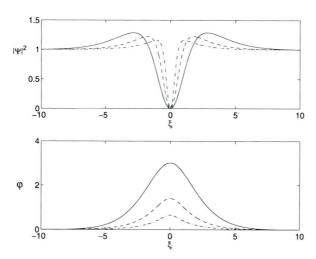

Fig. 1. The electron density $|\Psi|^2$ (the upper panel) and electrostatic potential φ (the lower panel) associated with a dark soliton supported by the system of equations (4) and (5), for $A = 5$ (solid lines), $A = 1$ (dashed lines), and $A = 0.2$ (dash-dotted line). After Ref. [4].

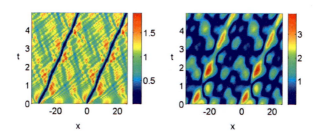

Fig. 2. The time-development of the electron density $|\Psi|^2$ (left-hand panel) and electrostatic potential φ (the right-hand panel), obtained from a simulation of the system of equations (1) and (2). The initial condition is $\Psi = 0.18 + \tanh[20\sin(x/10)]\exp(iKx)$, with $K = v_0/2A$, $A = 5$ and $v_0 = 5$. After Ref. [4].

We have solved (4) and (5) numerically and have presented the results in Fig. 1. Here we have plotted the profiles of W^2 and φ for a few values of A, where W was set to -1 on the left boundary and to $+1$ on the right boundary, i.e. the phase shift is 180 degrees between the two boundaries. We see that we have solutions in the form of a dark soliton, with a localized depletion of the electron density $N_e = |W|^2$, and where W has different sign on different sides of the solitary structure. The local depletion of the electron density is associated with a positive potential. Larger values of the parameter A give rise to larger-amplitude and wider dark solitons. Unlike a dark soliton associated with usual cubic Schrödinger equation in which the group dispersion and the nonlinearity coefficient have opposite sign, the modulus of the wave function in the present work has localized maxima on both sides of the density depletion. If the boundary conditions are shifted below 180 degrees (i.e. by a complex number), we have a "grey soliton" which is characterized by a non-zero density at the center of the soliton. In order to assess the dynamics and stability of the dark soliton, we have solved the time-dependent system of Eqs. (1) and (2) numerically, and have displayed the result in Fig. 2. The initial condition is $\Psi = 0.18 + \tanh[20\sin(x/10)]\exp(iKx)$, where $K = v_0/2A$, $A = 5$ and $v_0 = 5$. We clearly see oscillations and wave turbulence in the time-dependent solution presented in Fig. 2. Two very clear and long-lived dark solitons are visible, associated with a positive potential of $\varphi \approx 3$, which is consistent with the quasi-stationary solution of Fig. 1 for $A = 5$. Hence, the dark solitons seem to be robust structures that can withstand perturbations and turbulence during a considerable time.

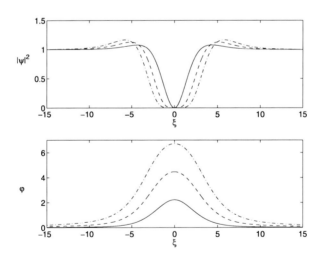

Fig. 3. The electron density $|\Psi|^2$ (upper panel) and electrostatic potential φ (lower panel) associated with a two-dimensional vortex supported by the system (7) and (8), for the charge states $n = 1$ (solid lines), $n = 2$ (dashed lines) and $n = 3$ (dash-dotted lines). We used $A = 5$ in all cases. After Ref. [4].

We next consider two-dimensional vortex structures of the form $\Psi = \psi(r)\exp(in\theta - i\Omega t)$, where r and θ are the polar coordinates defined via

Nonlinear Quantum Plasma Physics 197

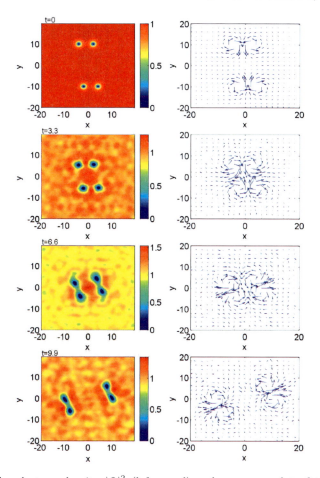

Fig. 4. The electron density $|\Psi|^2$ (left panel) and an arrow plot of the electron current $i(\Psi\nabla\Psi^* - \Psi^*\nabla\Psi)$ (right panel) associated with singly charged ($n=1$) two-dimensional vortices, obtained from a simulation of the time-dependent system of equations (1) and (2), at times $t = 0$, $t = 3.3$, $t = 6.6$ and $t = 9.9$ (upper to lower panels). We used $A = 5$. The singly charged vortices form pairs and keep their identities. After Ref. [4].

$x = r\cos(\theta)$ and $y = r\sin(\theta)$, Ω is a constant frequency shift, and $n = 0, \pm 1, \pm 2, \ldots$ for different excited states (charge states). With this, we can write Eqs. (1) and (2) in the form

$$\Omega\psi + A\left(\frac{d^2}{dr^2} + \frac{1}{r}\frac{d}{dr} - \frac{n^2}{r^2}\right)\psi + \varphi\psi - |\psi|^4\psi = 0, \quad (7)$$

and

$$\left(\frac{d^2}{dr^2} + \frac{1}{r}\frac{d}{dr}\right)\varphi = |\psi|^2 - 1, \quad (8)$$

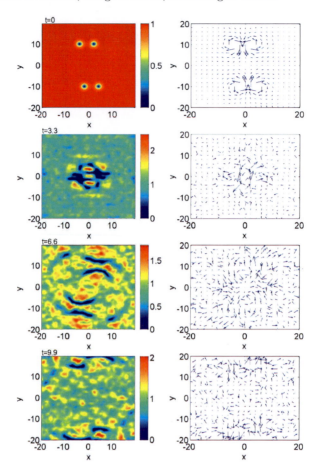

Fig. 5. The electron density $|\Psi|^2$ (left panel) and an arrow plot of the electron current $i(\Psi\nabla\Psi^* - \Psi^*\nabla\Psi)$ (right panel) associated with double charged ($n = 2$) two-dimensional vortices, obtained from a simulation of the time-dependent system of Eqs. (1) and (2), at times $t = 0$, $t = 3.3$, $t = 6.6$ and $t = 9.9$ (upper to lower panels). We used $A = 5$. The doubly charged vortices dissolve into nonlinear structures and wave turbulence. After Ref. [4].

respectively, where the boundary conditions $\psi = 1$ and $\varphi = d\psi/dr = 0$ at $r = \infty$ determine $\Omega = 1$. Different signs of n describe different rotation directions of the vortex. For $n \neq 0$, we must have $\psi = 0$ at $r = 0$, and from symmetry considerations we have $d\varphi/dr = 0$ at $r = 0$. In Fig. 3, we display numerical solutions of Eqs. (7) and (8) for different charge states n and for $A = 5$. We see that the vortex is characterized by a complete depletion of the electron density at the core of the vortex, and is associated with a positive electrostatic potential. In order to assess the stability of the vortices, we have numerically solved the time-dependent system of

Eqs. (1) and (2) in two-space dimensions for singly charged vortices and presented our results in Fig. 4. We have placed four vortex-like structures at some distance from each other, by the initial condition $\Psi = f_1 f_2 f_3 f_4$, where $f_j = \tanh[\sqrt{(x - x_j)^2 + (y - y_j)^2}] \exp[+in \arg(x - x_j, y - y_j)]$. Here $(x_1, y_1) = (-4, 10)$, $(x_2, y_2) = (2, 10)$, $(x_3, y_3) = (-2, -10)$, and $(x_4, y_4) = (4, -10)$. The function $\arg(x, y)$ denotes the angle between the x axis and the point (x, y), and it takes values between $-\pi$ and π. The initial conditions are such that the vortices are organized in two vortex pairs, as seen in the upper panels of Fig. 4. The vortices in the pairs have opposite polarity on the rotation, as seen in the electron fluid rotation direction in the upper right panel. The time-development of the system exhibits that the "partners" in the vortex pairs attract each other and propagate together with a constant velocity. When the two vortex pairs collide and interact (see the second and third pairs of panels in Fig. 4), the vortices keep their identities and change partners in a manner of asymptotic freedom, resulting into two new vortex pairs which propagate obliquely to the original propagation direction. For vortices that are multiply charged ($|n| > 1$), we have a breakup of the vortices and the formation of quasi one-dimensional dark solitons and pairs of vortices with single charge states. One such example is shown in Fig. 5, where we have simulated the system of Eqs. (1) and (2), with the same initial condition as the one in Fig. 4, except that we here have taken $n = 2$ to make the vortices doubly charged. The second row of panels in Fig. 5 reveals that the vortex pairs keep their identities for some time, while a quasi one-dimensional density cavity is formed between the two vortex pairs. At a later stage, the four vortices dissolve into complicated nonlinear structures and wave turbulence. Hence, the nonlinear dynamics is very different between singly and multiply charged solitons, where only singly charged vortices are long-lived and keep their identities. This is in line with previous results on the nonlinear Schrödinger equation, where it was noted that vortices with higher charge states are unstable [42]. In the numerical simulations of Eqs. (1) and (2), we used a pseudo-spectral method to approximate the x and y derivatives and a fourth-order Runge-Kutta scheme for the time-stepping. The numerical simulations confirmed the conservation laws of the electron number, momentum and energy up to the accuracy of the numerical scheme. The numerical solutions of the time-independent systems (4)–(5) and (7)–(8) were obtained by using the Newton method, where the ξ derivatives were approximated with a second-order centered difference scheme with appropriate boundary conditions on Ψ and φ.

3 Turbulence in quantum plasmas

In this Section, we use the coupled NLS and Poisson equations for investigating, by means of computer simulations, the properties of 2D electron fluid turbulence and associated electron transport in quantum plasmas [43]. We find that the nonlinear coupling between the EPOs of different scale sizes gives rise to small-scale electron density structures, while the electrostatic potential cascades towards large-scales. The total energy associated with our quantum electron plasma turbulence, nonetheless, processes a characteristic spectrum, which is a *non-* Kolmogorov-like. The electron diffusion caused by the electron fluid turbulence is consistent with the dynamical evolution of turbulent mode structures.

For our 2D turbulence studies, we use the nonlinear Schrödinger-Poisson equations [2, 4]

$$i\sqrt{2H}\frac{\partial \Psi}{\partial t} + H\nabla^2\Psi + \varphi\Psi - |\Psi|^2\Psi = 0, \tag{9}$$

and

$$\nabla^2\varphi = |\Psi|^2 - 1, \tag{10}$$

which are valid at zero electron temperature for the Fermi-Dirac equilibrium distribution, and which govern the dynamics of nonlinearly interacting EPOs of different wavelengths. In Eqs. (9) and (10) the wave function Ψ is normalized by $\sqrt{n_0}$, the electrostatic potential φ by T_F/e, the time t by the electron plasma period ω_{pe}^{-1}, and the space $\mathbf{r}$ by the Fermi Debye radius λ_D. We have introduced the notations $\lambda_D = (T_F/4\pi n_0 e^2)^{1/2} \equiv V_F/\omega_{pe}$ and $\sqrt{H} = \hbar\omega_{pe}/\sqrt{2}T_F$, where the Fermi electron temperature $T_F = (\hbar^2/2m_e)(3\pi^2)^{1/3}n_0^{2/3}$, e is magnitude of the electron charge, and $\omega_{pe} = (4\pi n_0 e^2/m_e)^{1/2}$ is the electron plasma frequency. The origin of the various terms in Eq. (9) is obvious. The first term is due to the electron inertia, the H-term in (9) is associated from the quantum tunneling involving the Bohm potential, $\varphi\Psi$ comes from the nonlinear coupling between the scalar potential (due to the space charge electric field) and the electron wave function, and the cubic nonlinear term is the contribution of the electron pressure [2] for the Fermi plasma that has a quantum statistical equation of state.

Equations (9) and (10) admit a set of conservation laws [44], including the number of electrons $N = \int \Psi^2 dxdy$, the electron momentum $\mathbf{P} = -i\int \Psi^*\nabla\Psi dxdy$, the electron angular momentum $\mathbf{L} = -i\int \Psi^*\mathbf{r} \times \nabla\Psi dxdy$, and the total energy $\mathcal{E} = \int[-\Psi^*H\nabla^2\Psi + |\nabla\varphi|^2/2 + |\Psi|^3/2]dxdy$. In obtaining the total energy $\mathcal{E}$, we have used the relation $\partial \mathbf{E}/\partial t = iH(\Psi\nabla\Psi^* - \Psi^*\nabla\Psi)$, where the electric field $\mathbf{E} = -\nabla\varphi$. The conservations laws are used to maintain the accuracy of the numerical integration of Eqs. (9) and (10), which hold for quantum electron-ion plasmas with fixed ion background. The assumption of immobile ions is valid, since the EPOs (given by the dispersion relation [2, 3] $\omega^2 = \omega_{pe}^2 + k^2 V_F^2 + \hbar^2 k^4/4m_e^2$) occur on the electron plasma period, which is much shorter than the ion plasma period ω_{pi}^{-1}. Here ω and k are the frequency and the wave-number, respectively. The ion dynamics, which may become important in the nonlinear phase on a longer timescale (say of the order of ω_{pi}^{-1}), in our investigation can easily be incorporated by replacing 1 in Eq. (10) by n_i, where the normalized (by n_0) ion density n_i is determined from $d_t n_i + n_i\nabla \cdot \mathbf{u}_i = 0$ and $d_t\mathbf{u}_i = -C_s^2\nabla\varphi$, where $d_t = (\partial/\partial t) + \mathbf{u}_i \cdot \nabla$, $\mathbf{u}_i$

Nonlinear Quantum Plasma Physics   201

is the ion velocity, $C_s = (T_F/m_i)^{1/2}$ is the ion sound speed, and m_i is the ion mass.

The nonlinear mode coupling interaction studies are performed to investigate the multi-scale evolution of a decaying 2D electron fluid turbulence, which is described by Eqs. (9) and (10). All the fluctuations are initialized isotropically (no mean fields are assumed) with random phases and amplitudes in Fourier space, and evolved further by the integration of Eqs. (9) and (10), using a fully de-aliased pseudospectral numerical scheme [45] based on the Fourier spectral methods. The spatial discretization in our 2D simulations uses a discrete Fourier representation of turbulent fluctuations. The numerical algorithm employed here conserves energy in terms of the dynamical fluid variables and not due to a separate energy equation written in a conservative form. The evolution variables use periodic boundary conditions. The initial isotropic turbulent spectrum was chosen close to k^{-2}, with random phases in all three directions. The choice of such (or even a flatter than -2) spectrum treats the turbulent fluctuations on an equal footing and avoids any influence on the dynamical evolution that may be due to the initial spectral non-symmetry. The equations are advanced in time using a second-order predictor-corrector scheme. The code is made stable by a proper de-aliasing of spurious Fourier modes, and by choosing a relatively small time step in the simulations. Our code is massively parallelized using Message Passing Interface (MPI) libraries to facilitate higher resolution in a 2D computational box, with a resolution of 512^2 grid points.

We study the properties of 2D fluid turbulence, composed of nonlinearly interacting EPOs, for two specific physical systems. These are the dense plasmas in the next generation laser-based plasma compression (LBPC) schemes [10] as well as in superdense astrophysical objects [14, 15, 26] (e.g. white dwarfs). It is expected that in LBPC schemes, the electron number density may reach 10^{27} cm^{-3} and beyond. Hence, we have $\omega_{pe} = 1.76 \times 10^{18}$ s^{-1}, $T_F = 1.7 \times 10^{-9}$ erg, $\hbar\omega_{pe} = 1.7 \times 10^{-9}$ erg, and $H = 1$. The Fermi Debye length $\lambda_D = 0.1$ Å. On the other hand, in the interior of white dwarfs, we typically have [46] $n_0 \sim 10^{30}$ cm^{-3} (such values are also common in dense neutron stars and supernovae), yielding $\omega_{pe} = 5.64 \times 10^{19}$ s^{-1}, $T_F = 1.7 \times 10^{-7}$ erg, $\hbar\omega_{pe} = 5.64 \times 10^{-8}$ erg, $H \approx 0.3$, and $\lambda_D = 0.025$ Å. The numerical solutions of Eqs. (9) and (10) for $H = 1$ and $H = 0.025$ (corresponding to $n_0 = 10^{27}$ cm^{-3} and $n_0 = 10^{30}$ cm^{-3}, respectively) are displayed in Figs. 6 and 7, respectively, which are the electron number density and electrostatic (ES) potential distributions in the (x, y)-plane.

Figures 6 and 7 reveal that the electron density distribution has a tendency to generate smaller length-scale structures, while the ES potential cascades towards larger scales. The co-existence of the small and larger scale structures in turbulence is a ubiquitous feature of various 2D turbulence systems. For example, in 2D hydrodynamic turbulence, the incompressible fluid admits two invariants, namely the energy and the mean squared vorticity. The two invariants, under the action of an external forcing, cascade simultaneously in tur-

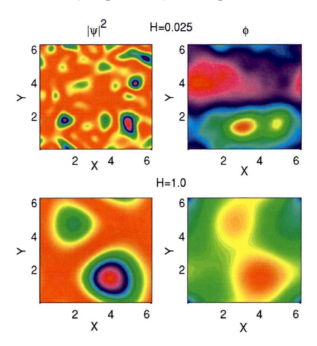

Fig. 6. Small scale fluctuations in the electron density resulted from a steady turbulence simulations of our 2D electron plasma. Forward cascades are responsible for the generation of small-scale fluctuations. Large scale structures are present in the electrostatic potential, essentially resulting from an inverse cascade. The 2D electron fluid turbulence interestingly relaxes towards an Iroshnikov-Kraichnan (IK) type $k^{-3/2}$ spectrum in a dense plasma for $H = 1$ as shown in the next figure. After Ref. [43].

bulence, thereby leading to a dual cascade phenomena. In these processes, the energy cascades towards longer length-scales, while the fluid vorticity transfers spectral power towards shorter length-scales. Usually, a dual cascade is observed in a driven turbulence simulation, in which certain modes are excited externally through random turbulent forces in spectral space. The randomly excited Fourier modes transfer the spectral energy by conserving the constants of motion in k-space. On the other hand, in freely decaying turbulence, the energy contained in the large-scale eddies is transferred to the smaller scales, leading to a statistically stationary inertial regime associated with the forward cascades of one of the invariants. Decaying turbulence often leads to the formation of coherent structures as turbulence relaxes, thus making the nonlinear interactions rather inefficient when they are saturated. The power spectrum exhibits an interesting feature in our 2D electron plasma system, unlike the 2D hydrodynamic turbulence [47, 48, 49]. The spectral slope in the 2D quantum electron fluid turbulence is close to the Iroshnikov-Kraichnan power law [50, 51] $k^{-3/2}$, rather than the usual Kolomogrov power law [47] $k^{-5/3}$. We

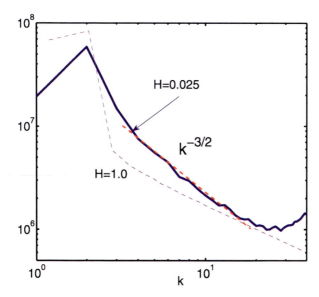

Fig. 7. The 2D electron fluid turbulence interestingly relaxes towards an Iroshnikov-Kraichnan (IK) type $k^{-3/2}$ spectrum in a dense plasma for $H = 1$. $H = 0.025$ results in a flat spectrum. After Ref. [43].

further find that this scaling is not universal and is determined critically by the quantum tunneling effect. For instance, for a higher value of H=1.0 the spectrum becomes more flat (see Fig 7). Physically, the flatness (or deviation from the $k^{-5/3}$), results from the short wavelength part of the EPOs spectrum which is controlled by the quantum tunneling effect associated with the Bohm potential. The peak in the energy spectrum can be attributed to the higher turbulent power residing in the EPO potential, which eventually leads to the generation of larger scale structures, as the total energy encompasses both the electrostatic potential and electron density components. In our dual cascade process, there is a delicate competition between the EPO dispersions caused by the statistical pressure law (giving the $k^2 V_F^2$ term, which dominates at longer scales) and the quantum Bohm potential (giving the $\hbar^2 k^4 / 4 m_e^2$ term, which dominates at shorter scales with respect to a source). Furthermore, it is interesting to note that exponents other than $k^{-5/3}$ have also been observed in numerical simulations [52, 53] of the Charney and 2D incompressible Navier-Stokes equations.

We finally estimate the electron diffusion coefficient in the presence of small and large scale turbulent EPOs in our quantum plasma. An effective electron diffusion coefficient caused by the momentum transfer can be calculated from $D_{eff} = \int_0^\infty \langle \mathbf{P}(\mathbf{r}, t) \cdot \mathbf{P}(\mathbf{r}, t + t') \rangle dt'$, where $\mathbf{P}$ is electron momentum and the angular bracket denotes spatial averages and the ensemble averages are normalized to unit mass. Since the 2D structures are confined

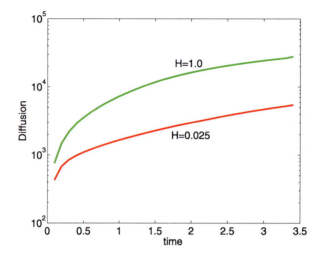

Fig. 8. Time evolution of an effective electron diffusion coefficient associated with the large-scale electrostatic potential and the small-scale electron density. Here a comparison between $H = 1$ and $H = 0.025$ is shown. After Ref. [43].

to a $x - y$ plane, the effective electron diffusion coefficient, D_{eff}, essentially relates the diffusion processes associated with random translational motions of the electrons in nonlinear plasmonic fields. We compute D_{eff} in our simulations, to measure the turbulent electron transport that is associated with the turbulent structures that we have reported herein. It is observed that the effective electron diffusion is lower when the field perturbations are Gaussian. On the other hand, the electron diffusion increases rapidly with the eventual formation of longer length-scale structures, as shown in Fig. 8. The electron diffusion due to large scale potential distributions in quantum plasmas dominates substantially, as depicted by the solid-curve in Fig. 8. Furthermore, in the steady-state, nonlinearly coupled EPOs form stationary structures, and D_{eff} saturates eventually. Thus, remarkably an enhanced electron diffusion results primarily due to the emergence of large-scale potential structures in our 2D quantum plasma.

4 Interaction between intense electromagnetic waves and quantum plasma oscillations

In this section, we discuss the nonlinear interaction between intense electromagnetic radiation and quantum plasma oscillations [54]. We consider a one-dimensional geometry of an unmagnetized dense electron-ion plasma, in which immobile ions form the neutralizing background. Thus, we are investigating the phenomena on a timescale shorter than the ion plasma period. Our

Nonlinear Quantum Plasma Physics 205

dense quantum plasma contains an intense circularly polarized electromagnetic (CPEM) plane wave that nonlinearly interacts with EPOs. The nonlinear interaction between intense CPEM waves and EPOs gives rise to an envelope of the CPEM vector potential $\mathbf{A}_\perp = A_\perp(\hat{\mathbf{x}} + i\hat{\mathbf{y}})\exp(-i\omega_0 t + ik_0 z)$, which obeys the nonlinear Schrödinger equation [40]

$$2i\Omega_0\left(\frac{\partial}{\partial t} + V_g\frac{\partial}{\partial z}\right)A_\perp + \frac{\partial^2 A_\perp}{\partial z^2} - \left(\frac{|\psi|^2}{\gamma} - 1\right)A_\perp = 0, \tag{11}$$

where the electron wave function ψ and the scalar potential are governed by, respectively,

$$iH_e\frac{\partial\psi}{\partial t} + \frac{H_e^2}{2}\frac{\partial^2\psi}{\partial z^2} + (\phi - \gamma + 1)\psi = 0, \tag{12}$$

and

$$\frac{\partial^2\phi}{\partial z^2} = |\psi|^2 - 1, \tag{13}$$

where $\Omega_0 = \omega_0/\omega_{pe}$, $V_g = v_g/c$, $H_e = \hbar\omega_{pe}/mc^2$, $v_g = k_0 c^2/\omega_0$ is the group velocity of the CPEM waves, and $\gamma = (1 + |A_\perp|^2)^{1/2}$ is the relativistic gamma factor due to the electron quiver velocity in the CPEM wave fields. Furthermore, $\omega_0 = (k_0^2 c^2 + \omega_{pe}^2)^{1/2}$ is the CPEM wave frequency, k_0 is the wavenumber, c is the speed of light in vacuum, $\omega_{pe} = (4\pi n_0 e^2/m)^{1/2}$ is the electron plasma frequency, e is the magnitude of the electron charge, n_0 is the equilibrium electron number density, and m is the electron rest mass. In (11)–(13) the time and space variables are normalized by the inverse electron plasma frequency ω_{pe}^{-1} and skin depth $\lambda_e = c/\omega_{pe}$, respectively, the scalar potential ϕ by mc^2/e, the vector potential $A_\perp$ by mc^2/e, and the electron wave function $\psi(z,t)$ by $n_0^{1/2}$. The nonlinear coupling between intense CPEM waves and EPOs comes about due to the nonlinear current density, which is represented by the term $|\psi|^2 A_\perp/\gamma$ in Eq. (11). The electron number density is defined as $n_e = \psi\psi^* = |\psi|^2$, where the asterisk denotes the complex conjugate. In Eq. (12), $1 - \gamma$ is the relativistic ponderomotive potential [40], which arises due to the cross-coupling between the CPEM wave-induced electron quiver velocity and the CPEM wave magnetic field. The second term in the left-hand side in (12) is associated with the quantum Bohm potential [5].

It is well known [55] that a relativistically strong electromagnetic wave in a classical electron plasma is subjected to the Raman scattering and modulational instabilities. At quantum scales, these instabilities will be modified by the dispersive effects caused by the tunnelling of the electrons. In order to investigate the quantum mechanical effects on the relativistic parametric instabilities in a dense quantum plasma in the presence of a relativistically strong CPEM pump wave, we let $\phi(z,t) = \phi_1(z,t)$, $A_\perp(z,t) = [A_0 + A_1(z,t)]\exp(-i\alpha_0 t)$ and $\psi(z,t) = [1 + \psi_1(z,t)]\exp(-i\beta_0 t)$, where A_0 is the large-amplitude CPEM pump and A_1 is the small-amplitude fluctuations of the CPEM wave amplitude due to the nonlinear coupling between CPEM waves and EPOs, i.e. $|A_1| \ll |A_0|$, and ψ_1 ($\ll 1$) is the small-amplitude

206   Padma K. Shukla, Bengt Eliasson, and Dastgeer Shaikh

perturbations in the electron wave function. The constant frequency shifts, determined from Eqs. (11) and (12), are $\alpha_0 = (1/\gamma_0 - 1)/(2\Omega_0)$ and $\beta_0 = (1 - \gamma_0)/H_e$, where $\gamma_0 = (1 + |A_0|^2)^{1/2}$. The first-order perturbations in the electromagnetic vector potential and the electron wave function are expanded into their respective sidebands as $A_1(z,t) = A_+ \exp(iKz - i\Omega t) + A_- \exp(-iKz + i\Omega t)$ and $\psi_1(z,t) = \psi_+ \exp(iKz - i\Omega t) + \psi_- \exp(-iKz + i\Omega t)$, while the potential is expanded as $\phi(z,t) = \widehat{\phi} \exp(iKz - i\Omega t) + \widehat{\phi}^* \exp(-iKz + i\Omega t)$, where Ω and K are the frequency and wave number of the electron plasma oscillations, respectively. Inserting the above mentioned Fourier ansatz into Eqs. (11)–(13), linearizing the resultant system of equations, and sorting into equations for different Fourier modes, we obtain the nonlinear dispersion relation

$$1 - \left(\frac{1}{D_+} + \frac{1}{D_-}\right)\left(1 + \frac{K^2}{D_L}\right)\frac{|A_0|^2}{2\gamma_0^3} = 0, \tag{14}$$

where $D_\pm = \mp 2\Omega_0(\Omega - V_g K) + K^2$ and $D_L = 1 + H_e^2 K^4/4 - \Omega^2$. We note that $D_L = 0$ yields the linear dispersion relation $\Omega^2 = 1 + H_e^2 K^4/4$ for the EPOs in a dense quantum plasma [1]. For $H_e \to 0$ we recover from (14) the nonlinear dispersion relation for relativistically large amplitude electromagnetic waves in a classical electron plasma [55]. The dispersion relation (14) governs the Raman backward and forward scattering instabilities, as well as the modulational instability. In the long wavelength limit $V_g \ll 1$, $\Omega_0 \approx 1$ we introduce the ansatz $\Omega = i\Gamma$, where the normalized (by ω_{pe}) growth rate $\Gamma \ll 1$, and obtain from Eq. (14) the growth rate $\Gamma = (1/2)|K|\{(|A_0|^2/\gamma_0^3)[1 + K^2/(1 + H_e^2 K^4/4)] - K^2\}^{1/2}$ of the modulational instability. For $|K| < 1$ and $H_e < 1$, the linear growth rate is only weakly depending on the quantum parameter H_e. However, possible nonlinear saturation of the modulational instability may lead to localized CPEM wave packets, which are trapped in a quantum electron hole. Such localized electromagnetic wavepackets would have length scales much shorter than those involved in the modulational instability process. Here quantum diffraction effects (associated with the quantum Bohm potential) become very important. In order to investigate the quantum diffraction effect on such localized electromagnetic pulses, we consider a steady state structure moving with a constant speed V_g. Inserting the ansatz $A_\perp = W(\xi)\exp(-i\Omega t)$, $\psi = P(\xi)\exp(ikx - i\omega t)$ and $\phi = \phi(\xi)$ into Eqs. (11)–(13), where $\xi = z - V_g t$, $k = V_g/H_e$ and $\omega = V_g^2/2H_e$, and where $W(\xi)$ and $P(\xi)$ are real, we obtain from (11)-(13) the coupled system of equations

$$\frac{\partial^2 W}{\partial \xi^2} + \left(\lambda - \frac{P^2}{\gamma} + 1\right)W = 0, \tag{15}$$

$$\frac{H_e^2}{2}\frac{\partial^2 P}{\partial \xi^2} + (\phi - \gamma + 1)P = 0, \tag{16}$$

where $\gamma = (1 + W^2)^{1/2}$, and

$$\frac{\partial^2 \phi}{\partial \xi^2} = P^2 - 1, \tag{17}$$

Nonlinear Quantum Plasma Physics   207

with the boundary conditions $W = \Phi = 0$ and $P^2 = 1$ at $|\xi| = \infty$. In Eq. (15), $\lambda = 2\Omega_0\Omega$ represents a nonlinear frequency shift of the CPEM wave. In the limit $H_e \to 0$, we have from (16) $\phi = \gamma - 1$, where $P \neq 0$, and we recover the classical (non-quantum) case of the relativistic solitary waves in a cold plasma [56]. We note that the system of equations (15)–(17) admits a Hamiltonian

$$Q_H = \frac{1}{2}\left(\frac{\partial W}{\partial \xi}\right)^2 + \frac{H_e^2}{2}\left(\frac{\partial P}{\partial \xi}\right)^2$$
$$-\frac{1}{2}\left(\frac{\partial \phi}{\partial \xi}\right)^2 + \frac{1}{2}(\lambda + 1)W^2 + P^2 - \gamma P^2 + \phi P^2 - \phi = 0, \quad (18)$$

where we have used the boundary conditions $\partial/\partial\xi = 0$, $W = \phi = 0$ and $|P| = 1$ at $|\xi| = \infty$.

In order to asses the importance of our investigation, we now present numerical solutions of (8)–(13) and (15)–(17), ensuring that (18) is conserved. We chose parameters that are representative of the next generation laser-based plasma compression (LBPC) schemes [10, 11]. The formula [40] $eA_\perp/mc^2 = 6 \times 10^{-10}\lambda_s\sqrt{I}$ will determine the normalized vector potential, provided that the CPEM wavelength λ_s (in microns) and intensity I (in W/cm^2) are known. It is expected that in LBPC schemes, the electron number density n_0 may reach 10^{27} cm^{-3} and beyond, and the peak values of $eA_\perp/mc^2$ may be in the range 1-2 (e.g. for focused EM pulses with $\lambda_s \sim 0.15$ nm and $I \sim 5 \times 10^{27}$ W/cm^2). For $\omega_{pe} = 1.76 \times 10^{18}$ s^{-1}, we have $\hbar\omega_{pe} = 1.76 \times 10^{-9}$ erg and $H_e = 0.002$, since $mc^2 = 8.1 \times 10^{-7}$ erg. The electron skin depth $\lambda_e \sim 1.7$ Å. On the other hand, a higher value of $H_e = 0.007$ is achieved for $\omega_{pe} = 5.64 \times 10^{18}$ s^{-1}. Thus, our numerical solutions below, based on these two values of H_e, have focused on scenarios that are relevant for the next generation intense laser-solid density plasma interaction experiments [10].

We first numerically solved Eqs. (15)–(17) for several values of H_e. Here, we solved the nonlinear boundary value problem with the boundary conditions $W = \phi = 0$ and $P = 1$ at the boundaries at $\xi = \pm 10$. We used centered second-order approximations for the second derivatives and solved the obtained nonlinear system of equations numerically by using the Newton method. The results are displayed in Figs. 9 and 10. We see that the solitary envelope pulse is composed of a single maximum of the localized vector potential W and a local depletion of the electron density P^2, and a localized positive potential ϕ at the center of the solitary pulse. The latter has a continuous spectrum in λ, where larger values of negative λ are associated with larger amplitude solitary EM pulses. At the center of the solitary EM pulse, the electron density is partially depleted, as in panels a) of Fig. 9, and for larger amplitudes of the EM waves we have stronger depletion of the electron density, as shown in panels b) and c) of Fig. 9. For cases where the electron density goes to almost zero in the classical case [56], one important quantum effect is that the electrons can tunnel into the depleted region. This is seen in Fig. 10, where the electron density remains nonzero for the larger value of H_e

in panels a), while the density shrinks to zero for the smaller value of H_e in panel b).

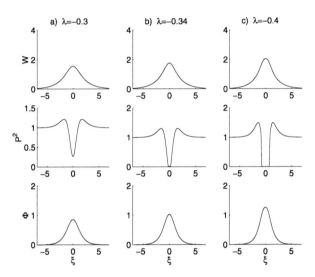

Fig. 9. The profiles of the CPEM vector potential $A_\perp$, the electron number density and the scalar potential (upper to lower rows of panels) for $\lambda = -0.3$, $\lambda = -3.4$ and $\lambda = -0.4$, with $H_e = 0.002$. After Ref. [54].

In order to investigate the quantum diffraction effects on the dynamics of localized CPEM wavepackets, we have solved the system of Eqs. (11)–(13) numerically. We considered the long-wavelength limit $\omega_0 \approx 1$ and $V_g \approx 0$. In the initial conditions, we use an EM pump with a constant amplitude $A_\perp = A_0 = 1$ and a uniform plasma density $\psi = 1$. A small amplitude noise (random numbers) of order 10^{-2} is added to $A_\perp$ to give a seed for any instability. The numerical results are displayed in Figs. 11 and 4 for $H_e = 0.002$ and $H_e = 0.007$, respectively. In both cases, we see an initial linear growth phase and a wave collapse at $t \approx 70$, in which almost all the CPEM wave energy is contracted into a few well separated localized CPEM wave pipes. These are characterized by a large bell-shaped amplitude of the CPEM wave, an almost complete depletion of the electron number density at the center of the CPEM wavepacket, and a large-amplitude positive electrostatic potential. Comparing Fig. 11 with Fig. 12, we see that there is a more complex dynamics in the interaction between the CPEM wavepackets for the larger $H_e = 0.007$, shown in Fig. 12, in comparison with $H_e = 0.002$, shown in Fig. 11, where the wavepackets are almost stationary when they are fully developed. We have here neglected the effects of the ion dynamics. The latter may be important for the development of expanding plasma bubbles (cavities) on longer timescales (e.g. the ion plasma period) [57].

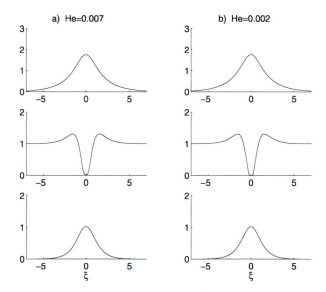

Fig. 10. The profiles of the CPEM vector potential $A_\perp$, the electron number density and the scalar potential (upper to lower rows of panels) for $H_e = 0.007$ and $H_e = 0.002$, with $\lambda = -0.34$. After Ref. [54].

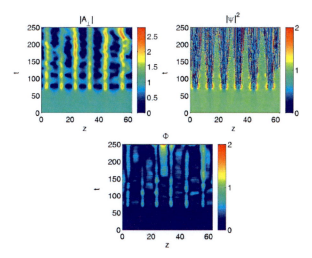

Fig. 11. The dynamics of the CPEM vector potential $A_\perp$ and the electron number density $|\psi|^2$ (upper panels) and of the electrostatic potential Φ (lower panel) for $H_e = 0.002$. After Ref. [54].

5 Conclusions

In summary, we have demonstrated the existence of localized nonlinear structures in quantum electron plasmas. The electron dynamics in the latter is

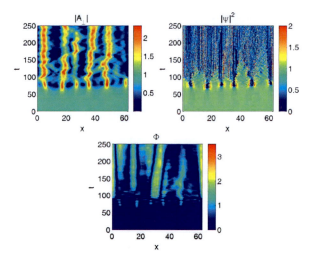

Fig. 12. The dynamics of the CPEM vector potential $A_\perp$ and the electron number density $|\psi|^2$ (upper panels) and the electrostatic potential ϕ (lower panel) for $H_e = 0.007$. After Ref. [54].

governed by a coupled nonlinear Schrödinger and Poisson system of equations, which admit a set of conserved quantities (the total number of electrons, the electron momentum, the electron angular momentum, and the electron energy). The latter were checked numerically. Quasi-stationary, localized structures in the form of one-dimensional dark solitons and two-dimensional vortices were found by solving the time-independent coupled system of equations numerically. These structures are associated with a local depletion of the electron density associated with positive electrostatic potential, and are parameterised by the quantum coupling parameter only. In the two-dimensional geometry, we have a class of vortices of different excited states (charge states) associated with a complete depletion of the electron density and an associated positive potential. The numerical simulation of the time-dependent system of equations shows the formation of stable dark solitons in one-space dimension with an amplitude consistent with the one found from the time-independent solutions. In two-space dimensions, the dark solitons of the first excited state were found to be stable and the preferred nonlinear state was in the form of vortex pairs of vortices with different polarities. One-dimensional dark solitons and singly charge two-dimensional vortices are thus long-lived nonlinear structures, which can transport information at quantum scales in micro-mechanical systems and dense laboratory plasmas. We have presented computer simulation studies of 2D fluid turbulence in a dense quantum plasma. Our simulations, for the parameters that are representative of the next generation intense laser-solid density plasma experiments as well as of the superdense astrophysical bodies, reveal new features of the dual cascade in a fully developed 2D electron fluid turbulence. Specifically, we find that the power spectrum associated with nonlinearly interacting EPOs in quantum plasmas follow a non-Kolmogorov-like spectrum. The deviation from a Kolmogorov-like spectrum resulting from the flattening of the spectrum is mediated essentially by the nonlinear EPOs interactions in the inertial range (basically controlled by the electron plasma wave dispersion effect represented by $\hbar^2 k^4/4m_e^2$), which impedes the spectral transfer of the turbulent power associated with the short scale Fourier modes. In the nonlinear regime, the inhibition of the spectral transfer is caused by short scale EPOs that are nonlinearly excited by the mode coupling of the EPOs in the forward cascade regime, which then grow, acquire nonlinear amplitudes, and eventually saturate in the nonlinear phase. We have also presented theoretical and computer simulation studies of nonlinearly interacting intense CPEM waves and EPOs in very dense quantum plasmas. The localized dark solitons, vortices, and CPEM wave structures, as discussed here, may be useful for information transfer as well as for electron acceleration in dense quantum plasmas.

Acknowledgements This work was partially supported by the Deutsche Forschungsgemeinschaft through the Sonderforschungsbereicht 591, as well as by the Swedish Research Council (VR).

References

1. D. Pines: J. Nucl. Energy: Part C: Plasma Phys. **2**, 5 (1961)
2. G. Manfredi, F. Haas: Phys. Rev. B **64**, 075316 (2001)
3. G. Manfredi: Fields Inst. Commun. **46**, 263 (2005)
4. P.K. Shukla, B. Eliasson: Phys. Rev. Lett. **96** 245001 (2006)
5. C.L. Gardner, C. Ringhofer: Phys. Rev. E **53**, 157 (1996)
6. M. Marklund, G. Brodin: Phys. Rev. Lett. **98**, 025001 (2007)
7. S.X. Hu, C.H. Keitel: Phys. Rev. Lett. **83**, 4709 (1999)
8. Y.A. Salamin et al: Phys. Rep. **427**, 41 (2006)
9. S.H. Glenzer et al: Phys. Rev. Lett. **98**, 065002 (2007)
10. V.M. Malkin et al: Phys. Rev. E **75**, 026404 (2007)
11. H. Azechi et al: Plasma Phys. Control. Fusion **48**, B267 (2006)
12. M. Opher et al: Phys. Plasmas **8**, 2454 (2001)
13. O.G. Benvenuto, M.A. De Vito: Mon. Not. R. Astron. Soc. **362**, 891 (2005)
14. G. Chabrier et al: J. Phys.: Condens. Matter **14**, 9133 (2002)
15. G. Chabrier et al: J. Phys. A: Math. Gen. **39**, 4411 (2006)
16. Y.Y. Lau et al: Phys. Rev. Lett. **66**, 1446 (1991)
17. L.K. Ang et al: Phys. Rev. Lett. **91**, 208303 (2003)
18. L.K. Ang, P. Zhang: Phys. Rev. Lett. **98**, 164802 (2007)
19. E.P. Wigner: Phys. Rev. **40**, 749 (1932)
20. M. Hillery et al: Phys. Rep. **106**, 121 (1984)
21. F. Haas, G. Manfredi, M. Feix: Phys. Rev. E **62**, 2763 (2000)
22. D. Anderson et al: Phys. Rev. E **65**, 046417 (2002)
23. F. Haas, L.G. Garcia, J. Goedert, G. Manfredi: Phys. Plasmas **10**, 3858 (2003)
24. F. Haas: Phys. Plasmas **12**, 062117 (2005)

25. G. Mourou et al: Rev. Mod. Phys. **78**, 309 (2006)
26. A.K. Harding, D. Lai: Rep. Prog. Phys. **69**, 2631 (2006)
27. G.V. Shpatakovskaya: JETP **102**, 466 (2006)
28. W.L. Barnes et al: Nature (London) **424**, 824 (2003)
29. D.E. Chang et al: Phys. Rev. Lett. **97**, 053002 (2006)
30. P.A. Markowich et al: *Semiconductor Equations* (Springer, Berlin 1990)
31. K.H. Becker, K.H. Schoenbach, J.G. Eden: J. Phys. D: Appl. Phys. **39**, R55 (2006)
32. M. Loffredo, L. Morato: Nuovo Cimento Soc. Ital Fis. B **108B**, 205 (1993)
33. R. Feynman: *Statistical Mechanics, A Set of of Lectures* (Benjamin, Reading, 1972)
34. A. Domps et al: Phys. Rev. Lett. **80**, 5520 (1998)
35. P. Hohenberg, W. Kohn: Phys. Rev. **136**, B864 (1964)
36. W. Kohn, L.J. Sham: Phys. Rev. **140**, A1133 (1965)
37. L. Brey et al: Phys. Rev. B **42**, 1240 (1990)
38. A.V. Andreev: JETP Lett. **72**, 238 (2000)
39. M. Marklund, P.K. Shukla: Rev. Mod. Phys. **78**, 591 (2006)
40. P.K. Shukla et al: Phys. Rep. **138**, 1 (1986)
41. E.B. Kolomeisky et al: Phys. Rev. Lett. **85**, 1146 (2000)
42. I.A. Ivonin, V.P. Pavlenko, H. Persson: Phys. Rev. E **60**, 492 (1999)
43. D. Shaikh, P.K. Shukla: Phys. Rev. Lett. **99**, 125002 (2007)
44. M. Marklund, G. Brodin: Phys. Rev. Lett. **98**, 025001 (2007)
45. D. Gottlieb, S.A. Orszag: *Numerical Analysis of Spectral Methods* (SIAM, Philadelphia 1977)
46. I. Iben Jr., A.V. Tutukov: Astrophys. J. **282**, 615 (1984)
47. A.N. Kolmogorov: C. R. Acad. Sci. USSR **30**, 301 (1941)
48. M. Lesieur: *Turbulence in Fluids* (Kluwer, Dordrecht 1990)
49. U. Frisch: *Turbulence* (Cambridge University Press, Cambridge, England 1995)
50. P. Iroshnikov: Sov. Astron. **7**, 566 (1963)
51. R.H. Kraichnan: Phys. Fluids **8**, 1385 (1965)
52. V.D. Larichev, J.C. McWilliams: Phys. Fluids A **3** 938 (1991)
53. R.K. Scott Phys. Rev. E **75**, 046301 (2007)
54. P.K. Shukla, B. Eliasson: Phys. Rev. Lett. **99**, 096401 (2007)
55. C.J. McKinstrie, R. Bingham: Phys. Fluids B **4**, 2626 (1992).
56. J.H. Marburger, R.F. Tooper: Phys. Rev. Lett. **35**, 1001 (1975).
57. M. Borghesi et al: Phys. Rev. Lett. **88**, 135002 (2002).

Dust Plasma Interactions in Space and Laboratory

Padma K. Shukla[1], Bengt Eliasson[2], and Dastgeer Shaikh[3]

[1] Theoretische Physik IV, Ruhr-Universität Bochum, D-44780 Bochum, Germany
`ps@tp4.rub.de`
[2] Theoretische Physik IV, Ruhr-Universität Bochum, D-44780 Bochum, Germany
`bengt@tp4.rub.de`
[3] Institute of Geophysics and Planetary Physics, University of Californina,
Riverside, CA 92521, USA `shaikh@ucr.edu`

Summary. Nonlinear aspects of coherent structures in dusty plasmas are discussed. These are the formation of envelope Langmuir solitons and shocks in a uniform dusty plasma as well as the self-organization of incompressible dust fluid in the form of vortical structures. We present conditions under which these excitations in dusty plasmas are possible. The relevance of our investigation to laboratory and space plasmas is discussed.

Key words: Dusty plasma: Vortices Dust – acoustic waves- Dust ion acoustic waves

1 Introduction

Nonlinear waves and structures in electron-ion plasmas have been thoroughly investigated [1, 2, 3]. The presence of charged dust grains in plasmas, which are ubiquitous in laboratory and space environments, brings a new dimension to collective dust-plasma interactions [4, 5, 6, 7]. Charged dust grains change the equilibrium quasineutrality conditions as well as introduce new features to collective phenomena when their dynamics is taken into consideration. The dust grain dynamics allows the possibility of new wave modes [8, 9, 10, 11], e.g. the dust acoustic waves, dust surface vortex modes, dust Coulomb waves, dust lattice waves, which are spectacularly observed in several laboratory experiments. When the waves are of large amplitudes, nonlinearities come into the picture [10, 11, 12]. Due to nonlinearities, wave fronts steepen and shocks are formed. If there is a delicate balance between nonlinearities and dissipation (e.g. due to dust charge fluctuations or dust fluid kinematic viscosity), one has the possibility of monotonic shock formation in dusty plasmas. In a collisionless system, a competition between nonlinearities and dispersion leads to the formation of bare solitons in plasmas. In a nonuniform plasma, the dust fluid motion can be incompressible and the two-dimensional dust flow can appear as eddies (vortices). In a driven dusty plasma system, say when electron or electromagnetic beams are fired into the plasma, we can have the trapping of plasmons and photons in dust density holes which are self-consistently created by the ponderomotive force of the high-frequency localized wave envelops. Such a structure is referred to as an envelope soliton. In this paper, we first present a fully nonlinear theory for Langmuir envelope solitons in a uniform plasma, taking into account the fully nonlinear dust acoustic wave response in the presence of the ponderomotive force of Langmuir waves [13]. Second, we consider the formation of structures associated with incompressible surface dust vortex modes (SDVMs) [14] and dust zonal flows (DZFs) in a nonuniform dusty plasma. The present results can help us to understand the salient features of localized excitations of Langmuir wave envelopes and coherent vortical structures in laboratory and space plasmas containing micron-sized charged dust particulates.

2 Langmuir envelope solitons

We here present a fully nonlinear theory for envelope solitons in a complex (dusty) plasma whose constituents are electrons, ions, and negatively charged massive dust grains [13]. At equilibrium, we have $n_{i0} = n_{e0} + Z_d n_{d0}$, where n_{j0} is the unperturbed number density of the particle species j (j equals i for ions, e for electrons, and d for dust grains) and Z_d is the number of electrons residing on the dust grain surface. Dust grains in laboratory plasmas are typically charged negatively due to collection of electrons from the ambient plasma [9, 15], but under UV radiation dust can also be charged positively [16]. We suppose that the presence of electron beams in a dusty plasma generates large amplitude Langmuir waves whose frequency is $\omega = (\omega_{pe}^2 + 3k^2 V_{Te}^2)^{1/2}$, where $\omega_{pe} = (4\pi n_e e^2/m_e)^{1/2}$ is the electron plasma frequency, n_e is number density of the electrons, e is the magnitude of the electron charge, m_e is the electron mass, k is the wavenumber, $V_{Te} = (T_e/m_e)^{1/2}$ is the electron thermal speed, and T_e is the electron temperature. Large amplitude Langmuir waves interacting nonlinearly with finite amplitude dust acoustic perturbations generate Langmuir wave electric field envelope whose electric field E evolves slowly in comparison with the electron plasma wave period according to a nonlinear Schrödinger equation

$$2i\omega_p \left(\frac{\partial}{\partial t} + v_g \frac{\partial}{\partial x} \right) E + 3V_{Te}^2 \frac{\partial^2 E}{\partial x^2} + \omega_p^2 \left(1 - \frac{n_e}{n_{e0}} \right) E = 0, \qquad (1)$$

where $\omega_p = (4\pi n_{e0} e^2/m_e)^{1/2}$ is the unperturbed electron plasma frequency and $v_g = 3k V_{Te}^2/\omega_p$ is the group velocity of the Langmuir waves. We note that Eq. (1) is derived by combining the electron continuity and momentum

Dust Plasma Interactions in Space and Laboratory 215

equations as well as by using Poisson's equation with fixed ions and retaining the arbitrary large electron number density perturbation n_{e1} associated with the dust acoustic waves in the presence of the Langmuir wave ponderomotive force. For our purposes, we have

$$n_e = n_{e0} \exp(\varphi - W^2), \tag{2}$$

where $\varphi = e\phi/T_e$, $W^2 = |E|^2/16\pi n_{e0}T_e$, and ϕ is the electrostatic potential of the DAWs whose phase speed is much smaller than the electron and ion thermal speeds. The ion number density perturbation associated with the DAWs is

$$n_i = n_{i0} \exp\left[-\tau(\varphi + \mu_i W^2)\right], \tag{3}$$

where $\tau = T_e/T_i$, T_i is the ion temperature, $\mu_i = m_e/m_i$, and m_i is the ion mass. We note that the W-terms in Eqs. (2) and (3) come from the averaging of the nonlinear term $m_j \mathbf{v}_{hj} \cdot \nabla \mathbf{v}_{hj}$ over the Langmuir wave period $2\pi/\omega_{pe}$, where $\mathbf{v}_{hj} \approx q_j E/m_j \omega_{pe}$ is the quiver velocity of the particle species j in the Langmuir wave electric field, $q_e = -e$, $q_i = e$, and $q_d = -Z_d e$. The dust dynamics is governed by the dust continuity and momentum equations

$$\frac{\partial n_d}{\partial t} + \frac{\partial(n_d u_d)}{\partial x} = 0, \tag{4}$$

and

$$m_d n_d \left(\frac{\partial u_d}{\partial t} + u_d \frac{\partial u_d}{\partial x}\right) = Z_d n_d T_e \frac{\partial(\varphi - \mu_d W^2)}{\partial x}, \tag{5}$$

where $\mu_d = Z_d m_e/m_d$, n_d is the dust number density, and u_d is the x component of the dust fluid velocity. The phase speed of the DAWs is assumed to be much larger than the dust thermal speed. The equations are closed with Poisson's equation

$$\lambda_{De}^2 \frac{\partial^2 \varphi}{\partial x^2} = \left(\frac{n_e}{n_{e0}} - \frac{n_i}{n_{e0}} + \frac{Z_d n_d}{n_{e0}}\right), \tag{6}$$

where $\lambda_{De} = (T_e/4\pi n_{e0}e^2)^{1/2}$ is the electron Debye radius. Thus, our nonlinear theory of cavitons accounts for arbitrary large amplitude density variations (and associated space charge potential ϕ) that are associated with fully nonlinear dispersive DAWs, contrary to the small amplitude cavitons based on the cubic nonlinear Schrödinger equation [4], which discards nonlinearities and the departure from the quasi-neutrality in the plasma slow motion.

We are interested in quasi-steady state solutions of Eqs. (1)–(6). Accordingly, we insert $E(x,t) = W(\xi) \exp\{i[X(x) + T(t)]\}$, $n_d(x,t) = n_d(\xi,t)$, $u_d(x,t) = u_d(\xi,t)$ and $\varphi(x) = \varphi(\xi)$, where $\xi = x - Vt$, and V is the constant speed of the soliton and $W(x)$, $X(x)$, $T(x)$ are assumed to be real. into Eqs. (1)–(6) we finally obtain the coupled set of nonlinear equations

216   Padma K. Shukla, Bengt Eliasson, and Dastgeer Shaikh

$$3\frac{\partial^2 W}{\partial \xi^2} - (\lambda - 1)W - W \exp(\varphi - W^2) = 0, \tag{7}$$

and

$$\frac{\partial^2 \varphi}{\partial \xi^2} - \exp(\varphi - W^2) + \alpha \exp\left[-\tau(\varphi + \mu_i W^2)\right] + (1-\alpha)\frac{M}{\sqrt{M^2 + 2(\varphi - \mu_d W^2)}} = 0, \tag{8}$$

where ξ is normalized by λ_{De}, $\lambda = 2\omega_p^{-1}(dT/dt) - 3k^2\lambda_{De}^2(1 - V^2/v_g^2)$ represents a nonlinear frequency shift, $\alpha = n_{i0}/n_{e0}$, and $M = V/C_D$ is the Mach number involving the dust acoustic speed $C_D = (Z_d T_e/m_d)^{1/2}$. Since in complex plasmas, we typically have $\mu_i \ll 1$ and $\mu_d \ll 1$, the contributions of ion and dust ponderomotive forces in Eq. (8) can be safely dropped. Thus, the electron ponderomotive force is transmitted to ions and dust via the ambipolar potential.

The system of Eqs. (7) and (8), without the $\mu_i W^2$ and $\mu_d W^2$ terms in Eq. (8), admits the first integral in the form of a Hamiltonian

$$H(W, \varphi, \lambda, M) = 3\left(\frac{\partial W}{\partial \xi}\right)^2 - \frac{1}{2}\left(\frac{\partial \varphi}{\partial \xi}\right)^2 - (\lambda - 1)W^2 + \exp(\varphi - W^2) - 1$$
$$+ \frac{\alpha}{\tau}\left[\exp(-\tau\varphi) - 1\right] + (\alpha - 1)M\left(\sqrt{M^2 + 2\varphi} - M\right) = 0, \tag{9}$$

where in the unperturbed state ($|\xi| = \infty$) we have used the boundary conditions $W = 0$, $\varphi = 0$, $\partial W/\partial \xi = 0$, $\partial \varphi/\partial \xi = 0$. Because we are interested in symmetric solutions defined by $W(\xi) = W(-\xi)$ and $\varphi(\xi) = \varphi(-\xi)$, the appropriate boundary conditions at $\xi = 0$ are $W = W_0$, $\varphi = \varphi_0$, $\partial W/\partial \xi = 0$, and $\partial \varphi/\partial \xi = 0$. Hence, from Eq. (9) we have $\exp(\varphi_0 - W_0^2) - 1 - (\lambda - 1)W_0^2 + (\alpha/\tau)\left[\exp(-\tau\varphi_0) - 1\right] + (\alpha - 1)M\left(\sqrt{M^2 + 2\varphi_0} - M\right) = 0$, which shows how the maximum values of W_0 and φ_0 are related with M and λ for given values of τ and α. It should be stressed that the static dust case has to be treated separately, where the last term in the left-hand side of Eq. (9) has to be replaced by $(\alpha - 1)\varphi$. Accordingly, the last term in the left-hand side of the simplified Hamiltonian above will be replaced by $(\alpha - 1)\varphi_0$.

In the absence of the Langmuir waves, the nonlinear DAWs are governed by the energy integral

$$\frac{1}{2}\left(\frac{\partial \varphi}{\partial \xi}\right)^2 + \Psi(\varphi, M) = 0, \tag{10}$$

where the Sagdeev potential is

$$\Psi(\varphi, M) = 1 - \exp(\varphi) + \frac{\alpha}{\tau}\left[1 - \exp(-\tau\varphi)\right] + (1 - \alpha)M\left(\sqrt{M^2 + 2\varphi} - M\right). \tag{11}$$

Equation (10), which is obtained from Eq. (9) in the limit of vanishing Langmuir wave electric fields, determines the profile of non-envelope (bare) dust acoustic solitary waves. The latter exist provided that $\Psi(\varphi)$ is negative between zero and $\pm\varphi_0$. Multivalued solutions of $\Psi(0)$ are ensured provided that $\partial^2\Psi/\partial^2\varphi = 0$, while at $\varphi = \varphi_0(-\varphi_0)$, we must have $\partial\Psi/\partial\varphi > 0(< 0)$. The condition $\Psi(\varphi_0, M) = 0$ gives a relation between φ_0 and M for given values of α and τ. It turns out that dust acoustic solitons have sub-dust acoustic speed, negative potential and dust density hump.

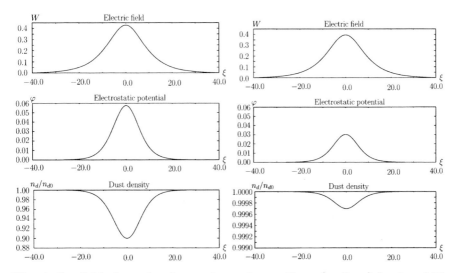

Fig. 1. Small Mach number Langmuir envelope solitons (cavitons) for $\lambda = 0.06$, $\tau = 2$, $\alpha = 2$, and $M = 0.7$ (left panel) and $M = 10$ (right panel). After Ref. [13].

The numerical solutions of Eqs. (7) and (8) are displayed in Figs. 1–2 for dusty plasmas containing micron-sized dust grains with $Z_d = 10^3$ and $m_d/m_i = 10^{12}$. The ion to electron mass ratio is typically 1836 and more depending upon ionized dusty gases. Figure 1 shows the profiles of a caviton for $M = 0.7$ and $M = 10$ with $\lambda = 0.06$ and $\alpha = 2$. We see that the envelope Langmuir solitons in complex plasmas are composed of a bell shaped Langmuir electric field and a dust density hole in association with a positive localized space charge electric potential. For higher Mach numbers, the influence of the potential on the dust dynamics becomes smaller; as can be seen from the dust density perturbations in Fig. 1. Thus, large and small amplitude cavitons move with sub and super dust acoustic speeds, respectively. By performing several numerical experiments we studied the influence of the electron to ion temperature ratio τ on the caviton. We found that the depth of the caviton decreases as τ increases, if the other parameters are kept constant.

In order to see the difference between the cavitons and the bare dust acoustic solitary waves [8], we integrated Eq. (10) numerically. The results

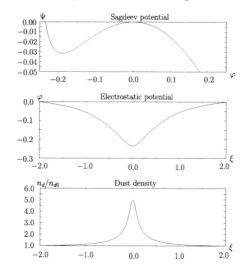

Fig. 2. Bare dust acoustic soliton ($W = 0$) for $\tau = 2$, $\alpha = 2$, and $M = 0.7$. After Ref. [13].

are displayed in Fig. 2. The numerical results shows that the dust acoustic solitary waves have sub–dust acoustic speed and they are associated with negative space charge potentials and the dust density hump. As can be seen from Fig. 2, the bare soliton also develops on a much smaller length scale than the envelope cavitons in Fig. 1. Evidently, the properties of cavitons are very different from the bare dust acoustic solitary waves. We suggest that new beam-plasma experimental and computer simulation studies should be conducted for verifying our theoretical predictions of dust cavitons. We also hope that forthcoming data from the Cassini mission will also reveal signatures of magnetic field aligned large amplitude localized Langmuir electric fields as well as an associated positive potential and finite amplitude dust density hole in Saturn's rings where $n_{e0} \sim 50$ cm^{-3}, $n_{i0} \sim 950$ cm^{-3}, $n_{d0} \sim 1$ cm^{-3}, $Z_d \sim 10^3$, the dust radius $r_d \sim 1$–5 μm, and $T_e \sim T_i \approx 100$ eV. In such a Saturn plasma environment, we expect an ambipolar potential of ten volts and a relative dust density depletion of ten percent if $E \sim 0.3$ millivolts/m and $M = 0.96$.

3 Self-steepening and shock formation

The formation of dust acoustic (DA) and dust ion-acoustic (DIA) shocks are are studied theoretically and numerically by means of simple-wave solutions and a comparison between fluid and kinetic model for DIA waves. A fluid model admits sharp discontinuities at the shock front while the kinetic model involves Landau-damping of the the shock front.

Dust Plasma Interactions in Space and Laboratory   219

3.1 Dust acoustic shocks

We present non-stationary solutions of fully nonlinear non-dispersive DAWs in an unmagnetized dusty plasma [22]. The inertialess electrons and ions are assumed to be Boltzmann distributed, $N_e(\phi) = \exp(\varphi)$ and $N_i(\phi) = \beta \exp(-\tau\varphi)$, where $N_e = n_e/n_{e0}$, $N_i = n_i/n_{i0}$, $\varphi = e\phi/T_e$, $\beta = n_{i0}/n_{e0}$, and $\tau = T_i/T_e$. At equilibrium, we have $n_{i0} = n_{e0} - \epsilon Z_d n_{d0}$, where n_{d0} is the unperturbed dust number density, Z_d is the dust charge state, and ϵ equals -1 ($+1$) for negatively (positively) charged dust grains. From the quasineutrality condition $n_e - n_i - \epsilon Z_d n_d = 0$, we have $N_d = [\beta \exp(-\tau\varphi) - \exp(\varphi)]/(\beta - 1)$ where $N_d = n_d/n_{d0}$. The continuity and momentum equation for the cold dust fluid is

$$\left(\frac{\partial}{\partial t} + u\frac{\partial}{\partial x}\right)[\beta \exp(-\tau\varphi) - \exp(\varphi)] + [\beta \exp(-\tau\varphi) - \exp(\varphi)]\frac{\partial u}{\partial x} = 0, \quad (12)$$

and

$$\frac{\partial u}{\partial t} + u\frac{\partial u}{\partial x} = -\epsilon\frac{\partial \varphi}{\partial x}, \quad (13)$$

where u is the normalized [by $C_d = (Z_d T_e/m_d)^{1/2}$] dust fluid velocity along the x axis, the time and space are normalized by the dust plasma period ω_{pd}^{-1} and the Debye radius C_d/ω_{pd}, respectively, where $\omega_{pd} = (4\pi Z_d^2 e^2 n_{d0}/m_d)^{1/2}$ is the dust plasma frequency. Simple waves solutions of Eqs. (12) and (13) can be found by rewriting them as

$$\frac{\partial}{\partial t}\begin{bmatrix} \varphi \\ u \end{bmatrix} + \begin{bmatrix} u & -\chi(\varphi) \\ \epsilon & u \end{bmatrix}\frac{\partial}{\partial x}\begin{bmatrix} \varphi \\ u \end{bmatrix} = 0, \quad (14)$$

where $\chi(\varphi) = [\beta \exp(-\tau\varphi) - \exp(\varphi)]/[\beta\tau \exp(-\tau\varphi) + \exp(\varphi)]$. The square matrix in the left-hand side of Eq. (14) is diagonalized by a diagonalizing matrix whose columns are eigenvectors of the matrix in Eq. (14), yielding

$$\frac{\partial\psi_\pm}{\partial t} + \lambda_\pm\frac{\partial\psi_\pm}{\partial x} = 0 \quad (15)$$

where the eigenvalues are $\lambda_\pm = u \pm \sqrt{-\epsilon\chi(\varphi)}$, and the new unknowns $\psi_\pm = u \mp F(\varphi)$ and $F(\varphi) = \int_0^\varphi [-\epsilon/\chi(s)]^{1/2} ds$. A simple-wave solution is found by setting ψ_- to zero so that $u = -F(\varphi)$, $\psi_+ = 2u$ and from Eq. (15) $\partial u/\partial t + \lambda_+(\varphi)\partial u/\partial x = 0$, where $\lambda_+(\varphi) = -F(\varphi) + \sqrt{-\epsilon\chi(\varphi)}$. Since u is a function of φ, we also have

$$\frac{\partial\varphi}{\partial t} + \lambda_+(\varphi)\frac{\partial\varphi}{\partial x} = 0, \quad (16)$$

which describes the self-steepening of the potential φ. The general solution of Eq. (16) for continuous φ is $\varphi = f_0[x - \lambda_+(\varphi)t]$, where f_0 is a function of one variable, obtained from the initial condition for φ at $t = 0$. The effective phase speed $\lambda_+(\varphi)$ is a function of the solution φ, and the solution may therefore self-steepen and develop shocks.

In recent experiments [23], it was observed that large-amplitude waves, $N_d = 2.2$, were associated with a ratio $u/v_{ph} \sim 0.5$ to 0.8 between the maximum particle particle (fluid) velocities and the phase speed of the waves. Linear theory gives $v_{ph}/u = N_d - 1 = 1.2$, which deviates from the experimental results [23]. Hence, we compare the experimental values with our theory. Using the dust density $N_d = 2.2$, the theoretical ratio between the particle and phase velocities, before shocks have developed, is $u/v_{ph} = (N_d-1)/N_d \approx 0.55$, and the ratio between the particle velocity and the speed of fully developed shock fronts is $u/v_{shock} = 2(N_d - 1)/(N_{d,left} - N_{d,right}) \approx 0.75$, where we used $N_{d,left} = 2.2$ and $N_{d,right} = 1$ ($N_{d,right}$ is the unperturbed dust density in front of the shock). The theoretical values are thus in excellent agreement with the experimental ones. In Fig 3, we have solved the simple-wave equation

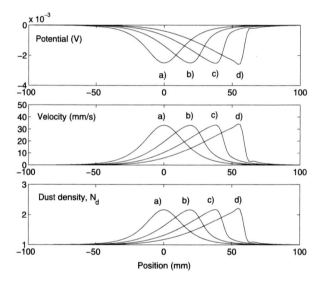

Fig. 3. The time evolution of the electrostatic potential, dust velocity and dust density with $\beta = 1.1$ and $\tau = 100$, for a) $t = 0$ s, b) $t = 0.24$ s, c) $t = 0.48$ s, and d) $t = 0.73$ s. After Ref. [22].

(16) numerically with parameters from the experiment in Ref. [23]. In order to convert the dimensionless units to dimensional ones used in the experiment, x was multiplied by 1.2, t by 1.1×10^{-3}, and u by 1200, and φ by 2.5. We see a clear signature of self-steepening of the DAWs and the creation of shocks [curves marked d) in Fig. 3], which can also be seen in some of the experiments of Ref [23].

3.2 Dust ion-acoustic shocks

The formation of dust ion-acoustic (DIA) shocks have been observed in recent experiments [24, 25, 26, 27]. We here discuss a theoretical model for DIA shocks in an unmagnetized dusty plasma [28], where the inertialess electrons are Boltzmann distributed, $N_e = \exp(\varphi)$, and the ions is governed by the normalized momentum and continuity equations [28]

$$\frac{\partial u}{\partial t} + u\frac{\partial u}{\partial z} + \left(\frac{1}{N+\alpha-1} + 3\tau N\right)\frac{\partial N}{\partial z} = 0, \tag{17}$$

and

$$\frac{\partial N}{\partial t} + N\frac{\partial u}{\partial z} + u\frac{\partial N}{\partial z} = 0, \tag{18}$$

where $\alpha = n_{e0}/n_{i0}$ and $\tau = T_{i0}/T_e$. The time is normalized by the ion plasma frequency and space by the electron Debye radius. Simple wave solutions of Eqs. (17) and (18) can be found by rewriting them in the matrix form

$$\frac{\partial}{\partial t}\begin{bmatrix} u \\ N \end{bmatrix} + \begin{bmatrix} u & \frac{1}{N+\alpha-1} + 3\tau N \\ N & u \end{bmatrix}\frac{\partial}{\partial z}\begin{bmatrix} u \\ N \end{bmatrix} = \begin{bmatrix} 0 \\ 0 \end{bmatrix}, \tag{19}$$

where the nonlinear wave speeds are given by the eigenvalues

$$\lambda_\pm = u \pm N^{1/2}\left(\frac{1}{N+\alpha-1} + 3\tau N\right)^{1/2} \tag{20}$$

of the square matrix multiplying the second term in Eq. (19). This matrix can be diagonalized by a diagonalizing matrix so that Eq. (19) takes the form [28]

$$\frac{\partial \psi_\pm}{\partial t} + \lambda_\pm\frac{\partial \psi_\pm}{\partial z} = 0, \tag{21}$$

where the new variables are

$$\psi_\pm = \frac{u}{2} \pm \frac{1}{2}\int_1^N \left(\frac{1}{N'(N'+\alpha-1)} + 3\tau\right)^{1/2} dN'. \tag{22}$$

Setting ψ_- to zero, we have $u(N) = \int_1^N [1/N'(N'+\alpha-1)+3\tau]^{1/2}\, dN'$, which inserted into Eq. (20) gives λ_+ as a function of N, and we have from Eq. (21) that

$$\frac{\partial N}{\partial t} + \lambda_+(N)\frac{\partial N}{\partial z} = 0, \tag{23}$$

with the general solution (for continuous N) $N = f_0(\xi)$, where $\xi = x - \lambda_+(N)t$ and f_0 is the initial condition for N. We have plotted λ_+ as a function of N in the left panel of Fig. 4. We see that λ_+ grows with increasing N in the two cases with $\alpha = 0.25$. For $\lambda = 0.05$, however, the phase speed first decreases for $N \approx 1$ while it increases for large N. In order to study the impact of kinetic effects, we have compared numerical solutions of Eqs. (17) and (18) with numerical

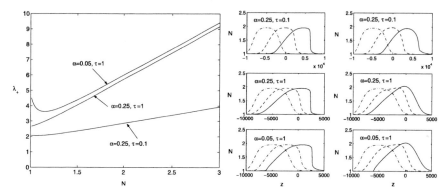

Fig. 4. The wave speed λ_+ as a function of N for different values on τ and α (left panel). The ion density N obtained from numerical solutions of the fluid and the Vlasov equation are displayed in the two right columns of panels, respectively. In the upper panels, N is shown at $t = 0$ (dash-dotted lines), $t = 1700$ (dashed lines) and $t = 3400$ (solid lines), while in the middle and lower panels N is shown at $t = 0$, $t = 500$ and $t = 1000$. After Ref. [28].

solutions of the Vlasov equation; see the right panels of Fig. 4. For $\tau = 0.1$ and $\alpha = 0.25$, we see that both the fluid and Vlasov solutions exhibit shocks. The shock front is distinct in the fluid solution and more diffuse in the Vlasov solution. For $\tau = 0.1$ and $\alpha = 0.25$, the fluid solution exhibits distinct shocks, while the Vlasov simulation shows a phase of self-steepening at $t = 500$, followed by an expansion of the diffuse shock at $t = 1000$. For $\alpha = 0.05$ and $\tau = 1$, the fluid solution shows a shock in the front end of the pulse, while the rear end of the pulse also steepens. The steepening of the pulse perturbations in the rear of the pulse can be explained by that the wave speed *decreases* for small-amplitude density perturbations ($1 < N < 1.1$), as seen in the left panel of Fig. 4, while it increases again for large-amplitude density perturbations. We finally note that the wave speed must be much larger than the sum of the ion thermal and fluid velocities [28], i.e. $\lambda_+ \gg V_T + u$, for the acceleration of particles and Landau damping of the shock to be negligible. Inserting the expressions for $u(N)$ and $\lambda_+(N)$ into this inequality, where the scaled ion thermal speed $V_T = (\tau T)^{1/2} \approx \tau^{1/2} N$, we obtain $N^{1/2}[1/(N+\alpha-1) + 3\tau N]^{1/2} \gg (\tau N^2)^{1/2}$, or $(1/[(N+\alpha-1)N\tau]+3)^{1/2} \gg 1$. This condition is fulfilled if $\tau \ll 1$ or/and if the electrons are evacuated due to the dust so that $\alpha \ll 1$, and at the same time $N = 1 + N_1$ where $|N_1| \ll 1$.

4 Surface dust vortices and zonal flows

Our investigations of surface dust vortices and zonal flows are based on a model put forward by Hasegawa and Shukla [14], who theoretically pointed out the existence of incompressible SDVMs in a nonuniform, unmagnetized dusty plasma. We first discuss the equilibrium state of our partially ionized laboratory dusty discharges in which collisions between stationary neutrals with electrons and ions are more frequent than those between electrons and ions. Here, the equilibrium electron and ion fluid velocities are $\mathbf{u}_{e0} \approx -(e\mathbf{E}_0/m_e\nu_{en}) - \nabla P_{e0}/n_{e0}m_e\nu_{en}$ and $\mathbf{u}_{i0} \approx e\mathbf{E}_0/m_i\nu_{in}$, respectively, where e is the magnitude of the electron charge, $\mathbf{E}_0$ is the DC electric field, m_e (m_i) is the electron (ion) mass, ν_{en} (ν_{in}) is the electron (ion)-neutral collision frequency, and P_{e0} is the unperturbed electron pressure which is much larger than the ion pressure. From the conservation of the unperturbed current density, when sources and sinks (ionization, recombination, etc) in steady state balance, we then deduce a vertically upward (along the z axis) electric field $E_{0z} = [1/n_{e0}e(1+\alpha)]\,\partial P_{e0}/\partial z$, which levitates a negatively charged dust grain in the plasma sheath due to a balance between the electric force ($= Q_d E_{0z}$) and the vertically downward gravity force $-m_d g$. Here, $\alpha = n_{i0}m_e\nu_{en}/n_{e0}m_i\nu_{in}$, $n_{i0} = n_{e0} - Q_d n_{d0}/e$, n_{j0} is the unperturbed density of the particle species j (j equals e for electrons, i for ions, and d for dust grains), $Q_d = -Z_d e$ is the dust charge, Z_d is the number of electrons residing on the dust, E_{0z} is the z component of $\mathbf{E}_0$, m_d is the dust mass, and g is the gravity constant. We suppose that the dust charge distribution and the equilibrium plasma densities are nonuniform along the z axis. When the equilibrium is perturbed, one has the possibility of nonlinear SDVMs in the form of a bipolar vortex or a chain of vortex [3, 4], which can be associated with coherent vortical structures in laboratory experiments [6]. When the phase speed (wavelength) of the SDVMs and DZFs is much smaller than the electron and ion thermal speeds (electron and ion collisional mean free paths $V_{Te,Ti}/\nu_{en,in}$, where $V_{Te}(V_{Ti})$ is the electron (ion) thermal speed) in dusty plasmas, the perturbed electrostatic forces ($q_j n_{e0,i0}\mathbf{E}_1$) acting on electrons and ions balance the corresponding pressure gradient $-\nabla P_{j1}$, where $q_e = -e$, $q_i = e$, and $\mathbf{E}_1$ (P_{j1}) is the perturbed electric field (perturbed pressure). The dynamics of incompressible ($\nabla \cdot \mathbf{v}_d = 0$) SDVMs and DZFs is then governed by the dust continuity and dust momentum equations, namely [14]

$$\frac{\partial \rho_d}{\partial t} + \nabla \cdot (\rho_d \mathbf{v}_d) = 0, \tag{24}$$

and

$$\rho_d \left(\frac{\partial}{\partial t} + \nu_d - \eta \nabla^2 + \mathbf{v}_d \cdot \nabla \right) \mathbf{v}_d = -\nabla P_1 + \rho_d g \hat{\mathbf{z}}, \tag{25}$$

where $\rho_d = m_d(n_{d0} + n_{d1})$, $n_{d1}(\ll n_{d0})$ is a small perturbation in the equilibrium dust number density, $\mathbf{v}_d$ is the perturbed dust fluid velocity, ν_d is the dust-neutral collision frequency, η represents the kinematic dust fluid viscosity (typically $\simeq 10^{-2} - 10^{-1}$ cm^2/s in laboratory, similar to that of water $\sim 10^{-2}$ cm^2/s), and $P_1 = P_{e1} + P_{i1} + P_{d1}$ is the perturbation in the equilibrium pressure. We stress that Eq. (25), in which the electric force on charged dust grains is eliminated by using $\mathbf{E}_1 = (e/Q_d n_d)\nabla(P_{e1} + P_{i1})$, is widely used in the investigation of collective processes in dusty plasmas [4].

224   Padma K. Shukla, Bengt Eliasson, and Dastgeer Shaikh

Two-dimensional incompressible SDVMs and DZFs are characterized by the velocity vectors $\mathbf{v}_s = \hat{\mathbf{x}} \times \nabla \psi_s(y, z)$ and $\mathbf{v}_z = \hat{\mathbf{x}} \times \nabla \psi_z(y, z)$, respectively. Here, $\hat{\mathbf{x}}$ is the unit vector along the x axis, which is perpendicular to the z axis, and ψ_s and ψ_z are the stream functions of the SDVMs and DZFs, respectively. Thus, there exist SDVM and DZF vorticities characterized by $\Omega_s = \nabla \times \mathbf{v}_s \equiv \hat{\mathbf{x}} \nabla_\perp^2 \psi_s(y, z)$ and $\Omega_z = \nabla \times \mathbf{v}_z \equiv \hat{\mathbf{x}} \nabla_\perp^2 \psi_z$. Letting $\rho_d = \rho_{d0} + \rho_{ds}$ and $\mathbf{v}_d = \mathbf{v}_s + \mathbf{v}_z$ in Eqs. (24) and (25), we obtain the governing equations for the SDVMs in the presence of DZFs. We have

$$\left(\frac{\partial}{\partial t} + \hat{\mathbf{x}} \times \nabla \psi_s \cdot \nabla \right) \rho_{ds} + \rho'_{d0} \frac{\partial \psi_s}{\partial y} + (\hat{\mathbf{x}} \times \nabla \psi_z \cdot \nabla) \rho_{ds} = 0, \tag{26}$$

and

$$\left(\frac{\partial}{\partial t} + \nu_d - \eta \nabla^2 + \hat{\mathbf{x}} \times \nabla \psi_s \cdot \nabla \right) \nabla_\perp^2 \psi_s + g \frac{\partial \rho_{ds}}{\partial y}$$
$$+ (\hat{\mathbf{x}} \times \nabla \psi_s \cdot \nabla) \nabla_\perp^2 \psi_z + (\hat{\mathbf{x}} \times \nabla \psi_z \cdot \nabla) \nabla_\perp^2 \psi_s = 0. \tag{27}$$

where $\rho'_{d0} = \partial \rho_{d0}/\partial z$ and $\rho_{d0}(z) \ll \rho_{ds}$. We note that $\rho_{dz} \ll \rho_{ds}$ due to insignificant variation of the dust ZF stream function along the y axis.

The dynamics of DZFs in the presence of the SDVMs is governed by

$$\left(\frac{\partial}{\partial t} + \nu_d - \eta \nabla^2 + \hat{\mathbf{x}} \times \nabla \psi_z \cdot \nabla \right) \nabla_\perp^2 \psi_z + \Omega_{ave} = 0, \tag{28}$$

where $\Omega_{ave} = \langle \hat{\mathbf{x}} \times \nabla \psi_s \cdot \nabla \nabla_\perp^2 \psi_s \rangle$ and the angular bracket denotes the averaging over the SDVM period. In the absence of the nonlinear interactions, the SDVMs and DZFs are decoupled. The corresponding dispersion relations, obtained from (26)–(28), are $\omega^2 + i\omega \left(\nu_d + \eta k_\perp^2 \right) - \Omega_B^2 k_y^2/(k_y^2 + k_z^2) = 0$ and $\Omega + i \left(\nu_d + \eta k_\perp^2 \right) = 0$, respectively. Here, $\omega(\Omega)$ is the frequency of the SDVMs (DZFs), k_y (k_z) is the component of the wavevector along the y (z) axis. The buoyancy frequency squared is denoted by $\Omega_B^2 = g \partial \ln \rho_{d0}/\partial z$. If the equilibrium dust density is proportional to $\exp(-z/L)$, where $L = \rho_{d0}/\rho'_{d0}$ is the dust density gradient scalesize, then $\Omega_B^2 = -g/L$. We note that the latter is a positive definite for $L < 0$. In the absence of dissipation, we observe the frequency condensation of SDVMs for $k_z \gg k_y \to 0$, indicating the possibility of SDVMs driven DZFs with short scale structures along the z axis, as discussed below.

We first consider the parametric excitation of DZFs by large amplitude SDVMs. For this purpose, we neglect the self-interaction mode couplings in (26)–(28) and let $\rho_{ds} = \rho_{0\pm} \exp(\pm i\mathbf{k}_0 \cdot \mathbf{r} \mp i\omega_0 t) + \sum_{+,-} \rho_\pm \exp(i\mathbf{k}_\pm \cdot \mathbf{r} - i\omega_\pm t)$, $\psi_s = \psi_{0\pm} \exp(\pm i\mathbf{k}_0 \cdot \mathbf{r} \mp i\omega_0 t) + \sum_{+,-} \rho_\pm \exp(i\mathbf{k}_\pm \cdot \mathbf{r} - i\omega_\pm t)$, and $\psi_z = \varphi \exp(i\mathbf{q} \cdot \mathbf{r} - i\Omega t)$, where the subscript $0\pm$ $(\pm)$ stands for the SDVM pump (sidebands), $\mathbf{k}_\pm = \mathbf{q} \pm \mathbf{k}_0$, and $\omega_\pm = \Omega \pm \omega_0$. Fourier transforming (26)–(28) and matching the phasers, we obtain

$$H_\pm \psi_\pm = \pm i \frac{\omega_0 \hat{\mathbf{x}} \times \mathbf{q} \cdot \mathbf{k}_0}{k_\pm^2} \left(g k_{oy} + 2 k_{\perp 0}^2 - q_\perp^2 \right) \psi_{0\pm} \varphi, \tag{29}$$

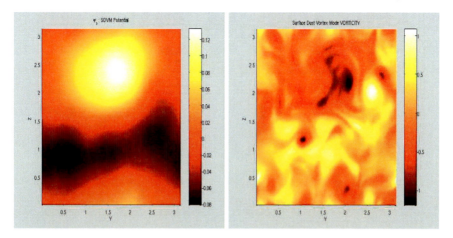

Fig. 5. Time evolution of the SDVMs yields long scale potential flows (left) and vorticity (right) for which k_z is finite and $k_y \approx 0$. The vorticity component $\nabla^2 \psi_s$ consequently possesses elongated small scale structures.

and

$$(\Omega + i\nu)\varphi = i\frac{\hat{\mathbf{x}} \times \mathbf{k}_0 \cdot \mathbf{q}}{q_\perp^2}\left(K_-^2 \psi_{0+}\psi_- - K_+^2 \psi_{0-}\psi_+\right), \quad (30)$$

where $H_\pm = \omega_\pm^2 + i(\nu_d + \eta k_\pm^2)\omega_\pm - \Omega_B^2 k_{0y}^2/k_\pm^2$, $k_\pm^2 = k_{0y}^2 + k_{z\pm}^2$, $K_\pm^2 = k_{\perp\pm}^2 - k_{0\perp}^2$, and $\nu = \nu_d + \eta q_\perp^2$. Eliminating $\psi_\pm$ from (30) by using (29), we have the nonlinear dispersion relation

$$\Omega + i\nu = \omega_0 \frac{|\hat{\mathbf{x}} \times \mathbf{k}_0 \cdot \mathbf{q}|^2}{q_\perp^2}\left(gk_{0y} + 2k_{\perp 0}^2 - q_\perp^2\right)\sum_{+,-}\frac{K_\pm^2 |\psi_0|^2}{H_\pm}, \quad (31)$$

where $|\psi_0|^2 = \psi_{0+}\psi_{0-}$. For $q_\perp \ll |\mathbf{k}_0|$ we obtain from (31) $\Omega(\Omega + i\nu) = \left(|\hat{\mathbf{x}} \times \mathbf{k}_0 \cdot \mathbf{q}|^2/q_\perp^2\right)(gk_{0y} + 2k_{\perp 0}^2)\mathbf{q} \cdot \mathbf{k}_{0\perp}|\psi_0|^2$, which predicts a purely growing instability ($\Omega = i\gamma$) if the growth rate $\gamma > \nu$ and $(gk_{0y} + 2k_{\perp 0}^2)\mathbf{q} \cdot \mathbf{k}_{0\perp} < 0$. For some typical values, viz $|\mathbf{v}_s| \sim C_D$, $|\mathbf{k}/\mathbf{k}_0| \sim 0.1$, we have $\gamma \sim$ one tenth of the dust acoustic wave frequency $k_0 C_D$, where C_D is the dust acoustic speed [4].

Next, in order to study the dynamics of nonlinearly coupled SDVMs and DZFs, we develop a spectral code to carry out high resolution computer simulations of (26)–(28) in a periodic box of length π in each directions. The time integration uses a second order predictor-corrector method. The spatial resolution is 1024 × 1024 Fourier modes. All fluctuations in the simulations are initialized with a Gaussian random number generator to ensure that the Fourier modes are all spatially uncorrelated and randomly phased. This ensures a nearly isotropic initial condition in the real space. We further make sure that no asymmetry is introduced in the dynamical evolution by the initial spectra and the boundary conditions. The normalization is as follows; ρ_{ds} by ρ_{d0}, $\psi_{s,z}$ by $C_D \lambda_D$, t by ω_{pd}^{-1}, and the space variable by $\lambda_D = C_D/\omega_{pd}$, where

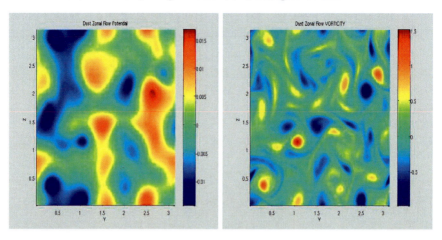

Fig. 6. The potential (left) and vorticity (right) of dust zonal flows (DZF), which tends to cascade towards shorter length scales that eventually align along the flow direction.

$\lambda_D = C_D/\omega_{pd}$ is the dusty plasma Debye radius [4]. Thus, the free parameters in our simulations are $\nu_d/\omega_{pd} \sim 10^{-2} - 10^{-3}$, $\eta/C_D\lambda_D \sim 10^{-2}$, $\lambda/L \sim 0.01$ and $g\lambda/C_D^2 \sim \lambda_D/L_p \sim 0.01$, where $L_p^{-1} = -P_0^{-1}\partial P_0/\partial z$. In Figs. 5 and 6 we show the evolutions of the stream functions and vorticities of the SDVMs and DZFs, respectively. We see that stream functions are coherent, contrary to the irregular vorticities. This is an indication of the dual cascade, in which energy from short scale SDVMs is transferred to large scale DZFs. It is pertinent to note from Eqs. (26)-(28) that the mode $k_y \approx 0$ cannot be excited in a linear regime in which this mode is entirely absent from the dynamics. Nevertheless, the $k_y \approx 0$ mode, essentially leading to an asymmetric large scale flows in a real space, is generated purely as a result of nonlinear interactions among SDVMs and DZFs in an inertial range spectral space across a large equilibrium dust density gradient. More precisely, in the SDVMs, the energy flows towards smaller k's, so that we expect large scale structures, similar to the modified NS turbulence. On the other hand, DZFs are zero-frequency limit of the SDVM mode; they have very small k_y and very large k_z, so that the frequency condensation occurs for short wavelengths (viz.$k_z \gg k_y$ structures. DZFs are excited due to the Reynolds stresses of the SDVMs. While the Reynolds stresses possess a tendency of typically generating the large-scale flows, their time average appearing in Eq. (28) causes a net nonlinear dissipation of zonal flows. This appears to be the primary reason why DZFs form short scales (see Fig. 6). The vorticity field of DZFs in Fig. 6 appears to be stretched across the equilibrium density gradient wrapping the small scale dust vortex structures around in the nonlinear saturated state. The vortices are trapped in the horizontal sheared flow and propagate along the self-consistent flow across the equilibrium gradient.

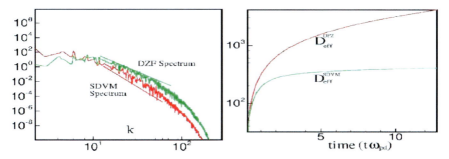

Fig. 7. The left panel displays Kolmogorov-like turbulent spectra of coupled DZF and SDVMs system. The spectrum of SDVMs in inertial range turbulence is more steep than that of DZF due to the presence of large scale flows that have $k_y \approx 0$. The numerical resolution is 1024×1024 Fourier modes in a two-dimensional box of $\pi \times \pi$. The right panel shows the effective diffusion coefficient associated with nonlinear turbulent transport in a coupled SDVM-DZF system. In agreement with the left panel, transport is suppressed due to the presence of large-scale flows in SDVM, while it has enhanced in short scales DZFs.

The energy spectrum decays due to dust-neutral collisions and dust kinematic viscosity at smaller scales. The left panel in Fig. 7 exhibits a high resolution Kolomogrov-like spectrum of a fully developed coupled SDVM-ZF turbulent system and the effective diffusion coefficient of the system. The spectrum of SDVMs is evidently steeper than that of DZFs in the inertial regime, thereby indicating the presence of large scale structures in its spectrum. This is further consistent with Fig. 5, which demonstrates $k_y \approx 0$ flows in the saturated SDVM-ZF turbulent state. It is to be noted that both the spectra in Fig. 7 are steeper compared to the fully developed 2D turbulent spectra for enstrophy or energy due to large-scale structures that condensate the lower Fourier modes because of inverse cascade processes. The dust density fluctuations also cascade towards long-scale structures due to short scale vortex merging. There also exist nonthermal transport associated with the effective turbulent diffusion ($D_{eff} = \int_0^{\tau} d\tau \langle v(y,t=0)v(y,t+\tau)\rangle$) of a dust particle in large scale ZF structures due to a random walk of the macroparticles in enhanced zonal flow fluctuations. As expected, emergence of large-scale coherent flows in SDVMs quench turbulent transport, in contrast to that involving random short scale DZFs, as shown in the right panel of Fig. 7. In nonuniform, nonlinear media without dust, 2D flows have been explained on the basis of the Navier-Stokes (NS) and the Charney-Hasegawa-Mima (CHM) equations [17, 18, 19, 20], which have two constants of motion, namely the energy and enstrophy (squared vorticity). The energy decays from a source to long wavelength, while enstrophy flows to shorter scales. Such a dual cascading, in agreement with the statistical quasi-equilibrium theory, ensures the formation of coherent vortical structures (eddies) [21], which are responsible for producing enhanced transport of fluids and plasma particles. For unbounded 2D NS turbulence, the conserved quantities are the kinetic energy

$E = \int_0^\infty E(k)dk$ and fluid enstrophy $Z = \int_0^\infty E(k)dk$, where E_k is the kinetic energy spectrum. For CHM turbulence, the conserved quantities are the total energy $E + \lambda^2 I$ and total enstrophy $Z + \lambda^2 E$, where $I = \int_0^\infty k^{-2}E(k)dk$ and λ is a positive constant.

5 Conclusions

In this paper, we have presented two new aspects of nonlinearities in dusty plasmas. First, we have shown that large-amplitude Langmuir waves interact with fully nonlinear dust acoustic waves via the ponderomotive force. Such a nonlinear interaction provides the possibility of Langmuir wave envelope trapping in a dust density cavity and associated positive space charge potential. There could exist the posibility of energization of charged particles in the localized potential of Langmuir wave envelope solitons. Second, we have considered the nonlinear dynamics of incompressible SDVMs and DZFs in a dusty plasma containing plasma inhomogeneities. We found that the Jacobian nonlinearity associated with dust fluid advection is responsible for the dual cascade leading to structuring of dust fluid in the form of vortical structures. The latter can cause non-thermal dust particle transport across the dust density gradient. Furthermore, various scale size dust vortices can be associated with dust hurricanes and dust devils that have been observed in laboratory experiments as well as in the Martian atmosphere.

Acknowledgements This work was partially supported by the Deutsche Forschungsgemeinschaft through the Sonderforschungsbereicht 591, as well as by the Swedish Research Council (VR).

References

1. R.Z. Sagdeev: In *Reviews of Plasma Physics*, vol 4, ed by M. A. Leontovich (Consultants Bureau, New York 1966) p 23
2. R.Z. Sagdeev: Rev. Mod. Phys. **51**, 11 (1979)
3. V.I. Petviashvilli, O.A. Pokhotelov: *Solitary Waves in Plasmas and in the Atmosphere* (Gordon Breach, Philadelphia, 1993)
4. P.K. Shukla, A.A. Mamun: *Introduction to Dusty Plasma Physics* (Institute of Physics, Bristol, 2002)
5. R. Merlino, J. Goree: Phys. Today **57**, 32 (2004)
6. V.E. Fortov et al: Phys. Uspekhi **47**, 447 (2004)
7. M. Horányi et al: Rev. Geophys. **42**, RG4002 (2004)
8. N.N. Rao, P.K. Shukla, M.Y. Yu: Planet. Space Sci. **38**, 543 (1990)
9. A. Barkan, N. D'Angelo, R.L. Merlino: Phys. Rev. Lett. **73**, 3093 (1994)
10. P.K. Shukla: Phys. Plasmas **8**, 1761 (2001)
11. P.K. Shukla: Phys. Plasmas **10**, 1619 (2003)
12. P.K. Shukla, A.A. Mamun: New J. Phys. **5**, 17 (2003)
13. P.K. Shukla, B. Eliasson, I. Sandberg: Phys. Rev. Lett. **91**, 075005 (2003)

Dust Plasma Interactions in Space and Laboratory

14. A. Hasegawa, P.K. Shukla: Phys. Lett. A **332**, 82 (2004)
15. Z. Sternovsky, M. Horányi, S. Robertson: J. Vac. Sci. Technol. **19**, 2533 (2001)
16. B. Walch, M. Horányi, S. Robertson: Phys. Rev. Lett. **75**, 838 (1995)
17. J.G. Charney: Geofys. Publ. **17**, 1–17 (1948)
18. A. Hasegawa, K. Mima: Phys. Rev. Lett. **39**, 205 (1977)
19. A. Hasegawa, K. Mima: Phys. Fluids **21**, 87 (1978)
20. A. Hasegawa: Ad. Physics **34**, 1 (1985)
21. W. Horton, Y.H. Ichikawa: *Chaos and Structures in Nonlinear Plasmas* (World Scientific, Singapore, 1996)
22. B. Eliasson, P.K. Shukla: Phys. Rev. E **69**, 067401 (2004)
23. E. Thomas Jr, R. Merlino: IEEE Trans. Plasma Sci. **29**, 152 (2000)
24. Y. Nakamura, H. Bailung, P.K. Shukla: Phys. Rev. Lett. **83**, 1602 (1999)
25. Q.-Z. Luo, R.L. Merlino: Phys. Plasmas **6**, 3455 (1999)
26. Q.-Z. Luo, N. D'Angelo, R.L. Merlino: Phys. Plasmas **7**, 2370 (2000)
27. Y. Nakamura: Phys. Plasmas **9**, 440 (2002)
28. B. Eliasson, P.K. Shukla: Phys. Plasmas **12**, 024502 (2005)

Part V

Accretion Disks

Magnetorotational Instability In Accretion Disks

V. Krishan[1] and S.M. Mahajan[2]

[1] Indian Institute of astrophysics, Bangalore-560034, India,
Raman Research Institute, Bangalore-560080, India
[2] Institute for Fusion Studies, The University of Texas at Austin, Austin,Texas
78712

Summary. Accretion disks are a basic and an ubiquitous construct of nature. We describe their formation, briefly and qualitatively, in diverse astrophysical situations. These systems, though controlled by the gravitational forces, have embedded in them a host of magnetohydrodynamic (MHD) and plasma-physical mechanisms without which many an observed phenomena would be hard to account for. The essential role of the MHD turbulence has been well emphasized. We give, here, a critical review of the ways and means of generating it through the magnetorotational instability and in the process correct the erroneous premise on which the local analysis continues to be carried out. The problem is complex and the present level of its understanding leaves much that needs novel techniques to fully capture its entire range of processes.

Key words: Accretion disks-magnetorotational instability-Hall effect

1 Introduction

Accretion disks form whenever a compact object such as a star or a black hole draws matter from its surroundings. More often than not this matter carries angular momentum and therefore cannot fall directly on to the compact object but instead settles in Keplerian motion around it. The accretion could take place via stellar winds as for example on to a black hole or through the Roche-Lobe filling process as in close binary stars. Accretion disks offer novel and efficient ways of extracting the gravitational energy from compact objects. A blob of gas would orbit a compact object with the Keplerian speed. It can spiral inwards only if its energy and angular momentum are removed by some kind of a dissipative process. With this mechanism, the binding energy of its innermost orbit can be extracted. Thus the inflow of the matter could be enhanced by the outflow of the angular momentum. It is well known that the Sun has the most of the mass and the planets the most of the angular momentum! It was realized quite early that the molecular viscous mechanism cannot produce the infall required for explaining the observed emissivities of these sources. Ever since, the turbulence assisted transport mechanisms have been at the centre of these investigations. The turbulence, ubiquitous as it is, needs instabilities for its generation and fluid shear more often than not fits the bill. The accretion disks in Keplerian motion are well endowed with shear. However, the shear of the Keplerian flow is stable since the specific angular momentum is an increasing function of the radial position and the Rayleigh criterion for the stability of the rotating fluids is well satisfied The frustrating search for a suitable hydrodynamic instability led eventually to the study of magnetohydrodynamic and plasma physical processes. The magnetic stresses produced by the stretching and shearing through dynamo type mechanisms could destabilize the system resulting in the outward transport of the angular momentum ([1], [2], [3], [4]). However, Balbus and Hawley [5] brought out the important and essential role of even a very weak magnetic field in exciting an interchange type linear instability in differentially rotating Keplerian disks. Since this work, there has been a flurry of activity to grasp the essence of the magnetorotational instability and its attendant consequences for the transport of the angular momentum deploying analytical and numerical methods ([6], [7]). The additional effects such as the Hall effect, the ambipolar diffusion and the dissipative processes have been found to be of particular importance in weakly ionized plasmas such as exist in the protoplanetary disks. The inclusion of the Hall effect and the electrical resistivity was initiated by wardle [8] wherein a formalism based on the conductivity tensor was adopted. Balbus and Terquem [9] handled the Hall effect in a dynamical way and determined the region of instability. They also claimed that the maximum growth rate continues to have the Oort value even with the inclusion of the Hall effect. We have examined this piece of work and find that both the neglect of the radial structure altogether as well as its inclusion in the so called local approximation leave much to be desired. For one the divergence free nature of the magnetic and the velocity perturbations was violated and it should not be. The so called local approximation under which the Fourier decomposition is carried out in the radial coordinate for a radially non autonomous system is another forbidden shortcut. In this paper after describing the basic properties of accretion disks in section 2, the nature of the magneto-rotational instability in a differentially rotating plasma is discussed in section 3. The mystery about the nonrecoverability of the Rayleigh criterion in the zero magnetic field limit is solved. The global formulation of the magnetorotational instability is presented in section 4 along with some exactly solvable cases. An approach based on being able to define an effective potential is briefly discussed in section 5. The specific case of the Keplerian rotation is presented in section 6. We end the paper with a mention of some new techniques which should be used in order to fully grasp the subtleties and the nuances of plasmas with flows, a system of great astrophysical interest.

2 Basic Characteristics of Accretion Disks

Accretion disks have been observed and postulated in three different classes of objects: 1) in star and planet forming regions, 2) in close binaries where the two stars orbit each other and 3)around black holes in active galactic nuclei. Planet formation has been known to be associated with accretion and evolution of gas and dust in disks around young stars. A molecular cloud of gas and dust or a nebula begins to collapse when the gravitational forces overcome the forces associated with gas pressure responsible for its expansion. The initial collapse might be triggered by a variety of perturbations such as a supernova blast wave or density waves in spiral galaxies. In the Nebular Hypothesis, a cloud of gas and dust collapsed by gravity begins to spin faster because of angular momentum conservation. Under the action of the competing forces associated with gravity, gas pressure, and rotation, the contracting nebula begins to flatten into a spinning disk with a bulge at the center. The instabilities in the collapsing and rotating cloud cause localized gravitational collapses. The central bulge becomes the star and condensations in the disk, the planets, as well as their moons and other debris resulting in the formation of a system such as the Solar System. Still undergoing condensation, the star without ignition and the planets are called the protostar and the protoplanets, respectively. Our Sun is a single star. Most stars occur in multiple or binary systems. In a binary system, the more massive star evolves faster and becomes a compact object - either a white dwarf star, a neutron star, or a black hole. The less massive star during its evolution into its expansion phase, may be so close to the compact star that its outer atmosphere may actually touch the surface of the compact star. If one star in a binary system is , say, a white dwarf and the other, a normal star like the sun, the white dwarf can pull gas off the normal star and accrete it onto itself. Since the stars are revolving around each other, the angular momentum conservation forbids the gas to fall directly onto the white dwarf and the gas settles in a disk around the compact star. The gas in the disk becomes very hot due to friction, eventually loses angular momentum and falls onto the white dwarf. Since this hot gas is being accelerated it radiates energy, usually in X-Rays which offer a good diagnostic for accretion disks. The evidence for the rotation of a disk is obtained from the observed double peaked line profiles. If the stars in a binary system are far from each other, the rotation period is long and given by Kepler's laws. If on the other hand the stars are nearly touching each other, as in close binaries, the orbital period is small, matter may stream from one star onto the other star. These are called accreting binaries. Accretion in such a system can be studied by plotting contours of equal gravitational potential. The contour through the inner Lagrange point where the net gravitational force vanishes defines two regions, one around each star, called the Roche lobes . Mass accretion can occur if one of the stars fills its Roche lobe, allowing matter to spill over the inner Lagrange point, in the form of a thin stream, onto the other star. Accretion in binary systems can also take the form of a wind from the surface of one star. Then the second star accumulates matter from the first star as it moves on its orbit through this wind. In complex situations, both winds and tidal accretion streams may be important. Accretion does not stop with the formation of the disk. The physical processes in the disk itself regulate further accretion. Accretion disks could be weakly ionized, cold and collisional, e.g., Protoplanetary Disks or they could be highly ionized, hot and collisionless e.g. those around black holes. All disks are expected to be magnetized and turbulent. The efficiency of conversion of the gravitational energy into electromagnetic energy (Hydroelectric Power!) is much higher than that for hydrogen or helium burning and varies from object to object. The emitted luminosity is $L = GM \odot MR^{-1}$ and the efficiency is $\eta = GM(Rc^2)^{-1}$ where M and R are the mass and the radius of the star and $\odot M$ is the mass accretion rate. For a white dwarf, $M = M_\odot$, the mass of the sun, R= 1000 Km, and $\eta = 0.1\%$; for a neutron star, $M = M_\odot$, R= 10 Km., and $\eta = 10\%$; for a black hole, the efficiency could be 10%. The range of particle densities and temperatures both within disks and from disk to disk is enormous. Disks occupy the broad density scale gap between the interstellar matter, which is at the most 10^6 cm^{-3} in molecular cloud cores, and the stellar interiors which have typically 10^{25} cm^{-3}. Disks in binary systems generally have interior densities above $10^{15} cm^{-3}$. The innermost regions of an accretion disk can be very hot. If 10^{37} ergs/s is emerging from a gas disk over a region of radial dimension of a neutron star 10^6cm and the gas is emitting as a blackbody, then its temperature should be of order 10^7 K , a good source of KeV photons. The surface temperature is expected to fall outwards in the disk. The local luminosity of a disk scales as $1/r$ and the radiated flux as $1/r^3$, implying an $r^{-3/4}$ scaling law for the surface temperature. Thus, on scales of 10^{10} cm., the fiducial disk cools to 10^4 K. Disks around white dwarfs are at 10^5K or so in their innermost orbits, and they are not expected to be powerful x-ray sources. This is the bare bones scenario. The accretion process becomes complex very near the stellar surface where boundary layer phenomena may dominate which may be the cause of flares and outbursts in white dwarf accretion. The equilibrium of the accretion disks is described by assuming a rotating mass of gas in a cylindrically symmetric potential well of a point mass at the origin, the centre of the disk. The radial component of the force balance, assuming radial velocity to be small and neglecting the vertical variation of the gravitational potential furnishes the Keplerian rotation law $\Omega = (GMr^{-3})^{0.5}$. The hydrostatic balance in the vertical direction gives the usual exponentially decreasing density profile with a scale height $H = \sqrt{2}c_s/\Omega$ where c_s is the sound speed. The azimuthal component of the force balance describes the angular momentum conservation in the absence of the viscous forces. The molecular diffusion produces an outward radial flow. Thus additional torque is required to transfer the angular momentum outwards and consequently the mass flow inwards ($V_r < 0$). Hence the search for shear stresses, anomalous viscosity coefficient ν, instabilities and turbulence! There are three basic timescales: 1) the shortest disk time scale $T_\Omega = 1/\Omega$

Magnetorotational Instability In Accretion Disks   237

is governed by the rotation velocity; 2) time scale over which the hydrostatic equilibrium is established in the vertical direction $T_z = H/c_s$ and 3) time scale over which surface density changes, the viscous time scale $T_\nu = r^2/\nu$ with $T_\nu \gg T_\Omega \sim T_z$.

3 Linear Study of Differentially Rotating Plasmas

We write the MHD equations in a dimensionless form: all lengths are normalized to an equilibrium length L, the magnetic and the velocity fields are respectively normalized to the uniform ambient field B_0, and the Alfvén speed $V_A = B_0/\sqrt{4\pi\rho}$, where ρ is the uniform mass density, and all frequencies are measured in terms of V_A/L. The induction and the force equations then read [10], [11]:

$$\frac{\partial \boldsymbol{B}}{\partial t} = \nabla \times [\boldsymbol{V} \times \boldsymbol{B}] \tag{1}$$

$$\frac{\partial}{\partial t} \boldsymbol{\nabla} \times \boldsymbol{V} = \nabla \times [\boldsymbol{V} \times (\boldsymbol{\nabla} \times \boldsymbol{V}) - \boldsymbol{B} \times (\boldsymbol{\nabla} \times \boldsymbol{B})]. \tag{2}$$

Equation (2) has been obtained by taking the curl of the equation of motion; the pressure term disappears because of the constant density assumption. Equations (1)-(2) allow the equilibrium $\boldsymbol{B}_0 = \widehat{e}_z, \boldsymbol{V}_0 = \frac{r\Omega}{V_A}\widehat{e}_\theta$. We split the fields into their ambient and fluctuating parts:

$$\boldsymbol{B} = \widehat{e}_z + \boldsymbol{b}$$
$$\boldsymbol{V} = \boldsymbol{V}_0 + \boldsymbol{v}.$$

Invoking the form of perturbations relevant to the cylindrical geometry

$$\boldsymbol{b} = \boldsymbol{b}(r)e^{-i\omega t + ikz + im\theta}$$

and using (prime denotes the differentiation with respect to r)

$$\boldsymbol{\nabla} \times \boldsymbol{V}_0 \equiv \widehat{e}_z g(r) = \widehat{e}_z \frac{1}{r}\frac{d}{dr}(r^2\Omega(r)) = \widehat{e}_z \left[2\Omega + r\Omega'(r)\right] \tag{3}$$

the linearized form of equations (1) and (2), in the incompressible limit ($\boldsymbol{\nabla}{\cdot}\boldsymbol{v} = 0$), turn out to be:

$$\omega_m \boldsymbol{b} = -k\boldsymbol{v} + ir\Omega'(r)b_r\widehat{e}_\theta, \tag{4}$$
$$\omega_m(\nabla \times \boldsymbol{v}) = -k(\nabla \times \boldsymbol{b}) + ir\Omega'(r)(\nabla \times \boldsymbol{v})_r\widehat{e}_\theta - g(r)k\boldsymbol{v} - ig'v_r\widehat{e}_z \tag{5}$$

where $\omega_m = \omega - m\Omega(r)$.

Equations (4) and (5), coupled with the divergence conditions ($\boldsymbol{\nabla} \cdot \boldsymbol{v} = 0$, $\boldsymbol{\nabla}{\cdot}\boldsymbol{b} = 0$), can be combined to obtain a single one dimensional second-order differential equation for the radial component of the velocity or the magnetic

238   V. Krishan and S.M. Mahajan

field perturbation for any arbitrary value of m. We do plan to analyze this general case in the near future. In this paper, however, we will consider only azimuthally symmetric perturbations ($m = 0$, $\omega_m = \omega$). By straightforward algebra, we derive the relatively compact mode equation ($\psi = v_r$ or b_r)

$$L\psi \equiv \frac{d}{dr}\frac{1}{r}\frac{d}{dr}r\psi - k^2\psi = -\frac{k^2}{F^2}\left[4\Omega^2 + \frac{2F}{\omega}r\Omega\Omega'(r)\right]\psi. \tag{6}$$

where $F = (\omega^2 - k^2)/\omega$, and the definition of $g(r) = 2\Omega(r) + r\Omega'(r)$ has been used. Equation (6), in which the mode frequency appears only through ω^2, is an exact linear consequence of Eqs. (1) and (2).

A Special Case

Let us now view the mode equation in a familiar setting by dwelling on two of its well-known limiting solutions. Let us begin with the simplest case of rigid rotation, $\Omega = \Omega_0$, $\Omega'(r) = 0$; the mode equation reduces to

$$\frac{d}{dr}\frac{1}{r}\frac{d}{dr}r\psi - k^2\psi = -\frac{4k^2}{F^2}\Omega_0^2\psi \tag{7}$$

allowing a solution

$$\psi = J_1(\mu r),$$

and the dispersion relation

$$\left(\mu^2 + k^2\right) = \frac{4k^2}{F^2}\Omega_0^2 \tag{8}$$

describing the Alfvén waves in a rotating plasma. Here μ is to be interpreted as an effective radial wavenumber. Notice that in Eq. (8), the left-hand represents the square of the total wavenumber comprised of the effective radial wavenumber μ, and the axial wavenumber k.

4 Differential Rotation — "The local lore"

Even when the rotation frequency varies with r, the conventional treatments set, in Eq. (6),

$$\frac{d}{dr}\frac{1}{r}\frac{d}{dr}r\psi = \frac{d^2\psi}{dr^2} + \frac{1}{r}\frac{d\psi}{dr} - \frac{\psi}{r^2} = 0,$$

to arrive at the following "local" dispersion relation

$$F^2 - 4\Omega^2 = 2r\Omega\Omega'\frac{F}{\omega}$$

or

$$\omega^4 - \omega^2\left[2k^2 + 4\Omega^2 + 2r\Omega\Omega'\right] + k^2\left[k^2 + 2r\Omega\Omega'\right] = 0. \tag{9}$$

It is straightforward to see that out of the two roots of the quartic, one reduces to the Rayleigh criterion for vanishing magnetic field. The second root represents a purely-growing mode, provided

$$k^2 + 2r\Omega\Omega' < 0 \qquad (10)$$

which is possible if $\Omega' < 0$, i.e., the rotation frequency decreases with distance. This purely-growing low-frequency root is called the Magnetorotational Instability (MRI). The theoretical and experimental pursuit of MRI is a major area of current research.

We contend that this "local" treatment is rather inadequate. In spite of its initial utility, its predictions cannot be relied on. In the nonautonomous differential equation, Eq. (6), the radial structure of the mode must be dictated by the radial variation of $\Omega(r)$ and cannot be chosen arbitrarily. In fact, different $\Omega(r)$ profiles with the same sign of Ω' could lead to vastly different mode structures with profound consequences for the existence and/or stability of the mode.

5 Magnetorotational Instability - "The Global Grope"

Global analyses of the magnetorotational instability have been carried out before. Knobloch [12] derived the governing eigenvalue equation and discussed the stability conditions for different magnetic field geometries and boundary conditions. The limited validity of the local analysis as well as the essential role of the azimuthal magnetic field were emphasized. Dubrulle and Knobloch [13] verified the WKB solutions numerically and assessed the relative role of the vertical and the toroidal magnetic fields in the excitation of the MRI. The growth rate was found to be of the order of the Alfven frequency in the limit of vanishing vertical field. In contrast Curry, Pudritz and Sutherland [14] studied the growth rates and the radial eigenfunctions of the MRI for a wide variety of physical conditions pertaining to rotation profiles and magnetic field strengths. They claimed to recover the local behaviour with radial eigenfunctions sinusoidal in the inner regions and exponentially decreasing in the outer regions. The unstable global mode in a Keplerian disk has been found to reside on the boundaries ([15]). Sano and Miyama [16] studied the global MRI including dissipation and the vertical structure by numerically solving the eigenvalue problem. A fully three dimensional numerical investigation of the linear MRI with boundaries in all directions has been carried out by Rudiger and Zhang [17].

More recently Coppi and Keyes [18] investigated the ballooning modes retaining the global character in the vertical direction but assuming sinusoidal variation in the radial direction. Kersale et al. [19] find that the wall modes have the properties of the local MRI for boundary conditions that constrain the total pressure at the radial boundaries. Thus the MRI has remained an

240   V. Krishan and S.M. Mahajan

active area of research over the years with its simulations ([20]), laboratory studies (e.g. Hantao, Goodman and Kageyama, [21]) and analytical studies under a host of assumptions. In spite of these many and varied approaches, the global nature of MRI is still far from delineation. The recovery of the local characteristics of the MRI depends crucially on the choice of boundary conditions. The typical boundary conditions, equivalent to walls with infinite potential, will surely define and confine the modes almost independent of the details of the rotation profile. We believe that it is of the utmost importance to learn about the nature of the instability from a boundary independent (the eigenmodes vanish at the boundaries) analysis; this kind of approach is , perhaps, the key to understanding the effects of different rotational profiles.

Even though the Keplerian accretion disks were the original inspiration for the work on MRI, we deal here with the stability of a generic magnetoro-tational system represented by Eq. (6). A general theoretical framework is needed since the $\Omega(r)$ profiles in laboratory experiments could be far from the Keplerian.

Fortunately, relatively straightforward and well known analytic methods are enough to extract the relevant physics quite rigorously. It is extremely convenient that the radial dependence of Eq. (6) (apart from the cylindrical operator) comes from Ω^2 and $r\Omega\Omega' = (r/2)(\Omega^2)'$. Thus a model for Ω^2 will specify the entire spatial dependence.

5.1 Exactly Solvable Profiles

We begin our search by investigating two nontrivial Ω^2 profiles for which the eigenvalue problem, Eq. (6), is exactly solvable in terms of the well-known special functions. For each of these examples, we will first manipulate Eq. (6) to derive an equivalent Schrödinger-like (self-adjoint) equation and then cal-culate the eigenspectrum; the effective eigenvalue condition will lead us to a dispersion relation involving ω^2.

I. The first of these profiles may be written as:

$$\Omega^2 = \Omega_0^2 \left[1 + \frac{\alpha}{r}\right]$$

$$r\left(\Omega^2\right)' = -\Omega_0^2 \frac{\alpha}{r}. \tag{11}$$

For this profile $\Omega' < 0$ if $\alpha > 0$. We have deliberately chosen this form so that the result for rigid rotation can be obtained by simply letting $\alpha \to 0$. We would see later that the problem associated with $\Omega^2 = \Omega_0^2/r$ is entirely equivalent, and by a simple transformation the results obtained for the former can be translated for the latter.

The relevant mode equation

Magnetorotational Instability In Accretion Disks 241

$$\frac{d}{dr}\frac{1}{r}\frac{d}{dr}r\psi - k^2\left[1 - \frac{4\Omega_0^2}{F^2}\right]\psi = -\frac{4k^2\Omega_0^2}{F^2}\frac{\alpha}{r}\left[1 - \frac{F}{4\omega}\right]\psi \tag{12}$$

is of the form

$$\frac{d}{dr}\frac{1}{r}\frac{d}{dr}r\psi - p_0\psi + \frac{q_0}{r}\psi = 0 \tag{13}$$

where

$$p_0 = k^2\left(1 - \frac{4\Omega_0^2}{F^2}\right), \qquad q_0 = \frac{4\alpha k^2\Omega_0^2}{F^2}\left(1 - \frac{F}{4\omega}\right). \tag{14}$$

By the substitutions

$$\psi = \frac{Q}{r^{1/2}}, \qquad z = 2\sqrt{p_0}\,r$$

we can derive the equivalent Schrödinger equation

$$\frac{d^2Q}{dz^2} + \left[-\frac{1}{4} + \frac{q_0}{2\sqrt{p_0}\,z} + \frac{\frac{1}{4} - 1}{z^2}\right]Q = 0 \tag{15}$$

which is the Whittaker equation with $\mu = \pm 1$, $\lambda = q_0/2\sqrt{p_0}$. Equation (15) has well-behaved solutions

$$\psi_n = \frac{Q_n}{r^{1/2}} = r\,e^{-\sqrt{p_0}\,r}\Phi\left[\frac{3}{2} - \frac{q_0}{2\sqrt{p_0}},\ 3,\ 2\sqrt{p_0}\,r\right] \tag{16}$$

where Φ is the Kummer function. The solution requires $\sqrt{p_0} > 0$ for its existence. The eigenvalue condition

$$\frac{3}{2} - \frac{q_0}{2\sqrt{p_0}} = -n, \tag{17}$$

then yields the dispersion relation,

$$q_0 = (2n + 3)\sqrt{p_0},\, or$$

$$\frac{4\alpha k^2\Omega_0^2}{F^2}\left(1 - \frac{F}{4\omega}\right) = |k|\sqrt{1 - \frac{4\Omega_0^2}{F^2}}\,(2n + 3). \tag{18}$$

One immediately notices that the dispersion relation, Eq. (18), looks very different from the "local" dispersion relation, Eq. (9); one also sees that the radial quantum number n appears in the spectral relation.

We will not attempt a general analysis of Eq. (18) for all the modes it allows; we will solve it primarily to obtain an approximate dispersion relation for the MRI-type of mode $\omega^2 < 0$, $-\omega^2 \ll k^2$. For these conditions

$$1 - \frac{F}{4\omega} = 1 - \frac{\omega^2 - k^2}{4\omega^2} = \frac{3}{4} + \frac{k^2}{4\omega^2} \simeq \frac{k^2}{4\omega^2}$$

and

242   V. Krishan and S.M. Mahajan

$$F^2 = \frac{(\omega^2 - k^2)^2}{\omega^2} \simeq \frac{k^4}{\omega^2}.$$

With these approximations, Eq. (18) leads to

$$\omega^2 = \left[k^2 - \frac{\alpha^2 \Omega_0^4}{(2n+3)^2} \right] \frac{k^2}{4\Omega_0^2}.$$

implying that for an instability we require $(\alpha > 0)$

$$\alpha \Omega_0^2 > |k| \, (2n+3)$$

which, for small $|k|$, is a much more severe constraint on Ω' (measured by α) than the local criterion would imply. The existence condition, $\sqrt{p_0} > 0$, is trivially satisfied for the unstable root.

As the radial mode number n rises, the instability criterion becomes harder to satisfy; thus $n = 0$ is the easiest to excite. In this case, the instability criterion though both qualitatively and quantitatively different from the "local" criterion, can indeed be satisfied for $\Omega' < 0$, and the eigenvalue problem is well posed; well-defined square integrable eigenfunctions are associated with the unstable mode. We shall see that it is not the case for the second exactly solvable case we present here.

II. Another analytically tractable problem emerges for the rotation profile:

$$\Omega^2 = \Omega_0^2 \left(1 + \beta r^2\right)$$

$$r(\Omega^2)' = \beta \Omega_0^2 2r^2.$$

(19)

where $\beta < 0 (> 0)$. for $\Omega' < 0 (> 0)$. For this profile Eq. (6) becomes

$$\frac{d}{dr} \frac{1}{r} \frac{d}{dr} r\psi - p_0 \, \psi - qr^2\psi = 0,$$

(20)

$$p_0 = k^2 \left[1 - \frac{4\Omega_0^2}{F^2}, \right]$$

$$q_0 = \frac{-4\beta \Omega_0^2 k^2}{F^2} \left[1 + \frac{F}{2\omega} \right].$$

(21)

With the substitutions, $P = r\psi$, and $r^2 = 2y$, we find the standard confluent hypergeometric form

$$\frac{d^2 P}{dy^2} + \left(-q_0 - \frac{p_0}{2y} \right) P = 0,$$

and another change of variable through $z^2 = 4q_0 y^2$ eventually furnishes

$$\frac{d^2 P}{dz^2} + \left[-\frac{1}{4} - \frac{p_0}{4\sqrt{q_0}} \frac{1}{z} \right] P = 0.$$

(22)

Equation (22) has the Whittaker form with $\mu = \pm\frac{1}{2}$, $\lambda = -p_0/4\sqrt{q_0}$ and leads to well-behaved eigensolutions

$$\psi_n = \frac{P_n}{r} = r \, e^{-\sqrt{q_0} \, r^2/2} \, \varPhi \left[1 + \frac{p_0}{4\sqrt{q_0}} \, , 2, \sqrt{q_0} \, r^2 \right], \tag{23}$$

provided Re $\sqrt{q_0} > 0$. The eigenvalue condition then leads to the dispersion relation

$$1 + \frac{p_0}{4\sqrt{q_0}} = -n,$$

$$\frac{p_0}{4\sqrt{q_0}} = -(n+1),$$

or

$$k^2 \left[1 - \frac{4\Omega^2}{F^2} \right] = -4(n+1) \left[-\frac{4\beta\Omega_0^2 k^2}{F^2} \left(1 + \frac{F}{2\omega} \right) \right]^{1/2}. \tag{24}$$

For the existence of the mode, the wave function ψ_n must fall at infinity. This can happen if Re $\sqrt{q_0} > 0$. However, for $\beta < 0 (\Omega' < 0)$ and $\omega^2 < 0$ both F^2 and $F\omega$ become negative and so does

$$q_0 = -\beta 4\Omega_0^2 k^2 \left[\frac{1}{F^2} + \frac{1}{2F\omega} \right].$$

Thus we are led to the contradiction that the right-hand side of Eq. (24) is purely imaginary, while the left-hand side is purely real. Therefore the profile $\Omega_0^2(1 + \beta r^2)$ for $\Omega' < 0$ does not admit the MRI-like mode for any value of Ω'.

In fact, although, for $\beta > 0$ ($\Omega' > 0$) q_0 is positive for $\omega^2 < 0$, Eq. (24) still cannot be satisfied, because the left-hand side is positive and the right-hand side [of Eq. (24)] is negative.

Here then is an example of a rotation profile which does not allow a mode with $\omega^2 < 0$ for either sign of Ω'.

6 Effective Potential Approach- General Profiles

The most important lesson one can learn from the two exactly solvable problems is that the eigenvalue problem associated with the magnetorotational system is rather subtle. Relatively crude methods based on the "local" or similar analyses can be seriously misleading. Operationally, the essence of the last section lies in the realization that the existence of a mode of the MRI-type, for instance, depends not just on the sign and magnitude of Ω', but on the full radial form of Ω. In addition, a given Ω may yield a perfectly acceptable solution for $\omega^2 > 0$ while no such solution may exist for $\omega^2 < 0$.

The dethronement of Ω' from its exalted position in the conventional treatment of the MRI is rather difficult to accept; the "local" criterion sounded so

244   V. Krishan and S.M. Mahajan

patently reasonable. In order to ensure that the exactly solvable cases have not lead us to conclusions that are somehow peculiar and belong to a set of measure zero, let us study the eigenvalue problems associated with classes of Ω profiles of which the Keplerian will be a particular case. We formulate the equivalent Schrödinger-like mode equation for an effective potential due to differential rotation.

Let us assume the rotation profile to have a general power law form:

$$\Omega^2 = \Omega_0^2 \left[1 + \alpha\, r^s\right], \ r\left(\Omega^2\right)' = s\alpha\Omega_0^2\, r^s \tag{25}$$

where s and α could be positive or negative. Straightforward algebraic manipulations yield the self- adjoint eigenvalue equation

$$\frac{d^2 Q}{dr^2} + \left[-p_0 - V(r)\right] Q = 0 \tag{26}$$

where

$$p_0 = k^2 \left[1 - 4\Omega_0^2 / F^2\right], \tag{27}$$

$$V(r) = \frac{3}{4r^2} - q_0\, r^s, \tag{28}$$

and

$$q_0 = 4k^2 \Omega_0^2 \left[\frac{\alpha}{F^2} + \frac{s\alpha}{4\omega F}\right]; \tag{29}$$

the sign of q_0 is fixed by the quantity in the square brackets.

The general problem of the magnetorotational modes is interesting in its own right but our current quest will be limited to a search for the MRI-type low-frequency instabilities. The eigenvalue problem, Eqs.(25-29), has the following interesting features:

1) The eigenvalue $(-p_0)$ is always negative for modes with $\omega^2 < 0$.

2) The potential has two terms: the first one represents the normal cylindrical centrifugal barrier and is always positive. The second term is due to differential rotation, and its sign will crucially determine the nature of $V(r)$.

3) The potential must have a minimum for finite and real r in order to yield a physically acceptable solution. In addition, the $V_{\min}$ must be negative and sufficiently deep so that the effective eigenvalue line $(-p_0$ with magnitude greater than k^2) intersects the potential; the intersection defines the turning points that trap the mode. These are still the necessary conditions. The condition of sufficiency will be satisfied if the potential is deep enough to accommodate the zero point energy associated with the lowest radial eigenmode. That requires an actual solution of the problem. Our attempt, here, is to explore if the potential for a given rotation profile can sustain even a single unstable mode ($\omega^2 < 0$).

4) In the range $-2 < s$, $V(r)$ approaches positive infinity as $r \to 0$. For this genre of potentials, the criteria for MRI can be succinctly summarized as:

a) If $V(r)$ (positive or negative) has no minimum, no mode is possible b) If $V(r)$ does have a minimum but $V_{\min}$ is positive, no mode is possible c) If $V(r)$ does have a minimum and $V_{\min}$ is negative, then the necessary condition for the existence of a mode is satisfied.

The second term in the potential can be considerably simplified for the low-frequency modes. For $\omega^2 < 0$, $(\omega^2 - k^2) < 0$ and $|\omega^2| \ll k^2$, we have

$$F^2 = \frac{(\omega^2 - k^2)^2}{\omega^2} \simeq \frac{k^4}{\omega^2}, \tag{30}$$

$$F\omega = \omega^2 - k^2 \simeq -k^2 \tag{31}$$

yielding

$$q_0 \simeq 4k^2 \, \Omega_0^2 \left(-\frac{\alpha s}{4k^2}\right) = -\alpha s \Omega_0^2. \tag{32}$$

We observe that for $s > 0$, $V(r)$ has no minimum for $\alpha < 0$ ($\Omega' < 0$). It does have a minimum for $\alpha > 0$ ($\Omega' > 0$), but then $V_{\min}$ is positive. In either case, therefore, no MRI-type mode is possible. This is in conformity with the conclusion of the exactly-solvable case I with rotation model [Eq. (19)] for $s = 2$. Thus we conclude that rotation profiles for $s > 0$ clearly contradict the standard lore that a sufficiently large $\Omega' < 0$, such that $(-2r\Omega\Omega' > k^2)$, can lead to an instability of the MRI type.

The situation for $-2 < s < 0$ turns out to be more favorable to the MRI. In this range of s, the potential $V(r)$ has a negative minimum for $q_0 > 0$ which is clearly possibly for $\alpha s < 0$ ($\Omega' < 0$) and the necessary conditions for an MRI-type mode are satisfied. This general result reinforces the conclusion of the exactly solvable case II with rotation profile given by Eq. (11). We also derived the sufficient condition for the instability for this particular case.

We are thus forced to admit that the mathematical physics behind MRI is a lot richer and subtler than it has been assumed to be so far. Two distinct Ω profiles with similar Ω' may lead to entirely different consequences; one of these may support an MRI-like unstable mode while the other may not.

We end this section by remarking that results for profiles of type $\Omega^2 = \Omega_0^2 r^s$ are readily obtained from the results for profiles $\Omega^2 = \Omega_0^2(1 + \alpha r^s)$ by putting $p_0 = k^2$ and $\alpha = 1$; there is no qualitative change.

7 Magnetorotational instability under Keplerian Rotation

We now take the tools of our analysis to the most important MRI related object — the Keplerian disk for which the rotation profile belongs to the class $\Omega^2 = Ar^{-(2+a)}, a > 0$. For the Keplerian disk proper, $a = 1$ and $A = GM/L^3$, where L is the normalizing distance. The potential in the now familiar eigenmode equation

246   V. Krishan and S.M. Mahajan

$$\frac{d^2Q}{dr^2} + \left[-k^2 - V(r)\right]Q = 0 \qquad (33)$$

has the form

$$V(r) = \frac{3}{4r^2} - \frac{4Ak^2}{r^{2+a}}\left[\frac{1}{F^2} - \frac{2+a}{4F\omega}\right]. \qquad (34)$$

For an MRI-type mode with $\omega^2 < 0$, the signs of the two terms in the square brackets are opposite. If the first term is dominant, then $V(r)$ is positive everywhere and has no minimum, and therefore cannot support a mode with an effective eigenvalue $(-k^2)$. If the second term is dominant, then the sign of the second term is negative, and the potential goes to negative infinity as $r \to 0$ (the Keplerian contribution diverges more rapidly than the centrifugal one for small r). The total potential $V(r)$ cannot have a minimum. The purely Keplerian potential then cannot support a bound state with $\omega^2 < 0$. The magnetorotational instability, considered as an eigenvalue problem on the infinite domain, cannot exist in a Keplerian accretion disk described within the magnetohydrodynamic model.

8 Conclusion

After describing some basic essentials of accretion disks the inadequacies of the local treatment of the magnetorotational instability have been emphasized. By deriving and analyzing the appropriate eigenvalue problem for a differentially-rotating ideal MHD plasma we have shown: 1) For a large class of rotation frequency profiles, including the Keplerian, the conventionally predicted low frequency, purely growing ($\omega^2 < 0$), axisymmetric, MRI-type instability is not possible even for $\Omega'(r) < 0$; the effective potential stemming from radial dependence of $(\Omega(r))$ is found to be inadequate for trapping even the lowest radial eigenmode, 2) For classes of $(\Omega(r))$ that do satisfy necessary conditions for the existence of MRI modes, the dispersion relation turns out to be qualitatively different from the one found by a "local" approximation. The local theory, therefore, can be often misleading, in addition to being inadequate. Future analyses as well as simulations of rotating magnetoplasmas must contend with this fact — the MRI they have been looking for may not be there. A nonautonomous system also known as a nonhermitian system such as a differentially rotating plasma does not easily submit to the standard methodology of investigating the linear stability. The eigenvalue problem of MRI can be cast into a Hermitian equation only for the modes with $m = 0$. When we analyze a general mode with a finite m, we encounter the difficulty of non-Hermitian operators. It is only for a static (no-flow) equilibrium that the stability is studied within the framework of the spectral analysis of Hermitian operators. The waves and instabilities in a flowing plasma, such as required for the excitation of the MRI ($m = 0$ modes are exceptional),are governed by the non-Hermitian operators.. Then, the standard notion of dispersion relation (i.e., replacement of ∂_t by $i\omega$ assuming a solution of the form of $e^{i\omega t} f(x)$), fails to capture the general nature of the dynamics. A well-known characteristic of a non-Hermitian system is the possibility of a complex valued ω (reality of the eigenvalues is a consequence of the Hermitian property). However, the problem is not so simple. For example, the system may be unstable even if all the eigenvalues (frequencies) are real, i.e., the dispersion relation can give only a necessary condition for stability. On the other hand, the energy principle can deduce only a sufficient condition for stability.

While the analysis of a general non-Hermitian system is still far beyond the range of present-day mathematical theories, stability of some class of equilibria may be analyzed with the help of the so-called Casimir invariants. For example the Beltrami equilibria are characterized by variational principles invoking some different constants of motion ([22]). The target functional of the variational principle consists of the energy (Hamiltonian) and helicities. The helicity is a Casimir invariant in an ideal vortex dynamics system. The addition of a Casimir invariant to the Hamiltonian does not change the dynamics. While the extremer of the original Hamiltonian is often trivial (for example, the absolute minimum, i.e., zero energy state), adding Casimirs can enrich the structure of such equilibria Beltrami equilibria are an example. When an equilibrium point of a dynamical system is characterized as an extremer of the Hamiltonian, we can derive a Lyapunov function from the transformed Hamiltonian, which often yields refined estimates of the sufficient condition of stability. Thus the study of plasmas with flows needs deployment of new techniques such as the effective potential approach and the determination of the spectrum of a non-Hermitian system. This paper does not pass a verdict against an instability in a rotating magnetoplasma; in particular, in a Keplerian disk. We have simply proved that in a minimal model (incompressible MHD, axisymmetric perturbations, uniform embedding field, purely Keplerian profiles extending all the way) one cannot find unstable eigenmodes defined over the infinite domain. We believe that a physics based modification of the rotation profile near the compact object is necessary for dependable stability prediction. In addition, the search for the origin of turbulence in accretion disks, must be carried to more comprehensible models which include, among other things, compressibility, nonideal and Hall Effects, and nonaxisymmetric fluctuations.

Acknowledgments

The authors gratefully acknowledge the insightful discussions with Prof. Zensho Yoshida of the University of Tokyo. The authors also express their deep gratitude to Dr. Baba Varghese for his help in the preparation of the manuscript.

References

1. D. Lynden-Bell, D., Nature, **223**, 690 (1969)
2. N. I. Shakura and R. A. Sunyaev, Astron. Astrophys, **24**, 337 (1973)
3. R. D. Blandford, 1976, MNRAS, **176**, 465 (1976)
4. J. E. Pringle, Ann. Rev. Astron. Astrophys., **19**, 137 (1981)
5. S. A. Balbus and J. F. Hawley, Ap.J., **376**, 214 (1991)
6. J. C. B. Papaloizou and D. N. C. Lin, Ann. Rev. Astron. Astrophys., **33**, 505 (1995)
7. S. A. Balbus and J. F. Hawley, Rev. Mod. Phys., **70**, 1 (1998)
8. M. Wardle, MNRAS, **279**, 767 (1999)
9. S. A. Balbus and C. Terquem, ApJ, **552**, 235 (2001)
10. V. Krishan and B.A. Varghese, Solar Phys., (2008)
11. S.M. Mahajan and V. Krishan, ApJ, In press, (2008)
12. E. Knobloch, MNRAS, **255**, 25P (1992)
13. B. Dubrulle and E. Knobloch, ASTRON. ASTROPHYS., **274**, 667 (1993)
14. C. Curry, R. E. Pudritz and G. Sutherland, Ap.J., **434**, 206 (1994)
15. G. I. Ogilvie and J. E. Pringle, MNRAS, **279**, 152 (1996)
16. T. Sano and S. M. Miyama, ApJ, **515**, 776 (1999)
17. G. Rudiger and Y. Zhang, Astron. Astrophys, **378**, 302 (2001)
18. B. Coppi and E. A. Keyes, Ap.J., **595**, 1000 (2003)
19. E. Kersale et al., Ap.J., **602**, 892 (2004)
20. J. F. Hawley, Ap.J., **528**, 462 (2001)
21. J. Hantao, J. Goodman and A. Kageyama, MNRAS, **325**, L1 (2001)
22. Z. Yoshida and S.M. Mahajan, Phys. Rev. Lett., **88**, 095001 (2002)

Hybrid Viscosity and Magnetoviscous Instability in Hot, Collisionless Accretion Disks

Prasad Subramanian[1], Peter A. Becker[2], and Menas Kafatos[2]

[1] Indian Institute of Astrophysics, Bangalore - 560034, India
psubrama@iiap.res.in
[2] College of Science, George Mason University, Fairfax, VA 22030, USA

Summary. We aim to illustrate the role of hot protons in enhancing the magnetorotational instability (MRI) via the "hybrid" viscosity, which is due to the redirection of protons interacting with static magnetic field perturbations, and to establish that it is the only relevant mechanism in this situation. It has recently been shown by Balbus [1] and Islam & Balbus [11] using a fluid approach that viscous momentum transport is key to the development of the MRI in accretion disks for a wide range of parameters. However, their results do not apply in hot, advection-dominated disks, which are collisionless. We develop a fluid picture using the hybrid viscosity mechanism, that applies in the collisionless limit. We demonstrate that viscous effects arising from this mechanism can significantly enhance the growth of the MRI as long as the plasma $\beta \gtrsim 80$. Our results facilitate for the first time a direct comparison between the MHD and quasi-kinetic treatments of the magnetoviscous instability in hot, collisionless disks.

Key words: MHD: Viscosity - Instability in Hot - Accretion Disks

1 Introduction

The microphysical source of viscosity in accretion disks has been a long-standing puzzle. Since the early 1990s, there has been a growing consensus that magnetic fields generated by the magnetorotational instability (MRI) are key to providing the required viscosity in cold accretion disks ([2] (and references therein), [4], [14]). The standard treatment of the MRI is valid only for collisional plasmas, which can be described in the MHD approximation. However, the plasmas comprising hot, two-temperature accretion flows, like those described in [8] and [19] (hereafter SLE) are clearly collisionless. This is also the case for the radiatively inefficient, advection-dominated accretion flows (ADAFs) treated in [15] and [16]. Filho ([9]), Kafatos ([12]) and Paczyński ([17]) had initially suggested that viscosity due to collisions between hot protons might be important in two-temperature accretion flows, although the effects of an embedded turbulent magnetic field were not included in their treatments. Subramanian, Becker & Kafatos ([23]; hereafter SBK96) proposed that a *hybrid* viscosity, due to protons colliding with magnetic scattering centers, might be the dominant viscosity mechanism in such accretion disks. In this paper we investigate the implications of the hybrid viscosity for the development of the MRI in hot disks. In particular, we show that this mechanism can be used to establish an interesting connection between the fluid models and the quasi-kinetic treatments used by previous authors to study the viscous enhancement of the growth rate during the early stages of the MRI.

2 MVI in hot accretion disks

Balbus ([1]) and Islam & Balbus ([11]) employed an MHD approach to study the effect of viscosity on the development of the MRI, and discovered a robust instability which they call the magnetoviscous instability (MVI). In the MVI, angular momentum is exchanged between fluid elements via viscous transport, which plays a central role in the development of the instability. Balbus ([1]) does not address the physical origin of the viscosity that is central to the development of the MVI, and therefore his results are stated in terms of an unspecified coefficient of dynamic viscosity, η. Islam & Balbus ([11]) assumed the plain Spitzer (collisional) viscosity due to proton-proton collisions in their treatment of the MVI, but this particular mechanism is not effective in hot, collisionless disks. There have been some recent attempts at quasi-kinetic treatments of MRI-like instabilities in collisionless plasmas (e.g., [18], [20], [21]). It is interesting to note that the pressure anisotropy concept discussed in these papers is somewhat similar to the idea embodied in the hybrid viscosity formalism of SBK96. This suggests that it may be possible to develop a "fluid" picture based on the hybrid viscosity that would be applicable in hot disks, hence bridging the gap between the two paradigms. The hybrid viscosity concept of SBK96 relies only on the momentum deposited by particles propagating along magnetic fields lines between adjacent annuli in the disk.

3 Applicability of the hybrid viscosity

Paczyński ([17]) and SBK96 noted that the presence of even a very weak magnetic field can effectively "tie" protons to magnetic field lines. Paczyński argued that in this situation the ion-ion collisional mean free path is much larger than the proton Larmor radius and therefore the effective mean free path is equal to the proton Larmor radius. This led him to conclude that the viscosity would effectively be quenched in such a plasma. However, the protons in hot accretion disks are typically super-Alfvénic, especially in the initial stages of a magnetic field-amplifying instability such as the MRI, when the plasma β parameter is quite large. This reflects the fact that the ratio of the proton thermal speed to the Alfvén speed is equal to $(3\beta/2)^{1/2}$. Since the magnetic field evolves on Alfvén timescales, it can be considered to be static for our purposes. The motion of collisionless, charged particles propagating through a static, tangled magnetic field has been explored extensively in the context of cosmic ray propagation (e.g., [5], [6], [10]). It has been conclusively established that the particle transport does not obey Bohm diffusion for a wide range of rigidities and turbulence levels (see, e.g., Fig. 4 of [5] and Fig. 4 of[6]). In particular, the low rigidity, low turbulence level case appropriate for our situation obeys the predictions of quasi-linear theory quite well, and the mean free paths are much larger than the Larmor radius as expected.

Under these conditions, SBK96 demonstrated the importance of a new kind of viscosity called the "hybrid viscosity," in which angular momentum is transported via collisions between protons and static irregularities ("kinks") in the magnetic field. In this picture, a proton spirals tightly along a magnetic field line until its gyro-averaged guiding center motion (and hence its gyro-averaged momentum) is changed via an encounter with a kink. During the encounter the proton therefore exchanges angular momentum with the field, which transfers the resulting torque to the plasma. The effective mean free path used in the computation of the viscosity is set equal to the distance between the magnetic kinks (i.e., the field coherence length). We express the hybrid viscosity mechanism in terms of a pressure anisotropy in § 6.1.

Here we examine the implications of the hybrid viscosity for the development of the MVI in hot, two-temperature accretion disks around underfed black holes. We assume that the accreting plasma is composed of fully ionized hydrogen. The physical picture involves the perturbation of an initially straight magnetic field line that eventually leads to the instability (see, e.g., Fig. 1 of [1]). Since the proton Larmor radius is negligible in comparison to a macroscopic length scale, we can effectively think of the proton as sliding along the field line like a bead on a wire. The proton is forced to change its direction upon encountering the kink associated with the initial field perturbation. In such a situation, the effective mean free path, λ, used in the description of the hybrid viscosity should be set equal to the wavelength of the initial perturbation. We demonstrate that the hybrid viscosity is the principle mediator of the MVI during the early stages of the instability.

4 Hybrid viscosity in hot accretion disks

The structure of hot, two-temperature accretion disks was first studied in detail by SLE, and later by Eilek & Kafatos ([8]) and SBK96. The closely related advection-dominated accretion flows were analyzed by Narayan & Yi ([15]), Narayan, Mahadevan, & Quataert ([16]), and many subsequent authors. In this section, we investigate the nature of the viscosity operative in hot, two-

252   Subramanian, Becker & Kafatos

temperature accretion disks based on a simplified set of model-independent relations that are applicable to both ADAF and SLE disks.

The gas in the disk can be considered collisionless with respect to the protons provided

$$\lambda_{ii} > H \ , \tag{1}$$

where H is the half-thickness of the disk, and the ion-ion Coulomb collisional mean free path, λ_{ii}, is given in cgs units by (SBK96)

$$\lambda_{ii} = 1.80 \times 10^5 \, \frac{T_i^2}{N_i \, \ln \Lambda} \ , \tag{2}$$

for a plasma with Coulomb logarithm $\ln \Lambda$ and ion temperature and number density T_i and N_i, respectively. We can combine equations (1) and (2) to obtain

$$\frac{\lambda_{ii}}{H} = 1.20 \times 10^{-19} \, \frac{T_i^2}{\tau_{es} \, \ln \Lambda} \ > 1 \ , \tag{3}$$

where the electron scattering optical thickness, τ_{es}, is given by

$$\tau_{es} = N_i \, \sigma_{\mathrm{T}} \, H \ , \tag{4}$$

and σ_{T} is the Thomson scattering cross section. Equation (3) can be rearranged to obtain a constraint on τ_{es} required for the disk to be collisionless, given by

$$\tau_{es} \ < \ \frac{1.20 \times 10^5 \, T_{12}^2}{\ln \Lambda} \ \sim \ 4 \times 10^3 \ , \tag{5}$$

where $T_{12} \equiv T/10^{12} \, \mathrm{K}$ and the final result holds for $\ln \Lambda = 29$ and $T_{12} \sim 1$. This confirms that tenuous, two-temperature disks with $T_i \sim 10^{11}$–$10^{12} \, \mathrm{K}$ will be collisionless for typical values of τ_{es}.

The collisionless nature of hot two-temperature accretion flows established by equation (5) strongly suggests that the plain Spitzer viscosity is not going to be relevant for such disks, although the answer will depend on the strength of the magnetic field. The hybrid viscosity will dominate over the Spitzer viscosity provided the ion-ion collisional mean free path λ_{ii} exceeds the Larmor radius, λ_{L}, so that the protons are effectively "tied" to magnetic field lines. We therefore have

$$\lambda_{ii} > \lambda_{\mathrm{L}} \ , \tag{6}$$

where the Larmor radius is given in cgs units by (SBK96)

$$\lambda_{\mathrm{L}} = 0.95 \, \frac{T_i^{1/2}}{B} \ , \tag{7}$$

where B is the magnetic field strength. Whether the disk is of the SLE or ADAF types, it is expected to be in vertical hydrostatic equilibrium, and therefore

$$H\Omega_{\mathrm{K}} = c_s = \sqrt{\frac{kT_i}{m_p}} \,, \tag{8}$$

where $\Omega_{\mathrm{K}} = (GM/r^3)^{1/2}$ is the Keplerian angular velocity at radius r around a black hole of mass M, c_s is the isothermal sound speed, and k and m_p denote Boltzmann's constant and the proton mass, respectively.

We can utilize equation (6) to derive a corresponding constraint on the plasma β parameter,

$$\beta \equiv \frac{8\pi N_i kT_i}{B^2} \,, \tag{9}$$

such that the hybrid viscosity dominates over the Spitzer viscosity. By combining equations (2), (4), (6), (7), (8), and (9), we find that

$$\beta \; < \; 3.71 \times 10^{32} \, \frac{T_{12}^{9/2} \, R^{3/2} \, M_8}{\tau_{es} \, (\ln \Lambda)^2} \,, \tag{10}$$

where $M_8 \equiv M/(10^8 M_\odot)$ and $R \equiv rc^2/(GM)$. The minimum possible value of the right-hand side in equation (10) is obtained for the maximum value of τ_{es}, which is given by equation (5). We therefore find that

$$\beta \; < \; 3.09 \times 10^{27} \, \frac{T_{12}^{5/2} \, R^{3/2} \, M_8}{\ln \Lambda} \,. \tag{11}$$

This relation is certainly satisfied in all cases involving the accretion of plasma onto a black hole, even in the presence of an infinitesimal magnetic field. We therefore conclude that the protons will be effectively tied to the magnetic field lines in two-temperature accretion disks around stellar mass and supermassive black holes, which implies that the hybrid viscosity dominates over the Spitzer viscosity in either SLE or ADAF disks.

The results of this section confirm that protons in two-temperature accretion disks rarely collide with each other, and are closely tied to magnetic field lines, even for very weak magnetic fields. If a field line is perturbed, a typical proton sliding along it will follow the perturbation, and will thus be effectively redirected. This is the basic premise of the hybrid viscosity concept, which we will now apply to the development of the MVI.

5 MVI driven by the hybrid viscosity

Figure 1 of [11] shows that magnetoviscous effects significantly enhance the MRI growth rates in the parameter regime

$$X \lesssim x \,, \quad Y \gtrsim y \,, \tag{12}$$

where $x \sim 1$, $y \sim 1$, and

$$X \equiv \frac{2.0\,(k_Z\,H)^2}{\beta}\,, \qquad Y \equiv \frac{1.5\,\eta\,k_\perp^2}{N_i\,m_p\,\Omega_{\rm K}}\,, \tag{13}$$

with η denoting the coefficient of dynamic viscosity and k_Z and $k_\perp$ representing the z and transverse components of the field perturbation wavenumber, respectively. The maximum MVI growth rate is $\sqrt{3}\,\Omega_{\rm K}$, which is $4/\sqrt{3} \sim 2.3$ times larger than the maximum MRI growth rate of $(3/4)\,\Omega_{\rm K}$. The conditions in equation (12) are derived from the dispersion relation given in equation (33) of [11], which is general enough to accommodate different prescriptions for the viscosity coefficient η. The condition $X \lesssim x$ implies a constraint on β given by

$$\beta \gtrsim \frac{2\,(k_Z\,H)^2}{x}\,. \tag{14}$$

As mentioned earlier, a proton sliding along a given field line is forced to change its direction when it encounters a kink/perturbation in the field line. The effective viscosity arises due to the momentum deposited in the fluid by the proton when it encounters the perturbation. In this picture, the perturbation wavelength plays the role of an effective mean free path. If we consider perturbations along an initially straight field line, as in Figure 1 of [1], then only the transverse component of the perturbation wavelength is relevant, and the effective mean free path for the proton is therefore

$$\lambda = \frac{2\,\pi}{k_\perp} \equiv \xi H\,, \tag{15}$$

where $\xi \leq 1$, since the perturbation wavelength λ cannot exceed the disk half-thickness H (SBK96).

In general, the Shakura-Sunyaev ([22]) viscosity parameter α is related to the coefficient of dynamic viscosity η via (SBK96)

$$\alpha P \equiv -\eta\,R\,\frac{d\Omega_{\rm K}}{dR} = \frac{3}{2}\,\eta\,\Omega_{\rm K}\,, \tag{16}$$

where $P = N_i\,k\,T_i$ is the pressure in a two-temperature disk with $T_i \gg T_e$. By combining equations (8), (13), and (16), we find that the condition $Y \gtrsim y$ can be rewritten as

$$(k_\perp\,H)^2\,\alpha \gtrsim y\,. \tag{17}$$

Following Islam & Balbus ([11]), we expect that $k_\perp \lesssim k_Z$. By combining equations (14) and (17), we therefore conclude that β must satisfy the condition

$$\beta \gtrsim \frac{2\,y}{\alpha x}\,. \tag{18}$$

We can also combine equations (14) and (15) to obtain the separate constraint

$$\beta \gtrsim \frac{79}{\xi^2 x}\,. \tag{19}$$

Equations (18) and (19) must *both* be satisfied if the MVI is to significantly enhance the MRI growth rates. Hence the combined condition for β is given by

$$\beta \gtrsim \text{Max} \left(\frac{79}{x\xi^2} , \frac{2\,y}{\alpha x} \right) . \tag{20}$$

We can use equation (16) to calculate the Shakura-Sunyaev parameter α_{hyb} describing the hybrid viscosity. The associated coefficient of dynamic viscosity is given by

$$\eta_{\text{hyb}} = \frac{\lambda}{\lambda_{ii}} \eta_{\text{s}} , \tag{21}$$

where λ_{ii} is computed using equation (2) and η_{s} is the standard Spitzer collisional viscosity, evaluated in cgs units using

$$\eta_{\text{s}} = 2.20 \times 10^{-15} \frac{T_i^{5/2}}{\ln \Lambda} . \tag{22}$$

The quantity η_{hyb} defined in equation (21) describes the effect of momentum deposition due to protons spiraling tightly along a magnetic field line over a mean free path λ. It differs from the expression given in equation (2.14) of SBK96 by a factor of $2/15$, because we do not consider tangled magnetic fields here. Setting $\eta = \eta_{\text{hyb}}$ in equation (16) and utilizing equations (2), (8), (15), (21), and (22), we find after some algebra that the expression for α_{hyb} reduces to the simple form

$$\alpha_{\text{hyb}} = 1.2\,\xi . \tag{23}$$

We can now combine equations (20) and (23) to conclude that in the case of the hybrid viscosity, the MVI is able to effectively enhance the MRI growth rates if

$$\beta \gtrsim \beta_{\text{crit}} \equiv \text{Max} \left(\frac{79}{x\xi^2} , \frac{1.7y}{x\xi} \right) . \tag{24}$$

In particular, we note that if $x \sim 1$ and $y \sim 1$, then equation (24) reduces to $\beta_{\text{crit}} = 79\,\xi^{-2}$, since $\xi \leq 1$. We therefore conclude that magnetoviscous effects driven by the hybrid viscosity will significantly enhance the growth rate (compared with the standard MRI growth rate) until the plasma β parameter reaches ~ 80, or, equivalently, until the field strength B reaches $\sim 10\%$ of the equipartition value. This assumes that the dominant perturbations have $\xi \sim 1$, which is expected to be the case during the early stages of the instability. Once the field exceeds this strength, the growth rate of the instability during the linear stage will be equal to the MRI rate.

6 Relation to previous work

It is interesting to contrast our result for the β constraint with those developed by previous authors using different theoretical frameworks.

256 Subramanian, Becker & Kafatos

6.1 Hybrid viscosity in terms of pressure anisotropy

Before proceeding on to discussing the result for the β constraint, we first cast the basic hybrid viscosity mechanism in terms of a pressure anisotropy. Several similar treatments appeal to a large-scale pressure anisotropy, rather than an explicit viscosity mechanism (e.g., [18], [20], [21]). It is therefore instructive to show that the hybrid viscosity mechanism we employ can be cast in these terms.

We follow the approach of SBK96 in considering a perturbation in the local magnetic field of an accretion disk. The pressure anisotropy due to the momentum flux carried by the particles can be analyzed in the local region using cartesian coordinates, with the $\hat{z}$-axis aligned in the azimuthal (orbital) direction, the $\hat{y}$-axis pointing in the outward radial direction, and the $\hat{x}$-axis oriented in the vertical direction. The unperturbed magnetic field is assumed to lie in the $\hat{z}$ direction, and the perturbed field makes an angle θ with respect to the $\hat{z}$-axis, and an azimuthal angle ϕ with respect to the $\hat{x}$-axis. In keeping with the hybrid viscosity scenario, we assume that the particles spiral tightly around the perturbed field line. In this situation, the component of the particle pressure in the direction *parallel* to the magnetic field, $P_{||}$, is equal to the $\hat{z}$-directed flux of the $\hat{z}$-component of momentum, P_{zz}. Likewise, the total particle pressure *perpendicular* to the field, $P_{\perp}$, is equal to the sum of the $\hat{x}$-directed momentum in the $\hat{x}$-direction and the $\hat{y}$-directed momentum in the $\hat{y}$-direction, denoted by P_{xx} and P_{yy}, respectively. Following the same approach that leads to equation (2.11) of SBK96, we obtain for the parallel pressure

$$P_{||} = P_{zz} = 2\,m_p\,N_i\,\cos^2\theta\,\times$$
$$\left[\frac{kT_i}{2m_p} - \left(\frac{2kT_i}{\pi m_p} \right)^{1/2} u'(0)\,\lambda\,\cos\theta\,\sin\theta\,\sin\phi \right], \qquad (25)$$

where $u(y)$ represents the shear velocity profile and the prime denotes differentiation with respect to y. Similarly, the total perpendicular pressure is given by

$$P_{\perp} = P_{xx} + P_{yy} = 2\,m_p\,N_i\,\sin^2\theta\,\times$$
$$\left[\frac{kT_i}{2m_p} - \left(\frac{2kT_i}{\pi m_p} \right)^{1/2} u'(0)\,\lambda\,\cos\theta\,\sin\theta\,\sin\phi \right]. \qquad (26)$$

Taken together, equations (25) and (26) imply that

$$\frac{P_{\perp}}{P_{||}} = \tan^2\theta \,. \qquad (27)$$

This result characterizes the pressure anisotropy associated with the hybrid viscosity mechanism. Equation (27) is strictly valid only in the limit of zero proton gyroradius, which is a reasonable approximation in hot advection-dominated disks. When the field line is unperturbed, so that it lies precisely along the $\hat{z}$-direction, then $\theta = 0$, and equation (27) indicates that the perpendicular pressure tends to zero; in reality, owing to finite gyroradius effects, the perpendicular pressure would actually be a small, but finite quantity even in this limit. Early in the instability, when the field line is slightly perturbed, θ has a small but non-zero value, and equation (27) predicts that the perpendicular pressure starts to increase in relation to the parallel pressure. We have cast the hybrid viscosity mechanism in terms of a pressure anisotropy in this section in order to make contact with that part of the literature in which viscous momentum transport is treated solely in this manner. The quasi-kinetic treatments of Quataert and co-workers rely on a Landau fluid closure scheme for deriving the perturbed pressure. The pressure anisotropy implied by the hybrid viscosity mechanism (eq. [27]) is much simpler than the corresponding result obtained using either the fluid closure scheme of Quataert et al., or the double adiabatic scheme ([7]) adopted by other authors.

6.2 Relation to MVI treatment

In their treatment of the MVI, Islam & Balbus ([11]) parametrized the viscous transport in terms of an unspecified proton-proton collision frequency, ν. Their estimates of the growth rates in *collisional* plasmas agree fairly well with those derived using quasi-kinetic treatments. Based on their formalism, they conclude that the β regime within which magnetoviscous effects can significantly impact the MRI growth rates in two-temperature accretion flows extends to $\beta_{crit} \sim 1$. However, as they point out, their approach breaks down in the collisionless limit $\nu \to 0$, which describes the ADAF disks of interest here. It is therefore not surprising that their constraint on β is significantly different from the one we have derived in equation (24).

6.3 Relation to quasi-kinetic treatment

Quataert, Dorland & Hammett ([18]) have treated the case of a strictly collisionless plasma using a fairly complex kinetic formalism. Their results suggest that, for the case with $B_\phi = B_z$ and $k_r = 0$ (which is the one considered by Islam & Balbus and ourselves), viscous effects will significantly impact the MRI growth rates for values of β that are several orders of magnitude larger than those predicted by our formalism. For example, their analysis predicts that a growth rate of $1.5\,\Omega_K$ can be achieved if $\beta \gtrsim \beta_{crit} \sim 10^4$ (see Fig. 4 of [18]) and Fig. 2 of [20]). On the other hand, Figure 1 of [11] indicates that a growth rate of $1.5\,\Omega_K$ can be achieved in the MHD model if $X \lesssim 0.35$, $Y \gtrsim 12$, which corresponds to $x = 0.35$, $y = 12$ in equation (12). Assuming that $\xi \sim 1$ as before, equation (24) yields in this case the condition $\beta \gtrsim \beta_{crit} \sim 225$. Hence our MHD model based on the hybrid viscosity predicts that viscous effects will enhance the MRI growth rates down to much lower values of β than those

258   Subramanian, Becker & Kafatos

obtained in the quasi-kinetic model. This difference reflects the differing role of the particle pressure in the two scenarios.

In our formulation, the viscosity is expressed by protons that deposit their momentum into the fluid upon encountering kinks in the magnetic field, which is anchored in the local gas. The importance of forces due to gas pressure relative to those due to the tension associated with the magnetic field thus scales as the plasma β. On the other hand, gas pressure forces are only $\sqrt{\beta}$ times as important as forces arising out of magnetic tension in the quasi-kinetic treatment of [18]. It follows that the value of $\beta_{\rm crit}$ computed using our MHD model based on the hybrid viscosity should be comparable to the square root of the $\beta_{\rm crit}$ value obtained using the quasi-kinetic model, and this is borne out by the numerical results cited above.

7 Conclusions

In this paper we have investigated the role of hot protons in influencing the magnetoviscous instability described in [1] and [11]. We have shown that the only relevant viscosity mechanism in this situation is the "hybrid" viscosity, which is due to the redirection of protons interacting with magnetic irregular-ities ("kinks") set up by the initial field perturbations. In particular, we have demonstrated in equation (24) that viscous effects associated with the hybrid viscosity will significantly augment the MRI growth rates for $\beta \gtrsim 80$, which corresponds to a magnetic field strength B below $\sim 10\%$ of the equipartition value. For smaller values of β, we expect the instability to grow at the MRI rate as long as it remains in the linear regime. This conclusion is expected to be valid in any hot, two-temperature accretion disk, including advection-dominated ones. We have obtained this result using a relatively simple fluid treatment, based upon the general dispersion relation obtained in [11]. Our use of the hybrid viscosity concept alleviates an important drawback in the fluid application made by Islam & Balbus ([11]), because their treatment of viscous transport breaks down in the collisionless plasmas of interest here. The new results we have obtained allow an interesting comparison between the MHD approach and the quasi-kinetic formalism used by other authors. We show that the differences between the predictions made by the two methodologies stem from the differing treatments of the particle pressure.

PS gratefully acknowledges the hospitality of the Jagannath Institute of Technology and Management, where part of this work was carried out.

References

1. Balbus, S. A. 2004, Astrophys. J., 616, 857
2. Balbus, S. A. 2003, Annu. Rev. Astron. Astrophys, 41, 555
3. Balbus, S. A. & Hawley, J. F. 1992, Rev. Mod. Phys., 70, 1

Hybrid viscosity and the MVI   259

4. Brandenburg, A., Nordlund, A., Stein, R. F., & Torkelsson, U. 1995, Astrophys. J., 446, 741
5. Candia, J. & Roulet, E. 2004, J. Cosmology & Astropart. Phys., 10, 7
6. Casse, F., Lemoine, M., & Pelletier, G. 2001, Phys. Rev. D, 65, 023002
7. Chew, G. F., Goldberger, M. L., & Low, F. E. 1956, Proc. R. Soc. London A, 236, 112
8. Eilek, J. A., & Kafatos, M. 1983, Astrophys. J., 271, 804
9. Filho, C. M. 1995, Astron. Astrophys., 294, 295
10. Giacalone, J. & Jokipii, J. R. 1999, Astrophys. J., 520, 204
11. Islam, T. & Balbus, S. A. 2005, Astrophys. J., 633, 328
12. Kafatos, M. 1988, in Supermassive Black Holes (Cambridge: Cambridge Univ. Press), 307
13. Mahadevan, R. 1997, Astrophys. J., 477, 585
14. Matsumoto, R., & Tajima, T. 1995, Astrophys. J., 445, 767
15. Narayan, R., & Yi, I. 1995, Astrophys. J., 452, 710
16. Narayan, R., Mahadevan, R. & Quataert, E. 1998, in Theory of Black Hole Accretion Disks, ed. M. Abramoqicz, G. Bjornsson, & J. E. Pringle (Cambridge: Cambridge Univ. Press), 148. astro-ph/9803141
17. Paczyński, B. 1978, Acta Astronomica, 28, 253
18. Quataert, E., Dorland, W., & Hammett, G. W. 2002, Astrophys. J., 577, 524
19. Shapiro, S. L., Lightman, A. L., & Eardley, D. M. 1976, Astrophys. J., 204, 187 (SLE)
20. Sharma, P., Hammett, G. W., & Quataert, E. 2003, Astrophys. J., 596, 1121
21. Sharma, P., Hammett, G. W., Quataert, E., & Stone, J. M. 2006, Astrophys. J., 637, 952
22. Shakura, N. I., & Sunyaev, R. A. 1973, Astron. Astrophys., 24, 337
23. Subramanian, P., Becker, P. A., & Kafatos, M. 1996, Astrophys. J., 469, 784 (SBK96)

Transonic Properties of Accretion Disk Around Compact Objects

Banibrata Mukhopadhyay

Astronomy and Astrophysics Programme, Department of Physics, Indian Institute of Science, Bangalore-560012, India `bm@physics.iisc.ernet.in`

Summary. An accretion flow is necessarily transonic around a black hole. However, around a neutron star it may or may not be transonic, depending on the inner disk boundary conditions influenced by the neutron star. I will discuss various transonic behavior of the disk fluid in general relativistic (or pseudo general relativistic) framework. I will address that there are four types of sonic/critical point possible to form in an accretion disk. It will be shown that how the fluid properties including location of sonic points vary with angular momentum of the compact object which controls the overall disk dynamics and outflows.

Key words: MHD: Accretion disk – compact objects

1 Introduction

Accretion disks serve to identify compact objects, mostly black holes and neutron stars, in the universe. The most common way to understand its formation is in a binary system where matter is pulled off a companion star and settles on to the compact object in the form of a disk. As practically black holes and neutron stars cannot be seen, by detecting and analyzing light rays off an accretion disk one can understand the properties of the central compact object. Other examples of the accretion disk are the protoplanetary disks, disks around active galactic nuclei, in star-forming systems, and in quiescent cataclysmic variables etc.

The molecular viscosity in an accretion disk is rather low. Shakura & Sunyaev [1] proposed that there is a significant turbulent viscosity explaining transport of matter inwards and angular momentum outwards with the Keplerian angular momentum profile [2]. However, until the work by Balbus & Hawley [3] origin of the instability and plausible turbulence was not understood. Applying the idea of Magneto-Rotational-Instability (MRI) described by Velikhov [4] and Chandrasekhar [5], they showed that the Keplerian disk flow may exhibit unstable modes under perturbation in the presence of a weak magnetic field. This plausibly generates turbulence which results transport. However, there are several accretion disk systems where gas is electrically neutral in charge and thus the magnetic field does not couple with the system. Examples of such system are protoplanetary and star-forming disks, disks around active galactic nuclei and in quiescent cataclysmic variables. Hence in these systems MRI is expected not to work to generate turbulence. While the origin of the transport in such systems is still an ill understood problem, some mechanisms have been proposed by various groups including the present author [6, 7, 8, 9, 10, 11, 12].

However, the problems of the transport and then the viscosity are not severe if the disk angular momentum profile deviates from the Keplerian type to sub-Keplerian type. In a sub-Keplerian disk, the gravitational force dominates over the centrifugal force resulting in a strong advective component of the infalling matter [13, 14, 15]. Due to the dominance of the gravity, the matter even with a constant angular momentum may transport inwards easily. An accretion disk with a strong advective component, namely the advective accretion disk, is hotter than the Keplerian disk which explains successfully several systems e.g. Sgr A^*, GRS 1515+105 which are expected to be consisting of hot disks. Therefore, such an accreting system may not be of pure Keplerian type. At least close to the compact object it is an advective accretion disk. In a sub-Keplerian disk, at far away off the black hole, the speed of the infalling matter is close to zero when it is practically out of the black hole's influence. However, temperature of the system is finite resulting in finite sound speed at that radius. On the other hand, matter speed at the black hole horizon reaches the speed of light (c), while, by causality condition, sound speed can be at the most $c/\sqrt{3}$. Therefore, there must be a location where matter speed crosses local sound speed making the flow transonic. However, the second condition namely the inner boundary condition not necessarily be satisfied in the flow around a neutron star. Therefore an accretion flow around a neutron star may or may not be transonic.

In the present article, I mainly concentrate on the sub-Keplerian accretion disk with possible multitransonic flow. If the flow is transonic, then the disk dynamics and corresponding outflow are influenced by the sonic locations/points [14, 16, 17, 18, 19]. Depending upon the flow energy (entropy) and angular momentum, a disk may exhibit one to four sonic points, through which all however, matter may not pass. Sonic points in accretion disks play role similar to the critical points in simple harmonic oscillators, particularly to control the dynamics of the system. Indeed an accretion disk can be looked upon as a damped harmonic oscillator [20].

I organize the paper as follows. In the next section, I discuss accretion flow, first the spherical one and then the disk flow, and formation of possible sonic points therein in the absence of any energy dissipation. In §3, I analyze how the rotation of the compact object affects the sonic points. Subsequently, for completeness, I present the generalized set of accretion disk equations

Transonic Properties of Accretion Disk Around Compact Objects   263

including possible dissipation effects in §4, without going into their solutions. Finally, I summarize in §5.

2 Accretion flow and formation of sonic points

First I discuss simple spherical accretion namely Bondi flow. Subsequently, I include angular momentum into the equations set describing disk accretion. I show that while the Bondi flow around a nonrotating compact star exhibits single sonic point the disk flow may exhibit multiple sonic points.

As gravity is expected to be very strong close to the compact object, in principle I should describe the flow system by the set of general relativistic equations. For a nonrotating compact object equations should be written in the Schwarzschild geometry and for a rotating compact object they should be in the Kerr geometry. However, that might hide transparency of the description, as the full general relativistic set of equations is so cumbersome that it is difficult to relate the terms with the physics they carry. Indeed under some occasions a full general relativistic description is not required. Therefore, I describe the system by pseudo-Newtonian approach. Here one uses Newtonian set of equations only along with a modified gravitational force/potential such that it mimics the general relativistic features approximately to describe the system.

2.1 Bondi flow

This happens for an isolated star when matter falls onto it from all directions, resulting in a spherical accretion. Therefore, to describe the system, we consider the spherical polar coordinate system where only nonzero component of velocity is $v_r = v$. Hence the equations describing the steady-state flow with negligible viscosity are given as:

$$v\frac{dv}{dx} + \frac{1}{\rho}\frac{dP}{dx} + F_g = 0, \tag{1}$$

$$\frac{1}{x^2}\frac{d}{dx}\left(\rho v x^2\right) = 0, \quad \text{hence} \quad \rho v x^2 = \text{constant} = \dot{M}_{ac}, \tag{2}$$

where all the variables are expressed in dimensionless units. v is velocity in unit of light speed, $x = r/r_g$ where r is radial coordinate and $r_g = GM_s/c^2$, M_s, c, G are respectively mass of the central star, speed of light, Newton's gravitation constant, P, ρ are respectively corresponding dimensionless pressure, density of the flow, $\dot{M}_{ac}$ is accretion rate and F_g is gravitational force given by

$$F_g = \frac{1}{x^2} \; : \; \text{Newtonian,} \tag{3}$$

$$= \frac{1}{(x-2)^2} \; : \; \text{pseudo} - \text{Newtonian, Schwarzschild geometry [21],} \tag{4}$$

$$= \frac{(x^2 - 2j\sqrt{x} + j^2)^2}{x^3(\sqrt{x}(x-2) + j)^2} \; : \; \text{pseudo} - \text{Newtonian, Kerr geometry [22], (5)}$$

where j is specific angular momentum of the compact object. Integrating equation (1) for an adiabatic flow to a nonrotating compact object I obtain

$$E = \frac{1}{2}v^2 + na^2 - \frac{1}{x-2}, \tag{6}$$

where polytropic index $n = 1/(\gamma - 1)$ and γ is defined as $\gamma = a^2\rho/P$ with $P = K\rho^\gamma$, K is gas constant and a is sound speed. If the Mach number of the flow is defined as $M = v/a$, then at a constant energy E, it is clearly understood from (6) that $M - x$ trajectory is always hyperbolic-type [23] with a sonic point at $M = 1$ shown in Fig. 1 [20]. The solution marked 'A' by an arrow indicates accretion and marked 'C'' as wind. None of the other solutions are physical to describe accretion and wind. However, for the flow around a rotating compact object, there might be more than one sonic point but I will not discuss them here, this can be checked with a pseudo-Newtonian potential proposed for the Kerr geometry [22].

2.2 Disk accretion flow

Now I consider a binary system when the compact object is closely associated with its binary companion star and pulling matter off it. The gravitational force acts on the donor star radially. On the other hand, the star is rotating and hence the matter detached off it due to gravitational pull has angular momentum. As a result the matter infalls towards the compact object in a spiral path forming a disk around it called the accretion disk. Therefore, the radial momentum balance equation describing disk dynamics in steady-state in the absence of any energy dissipation is [18]

$$v\frac{dv}{dx} + \frac{1}{\rho}\frac{dP}{dx} - \frac{\lambda^2}{x^3} + F_g = 0, \tag{7}$$

where λ is specific angular momentum of the infall which is a conserved quantity in a nondissipative flow. The corresponding vertically averaged mass conservation reads as

$$\frac{1}{x}\frac{d}{dx}(x\Sigma v) = 0, \; \text{hence} \; -4\pi x\Sigma v = \text{constant} = \dot{M}_{ac}, \tag{8}$$

where $\Sigma \sim h(x)\rho$ is the (vertically integrated) column density [24] with ρ is density at the equatorial plane, $h(x) = a\sqrt{x/F_g}$ is the half-thickness of

Transonic Properties of Accretion Disk Around Compact Objects   265

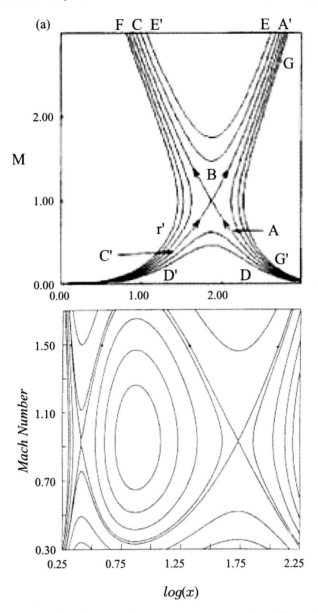

Fig. 1. Upper panel: Bondi flow: Mach number as a function of logarithmic radial coordinate for various energy of the flow. The point 'B' is sonic point and the solution marked by the arrow 'A' indicates accretion and 'C'' indicates wind. Lower panel: Disk flow: Same as Bondi flow.

the disk computed from the assumption of vertical equilibrium. Now integrating (7) for an adiabatic flow towards a nonrotating compact object I obtain specific energy of the flow

$$E = \frac{1}{2}v^2 + na^2 + \frac{\lambda^2}{2x^2} - \frac{1}{x-2},\tag{9}$$

which is a conserved quantity for the nondissipative system. It is seen from (9) that at far away from the black hole and close to its event horizon located at $x = 2$, last term (gravitational potential energy) on the right hand side dominates over the last but one term (centrifugal energy) resulting in a hyperbolic-type $M - x$ trajectory as shown in Fig. 1 [16, 20]. On the other hand, at an intermediate location, with an appropriate value of λ, centrifugal energy may dominate over gravitational potential energy resulting in an ellipse-type $M - x$ trajectory as shown in Fig. 1.

2.3 Formation of sonic points in disk accretion

Combining (7) and (8) I obtain

$$\frac{dv}{dx} = \frac{\frac{\lambda^2}{x^3} - F_g + \frac{a^2}{\gamma+1}\left(\frac{3}{x} - \frac{1}{F_g}\frac{dF_g}{dx}\right)}{v - \frac{2a^2}{(\gamma+1)v}} = \frac{N}{D}.\tag{10}$$

At the sonic point $x = x_c$, $D = 0$ and thus to have continuous dv/dx at the same location N must vanish. Hence, applying $N = D = 0$ at $x = x_c$ to (10), I obtain velocity, sound speed, specific energy at x_c

$$v_c = a_c\sqrt{\frac{2}{\gamma+1}}, \text{ hence Mach number } M_c = \sqrt{\frac{2}{\gamma+1}},$$

$$a_c = \sqrt{(\gamma+1)\left(\frac{\lambda^2}{x_c^2} - F_{gc}\right)\left[\frac{1}{F_{gc}}\left(\frac{dF_g}{dx}\right)_c - \frac{3}{x_c}\right]^{-1}},$$

$$E_c = \frac{2\gamma}{\gamma-1}\left[\frac{\frac{\lambda^2}{x_c^3} - F_{gc}}{\frac{1}{F_{gc}}\left(\frac{dF_g}{dx}\right)_c - \frac{3}{x_c}}\right].\tag{11}$$

I also compute a quantity that carries information of entropy

$$\dot{\mathcal{M}} = (\gamma K)^n \dot{M}_{ac}\tag{12}$$

which is a useful quantity to study the sonic point behavior. Now to obtain solutions for the accretion disk one has to integrate (10) with a proper boundary condition. As our interest is in transonic solutions, the matter necessarily passes through the sonic location and this I consider as our staring point of the integration. As the system under consideration is in the steady-state, integrating from the sonic point to outwards (inwards) and vice versa are equivalent.

However, dv/dx given in (10) is of $0/0$ form. Therefore, applying l'Hospital's rule, I obtain dv/dx at $x = x_c$

$$\left(\frac{dv}{dx}\right)_c = -\frac{\mathcal{B} + \sqrt{\mathcal{B}^2 - 4\mathcal{AC}}}{2\mathcal{A}}, \qquad (13)$$

where,

$$\mathcal{A} = 1 + \frac{2a_c^2}{(\gamma+1)v_c^2} + \frac{4a_c^2(\gamma-1)}{v_c^2(\gamma+1)^2},$$

$$\mathcal{B} = \frac{4a_c^2(\gamma-1)}{v_c(\gamma+1)^2}\left[\frac{3}{x_c} - \frac{1}{F_{gc}}\left(\frac{dF_g}{dx}\right)_c\right],$$

$$\mathcal{C} = \frac{a_c^2}{\gamma+1}\left[\left(\frac{1}{F_{gc}}\left(\frac{dF_g}{dx}\right)_c\right)^2 - \frac{1}{F_{gc}}\left(\frac{d^2F_g}{dx^2}\right)_c - \frac{3}{x_c^2}\right]$$
$$- \frac{3\lambda^2}{x_c^4} - \left(\frac{dF}{dx}\right)_c - \frac{a_c^2(\gamma-1)}{(\gamma+1)^2}\left[\frac{3}{x_c} - \frac{1}{F_{gc}}\left(\frac{dF_g}{dx}\right)_c\right]^2. \qquad (14)$$

Now the value of discriminant $\mathcal{D} = \mathcal{B}^2 - 4\mathcal{AC}$ determines the type of sonic point. Clearly $\mathcal{D}$ depends on the sonic location which can be computed for a given E_c from (11). In fact, the expression for E_c can be rewritten as a fourth order polynomial of x_c. Hence, E_c and λ are the input parameters determining flow behavior for a given system.

There are four different types of sonic point shown in Fig. 2 [25, 20] classified according to the trajectory around it. (1) When $\mathcal{D} < 0$ with $\mathcal{B} = 0$, sonic point is center-type (elliptical trajectories in Fig. 1). (2) $\mathcal{D} < 0$ with $\mathcal{B} \neq 0$ gives spiral-type sonic point. For $\mathcal{D} \geq 0$ with (3) $\mathcal{AC} > 0$ it is nodal-type, and (4) $\mathcal{AC} < 0$, saddle-type (hyperbolic trajectories in Figs. 1). When $\mathcal{AC} = 0$ for $\mathcal{D} \geq 0$, sonic points are called straight line.

Center **Spiral** **Saddle** **Nodal**

Fig. 2. Trajectories around various critical points where abscissa and ordinate are radial coordinate and Mach number respectively.

3 Sonic point analysis in disks around rotating compact objects

The energy and information of entropy at the sonic point E_c and $\dot{\mathcal{M}}_c$ as functions of sonic location x_c given by (11) and (12) respectively describe the loci of sonic point energy and entropy of the flow. Figure 3 [18] shows their behavior for various choice of specific angular momentum of the compact object. As I consider a nondissipative disk, the energy and entropy remain conserved throughout. The intersections of the curves by the constant energy and entropy line (which are the horizontal lines in the figure) indicate the sonic points of the accretion disk for that particular energy and entropy and specific angular momentum of the black hole (Kerr parameter). It is clearly seen that, at a particular energy and entropy, if the Kerr parameter increases, the region where sonic points form (as well as radii of marginally bound (x_b) and stable (x_s) orbits) shift(s) to a more inner region of the disk and possibility to have all sonic points in the disk outside event horizon increases. As an example, for $a = 0.998$, the inner edge of the accretion disk enlarges to such an extent that the fourth sonic point in the disk appears outside the event horizon. On the other hand, for retrograde orbits (counter rotating cases), all the sonic points come close to each other tending to overlap for a particular value of the Kerr parameter (when x_b and x_s move to greater radii). Therefore, as the value of the Kerr parameter decreases, the possibility of forming individual sonic points decreases and thus shrinking the region containing sonic points. This is very well-understood physically; as the Kerr parameter decreases, total angular momentum of the system decreases resulting in the disk tends to a *Bondi-like* flow, that has single saddle-type sonic point.

On the other hand, for a particular a, if E_c and $\dot{\mathcal{M}}_c$ decrease, then the possibility to have all the sonic points in the disk increases. This is understood from Fig. 3 that as E_c and $\dot{\mathcal{M}}_c$ decrease, the curve is more likely to intersect the horizontal line. If $E_c = 0.05$, then number of intersection of all the curves, except the dashed one, by the $E_c = 0.05$ line is one giving only sonic point. The dashed curve with $a = 0.998$ intersects twice. However, for $E_c = 0.0065$ (as considered in Fig. 3), the dashed curve intersects four times exhibiting four possible sonic points in the flow and the other curves intersect thrice. For a detailed description, see earlier works [18, 19].

Note that the sonic points occurring with a negative slope of the curve indicate the locations of 'saddle-type' sonic point and those with a positive slope indicate the 'center-type' sonic point. Thus the rotation of a black hole plays an important role in the formation of the sonic points which are related to the structure of accretion disks and presumably formation of outflows and jets.

Transonic Properties of Accretion Disk Around Compact Objects 269

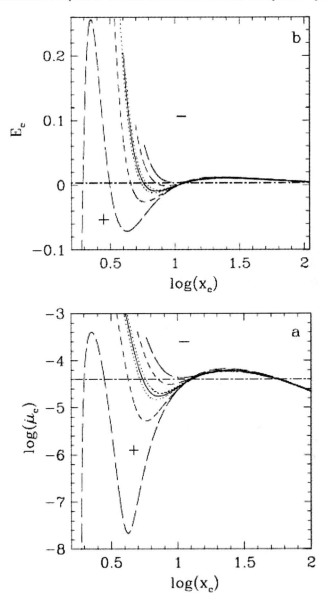

Fig. 3. Variation of energy (upper panel) and information of entropy (lower panel) as functions of logarithmic sonic radius when j is a parameter. Central solid curve is for nonrotating compact object ($j = 0$), while curves in regions of either side of it indicated by '+' and '−' are for prograde and and retrograde orbits respectively. Different curves from $j = 0$ line to downwards are for $j = 0.1, 0.5, 0.998$ and to upwards for $j = -0.1, -0.5, -0.998$. The horizontal line indicates the curve of constant energy of 0.0065 (upper panel) and of constant entropy of 5×10^{-5} (lower panel). Other parameters are $\lambda = 3.3$, $\gamma = 4/3$.

270   Banibrata Mukhopadhyay

4 Generalized set of equations describing a sub-Keplerian accretion disk

So far, for simplicity, I have described disks without any dissipative energy. In principle a disk must exhibit viscous dissipation making angular momentum varying with disk radius. The hot disk flow with ion temperature $T \geq 10^9$K is also expected to generate significant nuclear energy mostly via proton-capture reactions [26, 27]. On the other hand, significant energy is radiated out through the inverse-Compton, bremsstrahlung and synchrotron effects, cooling the disk. In addition, significant energy is expected to be absorbed through endothermic nuclear reactions, mostly via dissociation of elements [26, 27, 28], affecting disk dynamics significantly. Although the main purpose of the paper is to describe the basic mechanism to form sonic points in accretion disks and for that the inviscid set of equations without dissipation suffices, for completeness here I describe the generalized set of equations including effects of energy dissipation. However, I will not go for the solutions of such equations set which is beyond the scope of present paper.

Therefore, in general one should incorporate two more equations namely angular momentum balance and energy equation apart from (7) and (8) as given below. I express viscous dissipation q^+ in terms of shear stress $W_{x\phi}$ as $q^+ = W_{x\phi}^2/\eta$, where η is the coefficient of viscosity. Shakura & Sunyaev [1] parametrized shear stress in a Keplerian disk by gas pressure with a constant α, called Shakura-Sunyaev viscosity parameter, such that $W_{x\phi} = -\alpha P$. However, in an advective disk, there must be a significant contribution due to ram pressure and thus shear stress may be read as $W_{x\phi} = -\alpha(I_{n+1}P+I_n v^2\rho)h(x)$, where I consider vertically averaged disk and I_{n+1}, I_n appear due to integration in the vertical direction [24] when P and ρ are pressure and density respectively at the equatorial plane. However, if I use this expression of shear stress in describing q^+, it loses information of actual shear, effect due to variation of angular velocity in the disk. The proper way of writing shear stress should be $W_{x\phi} = \eta x \frac{d\Omega}{dx}$, where Ω is angular frequency of the flow. If I substitute this into q^+, then final equation to be solved will contain a nonlinear derivative term making it tedious to solve. Therefore, following Chakrabarti [29], I express $q^+ = -\alpha(I_{n+1}P+I_n v^2\rho)h(x)x\frac{d\Omega}{dx}$, where one $W_{x\phi}$ in the square is expressed by pressure and other by actual shear. Thus the angular momentum balance equation is

$$v\frac{d\lambda}{dx} = \frac{1}{\Sigma x}\frac{d}{dx}\left[x^2 q^+\right]. \tag{15}$$

Then the energy equation may read as

$$\Sigma v T\frac{ds}{dx} = \frac{vh(x)}{\Gamma_3 - 1}\left(\frac{dP}{dx} - \Gamma_1\frac{P}{\rho}\frac{d\rho}{dx}\right) = Q^+ - Q^-, \tag{16}$$

where s is entropy density, $Q^+ = q^+ + q_{\mathrm{nuc}}$ and $Q^- = q_{\mathrm{ic}} + q_{\mathrm{br}} + q_{\mathrm{syn}}$ when q_{nuc} is the nuclear energy released/absorbed in the disk and $q_{\mathrm{ic}}, q_{\mathrm{br}}, q_{\mathrm{syn}}$ are

Transonic Properties of Accretion Disk Around Compact Objects 271

respectively energy radiated out due to inverse-Compton, bremsstrahlung, synchrotron effect. Following Cox & Giuli [30], I define

$$\Gamma_3 = 1 + \frac{\Gamma_1 - \beta}{4 - 3\beta}, \Gamma_1 = \beta + \frac{(4 - 3\beta)^2(\gamma - 1)}{\beta + 12(\gamma - 1)(1 - \beta)}, \beta = \frac{\text{gas pressure}}{\text{total pressure}}. \quad (17)$$

To obtain generalized solutions of an accretion disk with a significant advective component, namely a sub-Keplerian accretion disk, one has to solve equations (7), (8), (15) and (16) simultaneously along with the set of equations generating nuclear energy through the reactions taking place in a hot flow among all the isotopes therein as given in earlier works in detail [26, 27, 28].

5 Summary

I have described how the various types of sonic point form in an accretion disk. While sonic points necessarily form in disks around a black hole, it need not be around a neutron star. I have discussed that upto four sonic points may exist outside event horizon of a black hole depending on the flow parameters and black hole's angular momentum. However, in a zero angular momentum Bondi flow around a nonrotating compact object, there is only one sonic point. When a disk has saddle and/or nodal -type sonic points, matter passes through them, while it does not for center and spiral -type sonic points. Therefore, the last two are not physical sonic points for accretion and wind because they may not be the part of solutions extending from infinity to the black hole horizon. However, the corresponding branches may be useful to explain certain scenarios in disks, as argued by some authors, e.g. formation of shock helping to understand observed truncation of disks [31, 32], launching of outflows and jets [17]. In these cases, as explained by previous authors [17, 18, 19, 20, 29], the infalling matter first becomes supersonic passing through the outer saddle-type sonic point, then the shock forms and it becomes subsonic around a center-type sonic point, and finally passing through the inner saddle-type sonic point it falls into the black hole. A detailed behavior of such solutions around a fast rotating compact object should be investigated for understanding jet physics in future.

References

1. N.I. Shakura, R.A. Sunyaev: A&A **24**, 337 (1973)
2. J.E. Pringle: ARA&A **19**, 137 (1981)
3. S.A. Balbus, J.F. Hawley: ApJ **376**, 214 (1991)
4. E. Velikhov: J. Exp. Theor. Phys. **36**, 1398 (1959)
5. S. Chandrasekhar: Proc. Natl. Acad. Sci. **46**, 253 (1960)
6. A. G. Tevzadze, G. D. Chagelishvili, J.-P. Zahn, R. Chanishvili, J. Lominadze: A&A **407**, 779 (2003)

272   Banibrata Mukhopadhyay

7. P. Yecko: A&A **425**, 385 (2004)
8. O. Umurhan, O. Regev: A&A **427**, 855 (2004)
9. N. Afshordi, B. Mukhopadhyay, R. Narayan: ApJ **629**, 373 (2005)
10. B. Mukhopadhyay, N. Afshordi, R. Narayan: ApJ **629**, 383 (2005)
11. B. Mukhopadhyay: ApJ **653**, 503 (2006)
12. A. G. Tevzadze, G. D. Chagelishvili, J.-P. Zahn: arXiv0710.3648
13. M. Abramowicz, W. Zurek: ApJ **246**, 314 (1981)
14. S. K. Chakrabarti: ApJ **347**, 365 (1989)
15. R. Narayan, I. Yi: ApJ **428**, L13 (1994)
16. S. K. Chakrabarti: Phys. Rep. **266**, 229 (1996)
17. S. K. Chakrabarti: A&A **351**, 185 (1999)
18. B. Mukhopadhyay: ApJ **586**, 1268 (2003)
19. B. Mukhopadhyay, S. Ghosh: MNRAS **342**, 274 (2003)
20. S. K. Chakrabarti: *Theory of transonic astrophysical flows*, (World Scientific, Singapore 1990)
21. B. Paczyński, P. J. Wiita: A&A **88**, 23 (1980)
22. B. Mukhopadhyay: ApJ **581**, 427 (2002)
23. H. Bondi: MNRAS **112**, 195 (1952)
24. R. Matsumoto, S. Kato, J. Fukue, A. T. Okazaki: PASJ **36**, 71 (1984)
25. J. M. T. Thompson, H. B. Stewart: *Nonlinear Dynamics and Chaos*, (John Willey and Sons Ltd. 1985)
26. S. K. Chakrabarti, B. Mukhopadhyay: A&A **344**, 105 (1999)
27. B. Mukhopadhyay, S. K. Chakrabarti: A&A **353**, 1029 (2000)
28. B. Mukhopadhyay, S. K. Chakrabarti: ApJ **555**, 816 (2001)
29. S. K. Chakrabarti: ApJ **464**, 664 (1996)
30. J. Cox, R. Giuli: *Principles of Stellar Structure*, (Gordon & Breach, New York 1968)
31. S. V. Vadawale, A. R. Rao, A. Nandi, S. K. Chakrabarti: A&A **370**, 17 (2001)
32. S. V. Vadawale, A. R. Rao, S. Naik, J. S. Yadav, C. H. Ishwara-Chandra, A. Pramesh Rao, G. G. Pooley: ApJ **597**, 1023 (2003)

Maximum Brightness Temperature for an Incoherent Synchrotron Radio Source

Ashok K. Singal

Astronomy & Astrophysics Division, Physical Research Laboratory, Navrangpura, Ahmedabad, India - 380009. `asingal@prl.res.in`

Summary. We discuss here a limit on the maximum brightness temperature achievable for an incoherent synchrotron radio source. This limit, commonly referred to in the literature as an inverse Compton limit, prescribes that the brightness temperature for an incoherent synchrotron radio source may not exceed $\sim 10^{12}$ K, a fact known from observations. However one gets a somewhat tighter limit on the brightness temperatures, $T_{\rm b} \lesssim 10^{11.5}$ K, independent of the inverse Compton effects, if one employs the condition of equipartition of energy in magnetic fields and relativistic particles in a synchrotron radio source. Pros and cons of the two brightness temperature limits are discussed.

Key words: Radio emission: non-thermal – inverse Compton scattering

1 Introduction

From the radio spectra combined with the VLBI (Very large Baseline Interferometry) observations it has been seen that the brightness temperatures for compact self-absorbed radio sources do not exceed about $\sim 10^{11-12}$ K. Kellermann and Pauliny-Toth [1] first time gave an explanation of this in terms of what has since then come to be known in the literature as an Inverse Compton limit. They argued that at brightness temperature $T_{\rm b} \gtrsim 10^{12}$ K energy losses of radiating electrons due to inverse Compton effects become so large that these result in a rapid cooling of the system, thereby bringing the synchrotron brightness temperature quickly below this limit. Of course much larger brightness temperatures have been inferred for the variable sources, however this excess in brightness temperatures has been explained in terms of a bulk relativistic motion of the emitting component [2, 3, 4]. The Doppler factors required to explain the excess in temperatures were initially thought to be $\sim 5-10$, similar to the ones required for explaining the superluminal velocities seen in some compact radio sources [5, 6, 7]. Singal and Gopal-krishna [8] pointed out that under the conditions of equipartition of energy between

274   Ashok K. Singal

magnetic fields and relativistic particles in a synchrotron radio source, much higher Doppler factors are needed to successfully explain the variability events. Singal [9], without taking recourse to any inverse Compton effects, derived a somewhat tighter upper limit $T_b \lesssim 10^{11.5}$ K, first time using the argument that due to the diamagnetic effects the energy in the magnetic fields cannot be less than a certain fraction of that in the relativistic particles and then an upper limit on brightness temperature follows naturally. However, it has to be noted that if one considers the drift currents at the boundaries, which may be present due to the non-uniformities in the magnetic fields [10], the above limit on the magnetic field energy gets modified. Later similar derivations [11] of the $T_b \lesssim 10^{11.5}$ K limit as well as of large Doppler factors, used essentially the same equipartition arguments as in [8, 9].

Here, we first derive the inverse Compton limit on T_b using the approach followed in [1], and then the equipartition limit as done in [9]. In fact we argue that even if we relax the condition of the equipartition of energy between magnetic fields and relativistic particles and let the energy in relativistic particles to be many orders of magnitude larger than that in the magnetic fields we still end up with a rather tight T_b limit. On the other hand if the energy in particles is *smaller* than that in magnetic fields, then in any case T_b *has to be lower than* $\sim 10^{11.5}$ K, as pointed out in [9].

2 Inverse Compton limit

We want to study the maximum brightness temperature limit in the rest frame of the source, therefore we assume that all quantities have been transformed to that frame. Hence we will not consider here any effects of the cosmological redshift or of the relativistic beaming due to a bulk motion of the radio source.

In inverse Compton interaction between relativistic electrons and their the synchrotron photons, the energy of the radio photons could get boosted to X-ray frequencies. The average energy of a photon in an inverse Compton interaction gets boosted by a factor γ_e^2 where γ_e is the Lorentz factor of the interacting relativistic electrons [12]. We confine our discussion to a single scattering case only.

A relativistic electron of Lorentz factor γ_e gyrating in a magnetic field B emits most of its radiation in a frequency band near its characteristic synchrotron frequency [13, 14]

$$\nu_c = 0.29 \frac{3}{4\pi} \frac{e B}{m_0 c} \gamma_e^2 \tag{1}$$

which gives us $\gamma_e^2 = 8.2 \times 10^2 \, \nu_c/B$ for ν_c in GHz and B in Gauss. From this we find that for $B \sim 10^{-3}$ Gauss as inferred in compact radio sources, $\nu_c = 0.1$ to 10 GHz yields $\gamma_e^2 \sim 10^5$ to $\sim 10^7$. Thus during an inverse Compton interaction while at the lower end the synchrotron photons can get boosted to infrared

Maximum Brightness Temperature   275

frequencies ($\sim 10^{13}$) Hz, at the higher end the synchrotron photons of radio frequencies $\sim 10^{10}$ Hz could get boosted to X-ray frequencies ($\sim 10^{17}$) Hz.

The power radiated by the inverse Compton process P_c as compared to that radiated by synchrotron mechanism, P_s is given as,

$$\frac{P_c}{P_s} = \frac{W_p}{W_b} \tag{2}$$

where W_p and W_b respectively are the photon energy density and magnetic field energy density within the source. This relation is true in the case where Thomson scattering cross-section is valid [12], which is true in our case as we have $\gamma_e h\nu << m_0 c^2$, for $\nu < 100$ GHz.

The photon energy density is related to the radiation intensity as [14]

$$W_p = \frac{4\pi}{c} \int_{\nu_1}^{\nu_2} I_\nu \, d\nu \tag{3}$$

where the specific intensity I_ν is defined as the flux density per unit solid angle, at frequency ν. In radio sources, the observed flux density in the optically thin part of the spectrum usually follows a power law, i.e., $I_\nu \propto \nu^{-\alpha}$, between the lower and upper cut off frequencies ν_1 and ν_2. In synchrotron theory this spectrum results from a power law energy distribution of radiating electrons $N(E) \propto E^{-\gamma}$ within some range E_1 and E_2, with $\gamma = 2\alpha + 1$ and E_1, E_2 related to ν_1, ν_2 by Eq. (1). In compact radio sources the source may become self-absorbed with flux density $\propto \nu^{2.5}$ below a turnover frequency ν_m. From Eq.(3) we can get

$$W_p = 2.27 \times 10^{-7} \frac{f(\alpha) \, F_m \, \nu_m^\alpha}{\Theta_x \, \Theta_y} \left[\frac{\nu_2^{1-\alpha} - \nu_1^{1-\alpha}}{1 - \alpha} \right] \text{erg cm}^{-3} \tag{4}$$

the expression to be evaluated in the 'limit' for $\alpha = 1$. Here F_m (Jy) is the flux density at frequency ν_m (GHz) corresponding to the point of spectral turnover, while Θ_x and Θ_y (mas) represent the angular size along the major and minor radio axes of the source component, assumed to be an ellipse. For calculating the photon energy density it may be appropriate to take the lower limit of the spectrum at ν_m itself, therefore we can put $\nu_1 = \nu_m$ in the above expression. Here it should be noted that F_ν is the flux density in the optically thin part of the synchrotron spectrum, accordingly F_{ν_m} obtained from an extrapolation up to $\nu = \nu_m$ using the straight slope α, is not the same as the actual peak flux density F_m at the turnover bend (see e.g, [15]). In Table 1 values of $f(\alpha) = F_{\nu_m}/F_m$ are listed, which as we see are of the order of unity.

We can express W_p in terms of the brightness temperature at the peak of the spectrum

$$T_m = 1.763 \times 10^{12} F_m \, \Theta_x^{-1} \Theta_y^{-1} \nu_m^{-2} \, \text{K} \tag{5}$$

to get

$$W_p = 1.29 \times 10^{-7} \frac{f(\alpha)}{1-\alpha} \left[\left(\frac{\nu_2}{\nu_m} \right)^{1-\alpha} - 1 \right] \nu_m^3 \left(\frac{T_m}{10^{12}} \right) \text{ erg cm}^{-3}. \quad (6)$$

On the other hand, from the synchrotron self-absorption we get,

$$B = 10^{-5} b(\alpha) F_m^{-2} \Theta_x^2 \Theta_y^2 \nu_m^5, \quad (7)$$

values of $b(\alpha)$[1] are given in Table 1.

Here a plausible assumption has been made that the direction of the magnetic field vector, with respect to the line of sight, changes randomly over regions small compared to a unit optical depth. For a uniform magnetic field direction throughout the source region, the co-efficients $a(\alpha)$ (see next section) and $b(\alpha)$ would be modified by factors of the order of unity.

Then the magnetic field energy density $W_b = B^2/8\pi$ can be written as,

$$W_b = 3.84 \times 10^{-11} b^2(\alpha) \nu_m^2 \left(\frac{T_m}{10^{12}} \right)^{-4} \text{ erg cm}^{-3}. \quad (8)$$

Equations (2), (6) and (8) lead us to,

$$\frac{P_c}{P_s} = \nu_m \left(\frac{T_m}{10^{11.3} p(\alpha)} \right)^5 \quad (9)$$

where the function

$$p(\alpha) = \left[\frac{f(\alpha)}{b^2(\alpha)(1-\alpha)} \left\{ \left(\frac{\nu_2}{\nu_m} \right)^{1-\alpha} - 1 \right\} \right]^{-1/5} \quad (10)$$

is of the order of unity (Table 1)[2].

3 Equipartition temperature limit

Energy density of the relativistic electrons in a synchrotron radio source component is given by [13, 16] [3]

$$W_e = \frac{8.22 \times 10^{-9}}{a(\alpha)(\alpha - 0.5)} \frac{F_\nu \nu^\alpha B^{-1.5}}{\Theta_x \Theta_y s} \left[\left(\frac{y_1(\alpha)}{\nu_1} \right)^{\alpha - 0.5} - \left(\frac{y_2(\alpha)}{\nu_2} \right)^{\alpha - 0.5} \right] \quad (11)$$

this expression to be evaluated in the 'limit' for $\alpha = 0.5$. Here W_e (erg cm^{-3}) is the energy density of radiating electrons; F_ν (Jy) is the flux density at frequency ν (GHz), with $\nu_1 < \nu < \nu_2$; and s(pc) is the characteristic depth of the component along the line of sight.

---

[1] Calculated from the tabulated functions in [14].

[2] In Table 1 and 2 we have taken ν_1 and ν_2 to be 0.01 and 100 GHz, and the turnover frequency ν_m is taken to be 1 GHz.

[3] Values of $a(\alpha)$ in Table 1, calculated by us from the tabulated functions given in [14], appear slightly different from the ones in [13] but are in agreement with those in [17].

Maximum Brightness Temperature

Table 1. Various functions of the spectral index α

α	γ	$a(\alpha)$	$b(\alpha)$	$f(\alpha)$	$p(\alpha)$	$t(\alpha)$	$y_1(\alpha)$	$y_2(\alpha)$
0.25	1.5	0.149	2.07	1.10	0.66	0.68	1.3	0.011
0.5	2.0	0.103	2.91	1.19	0.83	0.85	1.8	0.032
0.75	2.5	0.0831	2.85	1.27	0.94	0.80	2.2	0.10
1.0	3.0	0.0741	2.52	1.35	1.00	0.67	2.7	0.18
1.5	4.0	0.0726	1.79	1.50	1.03	0.43	3.4	0.38

In terms of the brightness temperature we can write,

$$
W_e = \frac{4.66 \times 10^{-9} \, f(\alpha)}{s \, a(\alpha) \, (\alpha - 0.5)} \left[\left(\frac{y_1(\alpha)}{\nu_1/\nu_m} \right)^{\alpha - 0.5} - \left(\frac{y_2(\alpha)}{\nu_2/\nu_m} \right)^{\alpha - 0.5} \right] \frac{\nu_m^{2.5}}{B^{1.5}} \left(\frac{T_m}{10^{12}} \right).
\tag{12}
$$

From the overall charge neutrality of the plasma we expect the electrons to be accompanied by an equal number of positive charges. Any positrons would have already been accounted for in our equation above, however presence of heavy particles will contribute additionally to the total particle energy density. Let the energy in the heavy particles be ξ times that in the lighter particles, then the total particle energy density is $W_k = (1 + \xi) \, W_e$.

Now if we assume the energy in particles to be related to that in magnetic fields by

$$
W_k = \eta \, W_b
\tag{13}
$$

then we get

$$
B^{7/2} = \frac{1.17 \times 10^{-7} f(\alpha)}{a(\alpha) \, (\alpha - 0.5)} \left[\left(\frac{y_1(\alpha)}{\nu_1/\nu_m} \right)^{\alpha - 0.5} - \left(\frac{y_2(\alpha)}{\nu_2/\nu_m} \right)^{\alpha - 0.5} \right] \times
$$
$$
\frac{(1 + \xi)}{s \, \eta} \, \nu_m^{2.5} \left(\frac{T_m}{10^{12}} \right).
\tag{14}
$$

Now substituting for magnetic field from Eq.(7) we get,

$$
\frac{\eta}{1 + \xi} = \frac{1}{s \, \nu_m} \left(\frac{T_m}{10^{10.9} \, t(\alpha)} \right)^8
\tag{15}
$$

where the function

$$
t(\alpha) = \left[\frac{f(\alpha)}{(\alpha - 0.5) \, a(\alpha) \, b^{3.5}(\alpha)} \left\{ \left(\frac{y_1(\alpha)}{\nu_1/\nu_m} \right)^{\alpha - 0.5} - \left(\frac{y_2(\alpha)}{\nu_2/\nu_m} \right)^{\alpha - 0.5} \right\} \right]^{-1/8}
\tag{16}
$$

is of the order of unity (Table 1).

278   Ashok K. Singal

4 A Correction to the derived T_m values

Actually T_m values have been calculated (both here as well as in the literature) for the turnover point in the synchrotron spectrum where the flux density peaks. However, the definition of the brightness temperature (Eq. 5) also involves ν^{-2}. Therefore a maxima of flux density is not necessarily a maxima for the brightness temperature also. In fact a zero slope for the flux density with respect to ν would imply for the brightness temperature a slope of -2. Therefore the peak of the brightness temperature will be at a point where flux density $\propto \nu^2$, so that $T_b \propto F \nu^{-2}$ has a zero slope with respect to ν. The peak of $F_\nu \nu^{-2}$, can be determined in the following manner. The specific intensity in a synchrotron self-absorbed source is given by [14],

$$I_\nu = \frac{c_5(\alpha)}{c_6(\alpha)} \left(\frac{\nu}{2c_1}\right)^{2.5} \left[1 - \exp\left\{-\left(\frac{\nu}{\nu_1}\right)^{-(\alpha+2.5)}\right\}\right] B_\perp^{-0.5} \tag{17}$$

where $c_1, c_5(\alpha), c_6(\alpha)$ are tabulated in [14]. The optical depth varies with frequency as $\tau = (\nu/\nu_1)^{-(\alpha+2.5)}$, ν_1 being the frequency at which τ is unity. The equivalent brightness temperature (in Rayleigh-Jeans limit) is then given by

$$T_\nu = \frac{c^2 c_5(\alpha)}{8 k c_1^2 c_6(\alpha)} \left(\frac{\nu}{2c_1}\right)^{0.5} \left[1 - \exp\left\{-\left(\frac{\nu}{\nu_1}\right)^{-(\alpha+2.5)}\right\}\right] B_\perp^{-0.5} \tag{18}$$

where k is the Boltzmann constant and c is the speed of light. We can maximize T_ν by differentiating it with ν and equating the result to zero. This way we get an equation for the optical depth τ_o, corresponding to the *peak brightness temperature* T_o, which is different from the one that is available in the literature for the optical depth τ_m at the *peak of the spectrum*. The equation that we get for τ_o is

$$\exp(\tau_o) = 1 + (2\alpha + 5)\,\tau_o, \tag{19}$$

solutions of this transcendental equation for different α values are given in Table 2. It is interesting to note that while the peak of the spectrum for the typical α values usually lies in the optically thin part of the spectrum ($\tau_m \lesssim 1$; Table 2), peak of the brightness temperature lies deep within the optically thick region ($\tau_o \sim 3$). Both the frequency and the intensity have to calculated for τ_o to get the maximum brightness temperature values. The correction factors are then given by,

$$\frac{\nu_o}{\nu_m} = \left(\frac{\tau_m}{\tau_o}\right)^{1/(\alpha+2.5)} \tag{20}$$

$$\frac{T_o}{T_m} = \left(\frac{\nu_o}{\nu_m}\right)^{0.5} \left[\frac{1 - \exp(-\tau_o)}{1 - \exp(-\tau_m)}\right] \tag{21}$$

Maximum Brightness Temperature   279

Table 2. Values for the temperature limits

α	γ	τ_m	τ_o	ν_o/ν_m	T_o/T_m	$\log(T_{ic})$	$\log(T_{eq})$
0.25	1.5	0.19	2.80	0.37	3.5	11.7	11.3
0.5	2.0	0.35	2.92	0.50	2.2	11.6	11.2
0.75	2.5	0.50	3.03	0.58	1.8	11.5	11.1
1.0	3.0	0.64	3.13	0.64	1.6	11.5	10.9
1.5	4.0	0.88	3.32	0.72	1.4	11.5	10.7

In Table 2 we have listed ν_o/ν_m and T_o/T_m, for different α values. Further we have also tabulated values of maximum brightness temperatures $\log(T_{ic})$ for $W_p = W_b$ and $\log(T_{eq})$ for $\eta = 1$ (with $\xi = 0$, $s = 1pc$) calculated from equations (9) and (15) respectively for different α values, incorporating the above corrections.

5 Discussion and Conclusions

From Table 2 we see that, depending upon the spectral index value α, theoretical maximum brightness temperature values for the inverse Compton limits are around $T_{ic} \sim 10^{11.6\pm0.1}$ K while for equipartition conditions the limits are lower by a factor of ~ 3 ($T_{eq} \sim 10^{11.0\pm0.3}$ K). From the observational data [18] it seems that the intrinsic T_b is $\lesssim 10^{11.3}$ K, which is broadly consistent with the either interpretation. It should be noted that though inverse Compton scattering increases the photon energy density, yet it does not increase the radio brightness as the scattered photons get boosted to much higher frequency bands. If anything, some photons get removed from the radio window, but the change in radio brightness due to that may not be very large. What could be important is the large energy losses by electrons which may cool the system rapidly. However, these inverse Compton losses become important only when $T_b > 10^{11.5}$ K, but the equipartition conditions may keep the temperatures well below this limit. This is not to say that inverse Compton effects cannot occur, it is only that conditions in synchrotron radio sources may not arise for inverse Compton losses to become very effective. Since we are considering the brightness temperature limit in the radio-band (after all that is where observationally such limits have been seen), then variations in ν_m that we may consider would at most be about an order of magnitude around say, 1 GHz. With the reasonable assumption that a self-absorbed radio source size may not be much larger than $\sim$ a pc, from Eq. (15) it follows that an order of magnitude higher T_m values would require η to increase by about a factor $\sim 10^8$, that is departure from equipartition will go up by about eight orders of magnitude. Actually for a given ν_m, the magnetic field energy density will go down by a factor $\sim 10^4$ (Eq. 8), while that in the relativistic particles will go up by a similar factor (Eq. 12, note the presence of $B^{1.5}$ in the denomi-

280   Ashok K. Singal

nator). This also implies that the total energy budget of the source will be higher by about $\sim 10^4$ than from the already stretched equipartition energy values. Further, it is not the photon energy density that really goes up drastically with higher brightness temperatures (as $W_p \propto T_{\mathrm{m}}$, Eq. 6), rather it is the drastic fall in the magnetic field energy ($W_b \propto T_{\mathrm{m}}^{-4}$, Eq. 8) that increases the ratio $P_{\mathrm{c}}/P_{\mathrm{s}}$ to go up as T_{m}^5. Therefore, it is still not clear whether the inverse Compton effects actually do play a significant role in maintaining the maximum brightness temperature limit in incoherent synchrotron radio sources.

References

1. K. I. Kellermann, I. I. K. Pauliny-Toth: ApJ **155**, L71 (1969)
2. M. J. Rees: Nature **211**, 468 (1966)
3. M. J. Rees: MNRAS **135**, 345 (1967)
4. L. Woltjer: ApJ **146**, 597 (1966)
5. M. H. Cohen, W. Cannon, G. H. Purcell et al: ApJ **170**, 207 (1971)
6. A. R. Whitney, I. I. Shapiro, A. E. E. Rogers et al: Science **173**, 225 (1971)
7. M. H. Cohen, R. P. Linfield, A. T. Moffet et al: Nature **268**, 405 (1977)
8. A. K. Singal, Gopal-krishna: MNRAS **215**, 383 (1985)
9. A. K. Singal: A&A **155**, 242 (1986)
10. G. Bodo, G. Ghisellini, E. Trussoni: MNRAS **255**, 694 (1992)
11. A. C. S. Readhead: ApJ **426**, 51 (1994)
12. G. B. Rybicki, A. P. Lightman: *Radiative Processes in Astrophysics*, (Wiley, New York 1979)
13. V. L. Ginzburg, S. I. Syrovatskii: ARAA **3**, 297 (1965)
14. A. G. Pacholczyk: *Radio Astrophysics*, (Freeman, San Francisco 1970)
15. M. A. Scott, A. C. S. Readhead: MNRAS **180**, 539 (1977)
16. V. L. Ginzburg: *Theoretical Physics and Astrophysics*, (Pergamon, Oxford 1979)
17. R. J. Gould: A&A **76**, 306 (1979)
18. D. C. Homan, Y. Y. Kovalev, M. L. Lister et al: ApJ **642**, L115 (2006)

Nonlinear Jeans Instability in an Uniformly Rotating Gas

Nikhil Chakrabarti[1], Barnana Pal[1] and Vinod Krishan[2]

[1] Saha Institute of Nuclear Physics, 1/AF Bidhannagar, Kolkata - 700064
[2] Indian Institute of Astrophysics, Bangalore – 560034, India
   Raman Research Institute, Bangalore - 560080, India

Summary. A nonlinear stability analysis of a uniformly rotating gas in a gravitational field has been performed. One dimensional non-linear equations have been solved by the double-Lagrangian transformation method. An explosive instability is shown to exist in contrast to the linear Jeans instability wherein a uniform rotation of the gas quenches the instability.

Key words: Jeans instability, Structure formation, Lagrangian technique, Explosive instability

1 Introduction

The problem of gravitational instability of large astrophysical system has been a subject of ongoing investigation in last few decades starting with the very early work of Jeans [1]. The Jeans instability can affect the gravitational collapse of the interstellar gas clouds. It is believed that the star formation mechanisms are due to the very high velocity cloud-cloud interaction in which a dense gaseous clump is formed. In these processes gaseous clump grows in mass, undergoes gravitational Jeans instability and fragments. The fragments in turn collapse further by self gravity and evolve into stars and /or clusters of stars.

The star formation mechanisms based on the linear description of the Jeans instability have been extensively discussed by various authors e.g.[2, 3, 5, 6]. Among these we mention the work of Chandrasekhar [2] in particular, where the system, an infinite homogeneous gas, rotating with a constant angular velocity, has been considered. It is shown that the rotation gives rise to a stabilizing effect that overcomes the Jeans instability in a direction perpendicular to the direction of rotation. In this report, we study the nonlinear stability of uniformly rotating gas in a gravitational field and demonstrate that there is a possibility of an explosive instability even in the presence of rotation.

282   Nikhil Chakrabarti, Barnana Pal and Vinod Krishan

The classical Jeans instability is obtained assuming a uniform density and pressure distribution throughout the gas. As is well known, in the presence of gravitational force, infinite homogeneous distribution of matter cannot occur. However, for the study of linear instability, such distribution, the so called "Jeans swindle" is assumed to prevail and the effect of the zero order gravitational field is neglected. Here we consider a more realistic model of equilibrium taking into account the effect of zeroth order gravitational field. Nonlinear effects are included using this model equilibrium system. Our purpose is to investigate the gravitational instability of nonuniform density distribution as well as nonuniform gravitational field. We exploit the extremely beautiful and powerful Lagrangian technique to solve the nonlinear equations [7, 8, 9, 10]. For simplicity the system with one spatial dimension is considered and this enables one to extend the investigation to a wider class of systems namely those with magnetic fields.

2 Basic equations

The model equations for non viscous fluid in a rotating frame may be written as, [4]

$$\rho \left[\frac{\partial \mathbf{u}}{\partial t} + (\mathbf{u} \cdot \nabla)\mathbf{u} \right] = -\nabla p + \rho \mathbf{g} + 2\rho(\mathbf{u} \times \mathbf{\Omega}), \tag{1}$$

where $\mathbf{g}$ is the effective gravity (after deduction of centrifugal term from the gravitational potential), ρ is the density, $\mathbf{u}$ is the velocity and $\mathbf{\Omega}$ is the angular velocity. For nonlinear analysis, we shall concentrate on one dimensional variation so that all variables are functions of (x, t) only. We take gravitational field in the negative x direction and the constant rotation $\mathbf{\Omega}$ in the z direction only. Using these conditions and relating pressure and density as $\partial p/\partial x = c_s^2 \partial \rho/\partial x$ equation (1) can be written in the component form as:

$$\rho \left(\frac{\partial}{\partial t} + u_x \frac{\partial}{\partial x} \right) u_x = -c_s^2 \frac{\partial \rho}{\partial x} - \rho g + 2\rho u_y \Omega \tag{2}$$

$$\rho \left(\frac{\partial}{\partial t} + u_x \frac{\partial}{\partial x} \right) u_y = -2\rho u_x \Omega, \tag{3}$$

where c_s is the sound speed The mass density conservation equation can be written as

$$\frac{\partial \rho}{\partial t} + \frac{\partial}{\partial x}(\rho u_x) = 0. \tag{4}$$

The Poisson equation for the gravitational field is

$$\frac{\partial g}{\partial x} = 4\pi G \rho, \tag{5}$$

Our aim is to solve these equations for the study of equilibrium and stability.

3 Local analysis

For the local analysis we recapitulate the results derived by Chandrasekhar[2] in infinite homogeneous medium. Writing $u_x = u_{x0} + u_{x1}$, $u_y = u_{y0} + u_{y1}$, $g = g_0 + g_1$, $\rho = \rho_0 + \rho_1$ and assuming $u_{x0} = u_{y0} = 0, \partial p_0 / \partial x = 0$, neglecting all nonlinear terms with g is in the negative x direction, linearized equations become,

$$\frac{\partial u_{x1}}{\partial t} = -g_1 - \frac{c_s^2}{\rho_0} \frac{\partial \rho_1}{\partial x} + 2\Omega u_{y1} \tag{6}$$

$$\frac{\partial u_{y1}}{\partial t} = -2\Omega u_{x1} \tag{7}$$

$$\frac{\partial \rho_1}{\partial t} + \rho_0 \frac{\partial u_{x1}}{\partial x} = 0 \tag{8}$$

$$\frac{\partial g_1}{\partial x} = 4\pi G \rho_1 \tag{9}$$

Variables with subscript one are the perturbed quantities and those with subscript zero are the equilibrium quantities. Taking the time derivative in Eq. (6) and using eq. (7 - 9) we get

$$\left[\frac{\partial^2}{\partial t^2} + (2\Omega)^2 \right] \left(-\frac{\partial \hat{\rho}_1}{\partial t} \right) = -\frac{\partial}{\partial t} \left[c_s^2 \frac{\partial^2 \hat{\rho}_1}{\partial x^2} + 4\pi G \rho_0 \hat{\rho}_1 \right]. \tag{10}$$

Assuming the perturbation $\hat{\rho}_1 \sim \exp(ikx - i\omega t)$ we have the dispersion relation as

$$\omega^2 = c_s^2 \left[k^2 - k_J^2 + \frac{4\Omega^2}{c_s^2} \right], \tag{11}$$

where $k_J^2 = 4\pi G \rho_0 / c_s^2$. In the absence of rotation ($\Omega = 0$), if $k_J^2 > k^2$ there is an instability known as the Jeans instability. Surely in the direction perpendicular to the direction of propagation of the wave a constant rotation acts as a stabilizing agent. As it is mentioned before that this result is derived in a homogeneous medium and with a gravitational field homogeneous distribution of mass is unrealistic. Therefore before going to the stability analysis it is necessary to study the inhomogeneous equilibrium. In the next section we shall study the inhomogeneous static equilibrium solution.

4 Equilibrium

Here we assume that in the equilibrium both, g_{eq} and ρ_{eq} depend only on the variable x and are denoted as $g_{eq}(x)$ and $\rho_{eq}(x)$ respectively. Therefore the static equilibrium $\mathbf{u}_0 = 0$ equations can be written from equations (2) and (5) as

$$- \rho_{eq} g_{eq} - c_s^2 \frac{\partial \rho_{eq}}{\partial x} = 0, \tag{12}$$

and
$$\frac{\partial g_{eq}}{\partial x} = 4\pi G \rho_{eq}, \tag{13}$$

Combining the above two equations we can write a single variable equation as
$$\frac{\partial^2}{\partial x^2} \ln \hat{\rho} = -\left(\frac{4\pi G \rho_0}{c_s^2}\right) \hat{\rho}, \tag{14}$$

where $\hat{\rho} = \rho_{eq}/\rho_0$, ρ_0 is a constant normalizing density. Equation (14) is an exactly solvable with a solution $\hat{\rho} = \rho_{eq}(x)/\rho_0 = \text{sech}^2(x/\lambda_J)$. Accordingly we can find the solution for the gravitational field as $g_{eq}(x) = -\hat{e}_x g_0 \tanh(x/\lambda_J)$ where $\lambda_J = c_s/\omega_J$, $\omega_J = 2\sqrt{\pi G \rho_0}$ is known as Jeans frequency and $g_0 = \sqrt{2} c_s \omega_J$. In stability analysis we will use $\rho_{eq}(x) \equiv \rho(\xi.0)$. The equilibrium solutions are shown in fig. (1) below

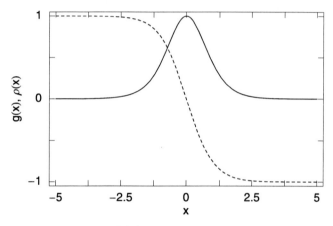

Fig. 1. Equilibrium density $\rho_{eq}(x)$ (solid line) and gravitational field $g_{eq}(x)$ (dotted line) as functions of x.

5 Stability Analysis

We formulate the nonlinear problem in the Lagrangian description by the Eulerian variables (x,t) to Lagrangian variables (ξ, τ) such that
$$\xi = x - \int u_x(\xi, \tau) d\tau, \quad \tau = t, \tag{15}$$

and define an auxiliary function ψ; where
$$\psi = \int u_x(\xi, \tau) d\tau, \quad \xi = x - \psi. \tag{16}$$

The co-ordinate ξ is in a frame moving with the fluid element. We simplify equations (2) - (5) using the transformation as:

$$\frac{\partial}{\partial \tau} \equiv \frac{\partial}{\partial t} + u_x \frac{\partial}{\partial x},$$

(17)

which implies that convective derivative reduces to partial time derivative and

$$\frac{\partial \psi}{\partial \tau} = u_x, \quad \frac{\partial x}{\partial \xi} = 1 + \frac{\partial \psi}{\partial \xi}.$$

In view of these transformations equations (4) and (5) simplify to

$$\frac{\partial}{\partial \tau} \left[\rho \left(1 + \frac{\partial \psi}{\partial \xi} \right) \right] = 0,$$

(18)

$$\frac{\partial g}{\partial \xi} = -4\pi G \rho(\xi, 0),$$

(19)

where $\rho(\xi, 0) \equiv \rho_{eq}$ is the value of ρ at $\tau = 0$. Using boundary conditions that at $\tau = 0$ both u_y and $\psi = 0$, momentum equation can be written as

$$\frac{\partial^2 \psi}{\partial \tau^2} + 4\Omega^2 \psi = -g - \frac{c_s^2}{\rho(\xi, 0)} \frac{\partial}{\partial \xi} \left[\frac{\rho(\xi, 0)}{1 + \frac{\partial \psi}{\partial \xi}} \right].$$

(20)

Now using boundary condition at $\xi = 0$, $\psi = 0$ and $\rho(0, 0) = \rho_0$ and normalizing the above equation we have

$$\frac{\partial^2 \hat{\psi}}{\partial \hat{\tau}^2} + \hat{\psi} = \frac{1}{4} \left(\frac{\omega_J^2}{\Omega^2} \right) \left(\frac{\lambda_J^2}{\lambda^2} \right) \frac{\partial^2 \hat{\psi}}{\partial \hat{\xi}^2}$$

(21)

where $\hat{\xi} = \xi/\lambda$, $\hat{\tau} = 2\Omega\tau$, $\hat{\psi} = \psi/\lambda_J$. For the large scale instability the characteristic scale length λ should be greater than Jeans scale length λ_J. Under this condition the coefficient on the right hand side of Eq. (21) is small and we denote it by ϵ. Removing all hats for simplicity of notation, we get,

$$\frac{\partial^2 \psi}{\partial \tau^2} + \psi = \epsilon \frac{\partial^2 \psi}{\partial \xi^2}$$

(22)

The above equation can be solved by the Bogoliuboff method [11]. In this method the solution is written as $\psi(\xi, \tau) = A(\xi, \tau) \sin[\tau + \chi(\xi, \tau)]$, where A and χ are determined form the following equations,

$$\frac{\partial A}{\partial \tau} \sin \theta + A \frac{\partial \chi}{\partial \tau} \cos \theta = 0$$

(23)

$$\frac{\partial A}{\partial \tau}\cos\theta - A\frac{\partial \chi}{\partial \tau}\sin\theta = \epsilon\left[\frac{\partial^2 A}{\partial \xi^2} - A\left(\frac{\partial \chi}{\partial \xi}\right)^2\right]\sin\theta + \epsilon\left[A\frac{\partial^2 \chi}{\partial \xi^2} + 2A\frac{\partial A}{\partial \xi}\frac{\partial \chi}{\partial \xi}\right]\cos\theta,$$

$$(24)$$

where $\theta = \tau + \chi$. Since $\partial A/\partial \tau$ and $\partial \chi/\partial \tau$ are proportional to the small parameter ϵ, A and χ are slowly varying functions of time τ. Substituting Eq. (23) in Eq. (24) and averaging over τ, the fast time scale over which A and χ do not change appreciably, we find,

$$\left(\frac{\partial}{\partial \tau_1} + \phi\frac{\partial}{\partial \xi}\right)\ln A = -\frac{1}{2}\frac{\partial \phi}{\partial \xi}$$

$$(25)$$

$$\left(\frac{\partial}{\partial \tau_1} + \phi\frac{\partial}{\partial \xi}\right)\phi = \frac{1}{2}\frac{\partial}{\partial \xi}\left(\frac{1}{A}\frac{\partial^2 A}{\partial \xi^2}\right),$$

$$(26)$$

where $\tau_1 = \epsilon\tau$, and $\phi = -\partial\chi/\partial\xi$.

To solve amplitude A and phase χ we now perform a second Lagrangian transformation from coordinates ξ, τ_1 to $\zeta, \bar{\tau}$, where $\zeta = \xi - \int_0^{\bar{\tau}} \phi(\zeta, \tau')d\tau'$, $\bar{\tau} = \tau_1$, to obtain

$$\frac{\partial}{\partial \bar{\tau}}\left[A^2\left(1 + \int^{\bar{\tau}}\frac{\partial \phi}{\partial \zeta}d\tau'\right)\right] = 0.$$

$$(27)$$

Expressing $\partial\phi/\partial\zeta$ in terms of A and substituting in Eq. (26) we obtain single variable equation for A,

$$\frac{\partial^2}{\partial \bar{\tau}^2}\left(\frac{1}{\hat{A}}\right)^2 = \frac{1}{2}\frac{\partial}{\partial \zeta}\left[\hat{A}^2\frac{\partial}{\partial \zeta}\left\{\frac{\hat{A}^2}{A}\frac{\partial}{\partial \zeta}\left(\hat{A}^2\frac{\partial A}{\partial \zeta}\right)\right\}\right],$$

$$(28)$$

where $\hat{A} = A(\zeta, \bar{\tau})/A(\zeta, 0)$. Equation (28) can be solved analytically by the method of separation of variables. Proposing $A(\zeta, \bar{\tau}) = P(\bar{\tau})Q(\zeta)$, Eq.(28) can be separated into two equations in space and time variables as

$$\frac{1}{2}\frac{d^2}{d\zeta^2}\left[\frac{1}{Q}\frac{d^2 Q}{d\zeta^2}\right] = \alpha^2,$$

$$(29)$$

$$\frac{1}{\hat{P}^6}\frac{d^2}{d\bar{\tau}^2}\left(\frac{1}{\hat{P}^2}\right) = \alpha^2,$$

$$(30)$$

where α is an arbitrary separation constant, $\hat{P} = P(\bar{\tau})/P(0)$ and we have used the relation $A(\zeta, 0) = P(0)Q(\zeta)$. From Eq. (30) together with the boundary conditions that at $\zeta = 0$, $Q(0) = 1/P(0)$ and $(dQ/d\zeta)_{\zeta=0} = 0$ we get the spatial solution for the amplitude as

$$Q(\xi) = Q(0)H_n(\alpha\zeta)\exp\left(-\frac{\alpha^2\zeta^2}{2}\right),$$

$$(31)$$

with the solvability condition $\alpha = 1/(1 + 2n)$, n being any positive integer and $H_n(\zeta)$, the Hermite functions. To obtain the evolution of this profile, we

Nonlinear Jeans Instability in an Uniformly Rotating Gas   287

solve Eq. (30) with the initial conditions $\hat{P}(0) = 1$ and $(d\hat{P}/d\tau)_0 = 0$. The solution is given by

$$P(\bar{\tau}) = \frac{P(0)}{(1 + \alpha^2 \bar{\tau}^2)^{1/4}}.$$ (32)

Therefore the complete solution for the amplitude A is found to be

$$A(\zeta, \bar{\tau}) = \frac{1}{(1 + \alpha^2 \bar{\tau}^2)^{1/4}} H_n(\alpha\zeta) \exp\left(-\frac{1}{2}\alpha^2\zeta^2\right),$$ (33)

Once the amplitude is known, the phase variable can be determined as $\phi(\zeta, \bar{\tau}) = \alpha\zeta\bar{\tau}\left(1 + \alpha^2\bar{\tau}^2\right)^{-1/2}$, assuming $\phi(\zeta, 0) = 0$. Knowing ϕ the relation between ζ and ξ is evident and the final solution for the density may be obtained from Eq. (19) as,

$$\rho(\xi, \tau) = \frac{\rho(\xi, 0)}{1 + f(\xi, \tau)}$$ (34)

where

$$f(\xi, \tau) = \frac{e^{-\bar{\alpha}\xi^2}}{(1 + \alpha^2\epsilon^2\tau^2)^{3/4}}\left[\sqrt{1 + \alpha^2\epsilon^2\tau^2}\{\bar{\alpha}H_n'(\bar{\alpha}\xi) - \bar{\alpha}^2\xi H_n(\bar{\alpha}\xi)\}\sin\theta\right.$$
$$\left. -\bar{\alpha}\epsilon\tau\xi H_n(\bar{\alpha}\xi)\cos\theta\right],$$ (35)

$$\rho(\xi, 0) = \rho_0 \mathrm{sech}^2\left(\frac{x}{\lambda_J}\right), \quad \bar{\alpha} = \frac{\alpha}{1 + \sqrt{1 + \alpha^2\epsilon^2\tau^2}}, \quad \theta = \tau\left(1 - \frac{\bar{\alpha}\epsilon\xi^2}{2\sqrt{1 + \alpha^2\epsilon^2\tau^2}}\right).$$

Once the solution for auxiliary function ψ is known we can easily find out the relation between ξ and x. Thus, the transformation from Lagrangian to Eulerian variables i.e. determination of ξ as a function of x and t results in the following relation

$$x = \xi + \frac{1}{(1 + \alpha^2\epsilon^2t^2)^{1/4}} H_n(\bar{\alpha}\xi) e^{-\frac{\bar{\alpha}^2\xi^2}{2}} \sin\theta.$$ (36)

Equation (34) together with Eq. (36) gives the complete solution. Equation (34) indicates that when $f(\xi, \tau) \to -1$, the system undergoes an explosive instability. For large τ, $f(\xi, \tau)$ asymptotically approaches

$$f(\xi, \tau) \sim \frac{e^{-\bar{\alpha}\xi^2}}{\sqrt{\alpha\epsilon\tau}}\left[\{2nH_{n-1}(\bar{\alpha}\xi) - \bar{\alpha}^2\xi H_n(\bar{\alpha}\xi)\}\sin\theta - \frac{\bar{\alpha}}{\alpha}\xi H_n(\bar{\alpha}\xi)\cos\theta\right]$$

which suggests that for large τ, $f(\xi.\tau) \to 0$ and the system approaches equilibrium For lower values of $n(= 0 - 2)$, the magnitude of $f(\xi, \tau)$ is small compared to 1. As a result $\rho(\xi, \tau)$ fluctuates a little and finally settles to the

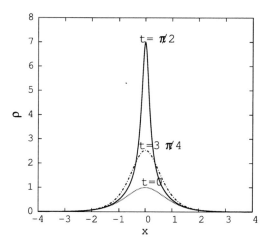

Fig. 2. Density amplitude $\rho(x,t)$ for $n = 3$ with time as a parameter.

value of $\rho(\xi, 0)$. A significant change in the density amplitude occurs at $n = 3$ as shown in figure (2). It indicates that nonlinearity feeds energy to the system by virtue of which the density amplitude gets enhanced considerably, reaching a maximum $\rho(x,t) \approx 7$ at $x = 0$, $t = \pi/2$ and subsequently returns back to its initial value $\rho(\xi, 0)$ at $t \to \infty$. For $n = 5$ onwards $f(\xi, \tau) \to -1$ as τ increases from zero giving rise to explosive instability. This is clearly demonstrated in figure (3). Thus we see that although rotation acts as a stabilizing effect for linear systems, the possibility of explosive instability is not ruled out in the presence of nonlinear effects.

6 Summary and conclusion

To summarize, we like to emphasize that the present nonlinear analysis leads to an explosive instability which can not be quenched by rotation. Non-relativistic gravitohydrodynamic equations are solved and the solutions indicate the existence of density bursts even in the presence of rotation. From the mathematical point of view it is interesting to note that the Lagrangian transformation is introduced twice to reduce the convective derivative to a partial derivative in some space. We have considered here only one dimensional case for simplicity of calculation. The solution obtained here may be relevant and applicable for other similar physical situations. To the best of our knowledge such time dependent solutions of gravitating system have not been studied earlier. We believe that more investigations of this nature would enrich our understanding of the structure formation in the universe.

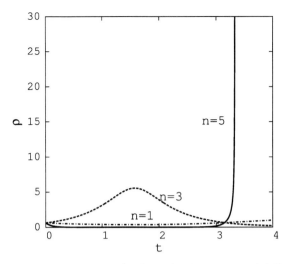

Fig. 3. Density amplitude $\rho(x,t)$ for fixed $\xi = 0.7$ and different n.

References

1. Jeans, J. H., 1902, Phillosophical. Trans., 199, 1.
2. Chandrasekhar, S., 1961, Hydrodynamics and Hydromagnetic Stability, (Oxford Univ. Press) p-588.
3. Kolb, E. W., & Turner, M. S. 1990, The Early Universe, (Addision-Wesley Publishing Co.).
4. Krishan, V., 1999, Astrophysical Plasmas and fluids(Kluwer Academic Publisher) p-270
5. Monte-Lima, I. & Ortega, V. G. 2004, New Astronomy, 9, 365.
6. Griv, E., Gedalim, M. & Yuan, C. 2002, Astronomy and Astrophysics, 383, 338.
7. Dawson, J. M. 1959, Physical Review, 113, 383.
8. Davidson, R. C. & Schram, P. P. 1968, Nuclear Fusion, 8, 183.
9. Infeld, E. & Rowlands, G. 1987, Phys. Rev. Letters, 58, 2063.
10. Davidson, R. C. 1972, Methods in Nonlinear Plasma Theory, (Academic, New York).
11. Pipes, L. A., & Harvill, L. R. 1971, Applied mathematics for Engineers and Physicists, p-560.

Part VI

Solar and Space Plasmas

An Overview of the Magnetosphere, Substorms and Geomagnetic Storms

G. S. Lakhina[1], S. Alex[1] and R. Rawat[1]

Indian Institute of Geomagnetism, Plot no. 5, Sector-18, Kalamboli Highway, Panvel (W), Navi Mumbai-410 218
`lakhina@iigs.iigm.res.in;salex@iigs.iigm.res.in;`
`rashmir@iigs.iigm.res.in`

Summary. The magnetosphere is the region of space to which the Earth's magnetic field is confined by the solar wind plasma which is continuously being blown outward from the Sun. The magnetosphere of the Earth extends to distances in excess of 60,000 kilometers on the Sunward side and to about million kilometers from Earth on the anti-sunward side, respectively. Much has been learned about this dynamic plasma region over the past 40 years from the direct measurements by various spacecrafts. This review first gives a brief introduction to this dynamic region of Earth's near space environment and then discusses important characteristics of magnetospheric substorms and storms, and their role in controlling the space weather processes.

Key words: Magnetosphere – Substorms: Geomagnetic storms

1 Introduction

Two Essential ingredients for the formation of the magnetosphere are the Earth's magnetic field and the solar wind. The major part of the geomagnetic field is generated in the earth's outer liquid core by a complex process called Geodynamo which is not fully understood till now. This part constitutes the Main Field which varies over long time scales of several decades to million of years. Outside the Earth, the main geomagnetic field has the same form as that of a bar magnet, a dipole field, aligned approximately at an angle of $11.5°$ with the Earth's spin axis. The solar wind is an ionized and highly conducting gas, consisting mainly of electrons and protons with a small fraction of alpha particles and other heavier ions, emitted continuously from the Sun. The solar wind flow drags the frozen-in solar magnetic field along with it into the interplanetary medium, the so called interplanetary magnetic field (IMF) [1, 2]. The solar wind interaction with the geomagnetic field gives rise to the Earth's Magnetosphere. The fast varying part of the geomagnetic field is controlled by the solar activity and it arises due to the interaction of solar ejecta with Earth's magnetosphere. Magnetospheric storms and substorms are indicators of geomagnetic activity. Magnetic Storms are driven directly by solar drivers like coronal mass ejections (CMEs), solar flares, fast solar wind streams, etc. The intensity of a geomagnetic storm is characterized by Dst (disturbance storm time) index. Substorms, in simplest terms, are disturbances occurring within the magnetosphere that are ultimately caused by the solar wind. They are delineated by geomagnetic AE (auroral electrojet) index. Magnetic reconnection plays an important role in energy transfer from solar wind to the magnetosphere [3] . Magnetic reconnection is very effective when the interplanetary magnetic field is directed southwards leading to strong plasma injection from the tail towards the inner magnetosphere causing in-tense auroras at high-latitude nightside regions. The solar wind input power into the magnetosphere is about $\sim 10^{11}$ W during a typical sub-storms, and about $\sim 10^{13}$ W during a moderate magnetic storm. Here, we give a brief overview of the magnetosphere, magnetospheric substorms and storms.

2 Magnetosphere

The Magnetosphere is the region of space, in the vicinity of the Earth, where the geomagnetic field is confined by the solar wind plasma emanating from the Sun. The magnetosphere of the Earth extends to distances in excess of 60,000 kilometers on the Sunward side and to about million kilometers from the Earth on the anti-sunward side, respectively. Much has been learned about this dynamic plasma region over the past 40 years from the direct measurements by various spacecrafts.

A schematic of the three dimensional view of the magnetosphere is shown in Figure 1. It shows various important plasma regions as well as the current systems.

2.1 Some Important Magnetospheric Plasma Regions

The magnetopause forms a boundary between the solar wind and the magnetosphere. Here the solar wind dynamic pressure is balanced by the geomagnetic field pressure [4]. The Earth's magnetic field is more or less confined inside the magnetopause boundary. This boundary layer is the site where solar wind energy and momentum are transferred into the magnetosphere [5]. Two main processes by which the solar wind plasma can cross the magnetopause are (1) direct entry involving magnetic reconnection [3, 6], and (2) the cross-field transport due to the scattering of particles by the waves across the closed magnetopause field lines [7, 8, 9], or the so called viscous interaction [10, 11, 12].

The plasma sheet is a thick layer of hot plasma centered on the magnetotail's equator, with a typical thickness 3-7 R_E and density of 0.3-0.5 ions

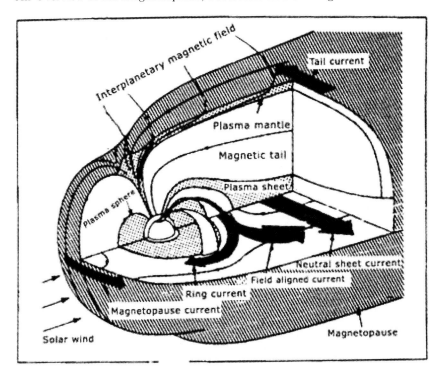

Fig. 1. Schematic 3-dimensional view of the earth's magnetosphere formed by the interaction of solar wind with the geomagnetic field. Small arrows indicate the direction of the magnetic field lines. Thick arrows show the direction of electric currents. Various current systems present in the magnetosphere are shown.

cm^{-3} and typical ion energy of 2-5 keV during geomagnetically quiet periods. Strong cross-tail currents flow across the width of the plasma sheet, from edge to edge. Once the current reaches the magnetopause, it divides into 2 parts which close around the lobes, over the top and under the bottom of the magnetotail. The tail lobes are two regions of relatively strong magnetic field, above the plasma sheet in both the north and the south hemisphere. Field lines of the lobes are smooth, and maintain roughly the same direction until they converge above the poles. They point towards the Earth north of the equator and away from the Earth south of it. The boundary between the lobes and the plasma sheet is known as plasma sheet boundary layer.

The plasmasphere contains essentially the high density cold plasma of the ionospheric origin. The outer boundary of the plasmasphere is called the plasmapause. Outside the plasmapause, the cold plasma density drops to much lower values than that inside the plasmasphere. Instead this region is characterized by the presence of hot and tenuous plasma that is accelerated in the tail on closed field lines downstream from the tail reconnection site, and is then transported inward toward the Earth. The size of the plasmasphere is not constant, and it changes with the geomagnetic activity and interplanetary conditions.

The Ring Current is formed due to the motion of trapped energetic particles (both electrons and protons) injected towards the earth from the plasma sheet. These particles undergo three types of basic motions in the dipole geometry of the geomagnetic field as shown in Figure 2. The fastest of the three

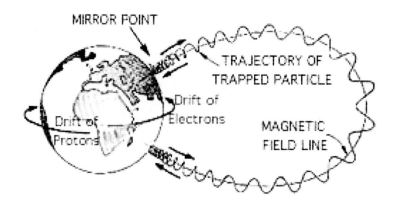

Fig. 2. schematics of 3 types of motions exhibited by charged particles (electrons and ions) in the geomagnetic field. The fast motion is the gyration around the magnetic field line, the so called cyclotron motion. The second type is sliding along the field line or the bouncing motion . The third type, the slowest type, is the drift of charged particles across the magnetic field due to the magnetic gradients or the gradient drift motion. The electrons gradient drift towards the east and the protons towards the west around the Earth, thus, constituting a westward current, called the Ring Current which encircles the Earth.

is known as the gyration or cyclotron motion where the charge particles go around a magnetic field line, typically thousands of times each second, in a plane perpendicular to the magnetic field. This is followed by a slower back-and-forth bounce motion along the field line, typically lasting 1/10 second. The third and the slowest type, is a slow drift around the magnetic axis of the Earth, from the current field line to its neighbor, staying roughly at the same distance from the axis. Typical time to circle the Earth is a few minutes or so. Due to drifting of positive ions and negative electrons in opposite directions (see Figure 2), an electric current is generated which circulates clockwise around the Earth when viewed from the north. The current is known as the ring current. The ring current energy is mainly carried by the ions, most of which are protons. However, during geomagnetically active periods significant energy is carried by He^{++} ions of solar wind origin and O^+ ions of ionospheric origin. In fact during intense geomagnetic storms, the O^+ ions of ionospheric origin are found to be the most prominent ring current energy carriers.

2.2 Current Systems

There are several current systems in the magnetosphere. The strong currents are usually encountered in the boundary layers. Some important cur-rent systems are magnetopause current or Chapman-Ferraro currents which flows in the magnetopause boundary layer, cross-tail or neutral sheet current flowing at the center of the plasma sheet, and field-aligned current or Birkeland current flowing along the magnetic field lines which couples the ionosphere and magnetosphere, and are linked with auroras. There is another important current system in the ionosphere called the Solar quiet (Sq) current which produces daily variations in the geomagnetic field. The fluctuations in these current systems produce variations in the geomagnetic field at various scales in time and space, for example, magnetic pulsations, magnetic bays, daily variations, etc. These current systems help to dissipate the solar wind energy transferred to the magnetosphere.

3 Magnetospheric Substorms

Magnetospheric substorms last for a period of $\sim$ one to a few hours, and are characterized by an explosive energy released from the magnetotail in the form of energetic particles ($\sim$ 5-50 KeV) and strong plasma flows ($\sim$ 100-1000 km s^{-1} or so). The energy is dissipated in the near-Earth nightside region. Auroras become widespread and intense, also much more agitated, and the Earth's magnetic field is disturbed mainly at high latitudes. During the substorm activity, the cross-tail current is disrupted and diverted towards the ionosphere as field-aligned current, and there is an enhancement in the ring current and the westward electrojet current. Magnetospheric convection increases, the plasma sheet tends to move earthward and a part of the plasma sheet is severed from the earth, forming a "plasmoid" that flows out tailwards. When the substorms occur frequently they may give rise to geomagnetic storms.

An isolated substorm has typically three phases. The growth phase usually begins with the start of southward turning of the interplanetary magnetic field (IMF). The size of the polar cap and the cross-section of the magnetotail increase, the near-earth plasma sheet starts thinning and the dipolar magnetic field is stretched into a tail-like field. The expansion phase starts with a sudden brightening of the discrete auroral arcs in the midnight sector and their rapid poleward advancement. The earthward injection of energetic particles intensifies the ring current, and the plasmoids are formed flowing in the anti-earthward direction. The recovery phase is manifested when the auroras start receding equatorward and the plasma sheet thickens [13].

3.1 Substorm Models

Several models have been proposed for the substorm phenomena based on the nature of solar wind energy input and that of the stored energy in the magnetosphere [14, 15]. Several statistical studies have shown that substorms clearly have both directly driven and loading - unloading components [16, 17]. Studies based on linear prediction filtering technique indicate that the typical time scales for the driven and loading-unloading processes are about 20 minutes and 1 hour respectively [18]. Two main models for the substorm are reconnection model, also known as near earth neutral line (NENL) model, and the current disruption model.

Reconnection Model

The reconnection model or the NENL model [19] treats the sub-storm phenomenon as a loading-unloading process. The growth phase is due to enhanced dayside magnetic reconnection which leads to a larger tail size and thinning of the plasma sheet. The magnetic energy stored in the magnetotail is released explosively through a magnetic reconnection process in the vicinity of a newly formed neutral line in the near-earth tail region (at about 10 - 20 R_E downstream). The neutral line formation leads to the disruption (actual mechanism some what uncertain) of the cross-tail current in the vicinity of the neutral line, and to the severance of the plasmasheet to produce a plasmoid.

Current Disruption Model

The current disruption model treats the substorm as a process directly driven by the solar wind [20]. According to this model, the solar wind energy input, equivalent to the power of solar wind-magnetosphere (SM) dynamo, directly controls the response of the magnetosphere such as the magnetospheric energy dissipation rate. When the solar wind energy input exceeds some critical value $\sim 10^{18}$ ergs s^{-1}, magnetospheric convection alone is unable to dissipate this much energy. The cross-tail current is disrupted and diverted to the ionosphere, along the magnetic field lines where it is dissipated efficiently. The resulting current circuit forms the substorm current wedge [21]. This is identified as substorm onset. A sudden increase in the anomalous resistivity due to a cross-field current instability is the likely cause for the current disruption [22, 23]. Particle injection results from the collapse of stretched field lines as a result of current disruption. The deflation of the plasma sheet is communicated downstream by the launching of a rarefaction wave, causing plasma sheet thinning further downstream of the current disruption region. Figure 3 shows the highlights of the current disruption model and the NENL model [24].

Two Substorm Onset Models

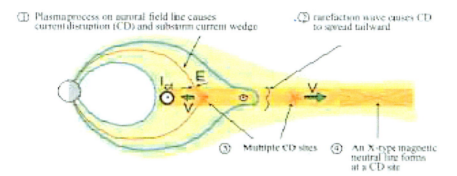

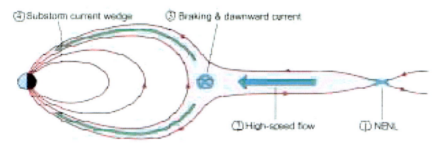

Fig. 3. A schematic diagram describing highlights of two substorm on-set models. Top panel shows the current disruption model and the bottom panel shows the near earth neutral line (NENL) model [after Lui [24]].

4 Geomagnetic storms

A geomagnetic storm is characterized by a Main Phase during which the horizontal component of the Earth's low-latitude magnetic field is significantly depressed over a time span of one to a few hours followed by its recovery which may extend over several days [25]. The main interplanetary causes of geomagnetic storms are the solar ejecta (due to CMEs and solar flares) having unusually intense magnetic fields and high solar wind speeds near the Earth, the corotating interaction regions (CIRs) and the fast streams from coronal holes. Geomagnetic storms occur when the solar wind-magnetosphere coupling becomes intensified during the arrival of fast moving solar ejecta accompanied by long intervals of south-ward interplanetary magnetic field (IMF), e.g., such as in a "magnetic cloud" [26]. Under such conditions, the efficiency of the reconnection process is considerably enhanced during southward IMF intervals

300   G. S. Lakhina, S. Alex and R. Rawat

[27, 9], leading to strong plasma injection from the magnetotail towards the inner magnetosphere. This leads to intense auroras at high-latitude nightside regions and at the same time intensifies the ring current which causes a diamagnetic decrease in the Earth's magnetic field measured at near-equatorial magnetic stations. The decrease in the equatorial magnetic field strength, measured by the Dst index, is directly related to the total kinetic energy of the ring current particles; thus the Dst index is a good measure of the energetics of the magnetic storm.

4.1 Geomagnetic storms types

The Dst index profile of geomagnetic storms is used to characterize the geomagnetic storms into various types and categories. The sudden commencement (SC) storms are characterized by a sudden increase in the horizontal magnetic field intensity shortly before the main phase [see top panel of Figure 4]. This sudden increase in magnetic field strength is caused by the interplanetary shock compression of the magnetosphere. The period between the sudden commencement and the main phase is called the initial phase. However, all magnetic storms do not have the initial phase. A geomagnetic storm not accompanied by a SC is called a gradual geomagnetic storm (SG) type [bottom panel of Figure 4]. Magnetic storms having a single main phase, wherein the Dst decreases more or less continuously to a minimum value and then starts to recover, are called Type 1 or one-step storms. In Type 2 or two-step storms the main phase undergoes a two-step growth in the ring current [28] in a way that before the ring current had decayed to a significant pre-storm level, a new major particle injection occurs, leading to further build-up of the ring current and further decrease of Dst. Hence, there is a possibility of multi-step storms depending on the ring current injection events caused by the interplanetary conditions.

The intensity of the magnetic storm is measured by the Dst index at peak of the main phase. The magnetic storms are called weak when Dst > -50 nT, moderate when -50> Dst > -100 nT, and intense when Dst < -100 nT [28] and super intense when Dst < -500 nT [29].

4.2 Some important Characteristics of Magnetic Storms

Taylor et al. [30] have shown that the SC type magnetic storms result from interplanetary shocks associated with coronal mass ejections (CMEs) while the gradual storms (SG type) are caused by corotating interaction regions (CIRs). From the superposed epoch analysis of more than 300 storms, Yokoyama and Kamide citeYK97 concluded that magnetic storm intensity depends on the duration of main phase; the more intense storms have longer main phase. The southward component of IMF plays a crucial role in both triggering the main phase and in determining the magnetic storm intensity. Tsurutani et al. [32] have shown that the number of major (Dst < -100 nT) storms follow the solar

Sudden Commencement (SC) Type Storm

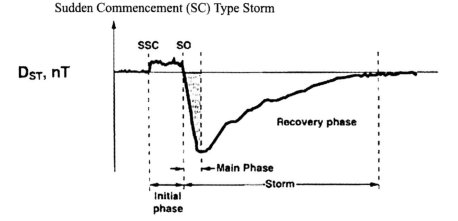

Gradual (SG) Type Storm

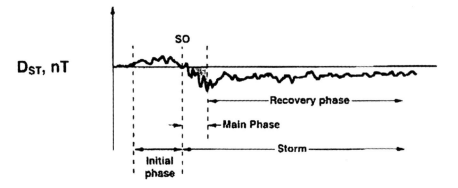

Fig. 4. Schematics of magnetic storms (top) sudden commencement (SC) type driven by ICMEs and (bottom) gradual (SG) type caused by CIRs. All storms may not have initial phases (taken from Tsurutani et al. [32] and slightly modified)

cycle sunspot number. For weak to moderate storms, there is a much smaller solar cycle dependence (see Figure 5).

They concluded that CIRs and high speed streams are presumably responsible for most of the weaker storms. They further showed that the CIR-generated magnetic storms appear to have very long recovery phases as compared to those driven by interplanetary coronal mass ejections (ICMEs). Relativistic "killer" electrons appear during "recovery" phase of the magnetic storms. These electrons pose great danger for the spacecraft.

From the study of 9 intense magnetic storms (Dst < - 175 nT) that occurred during 1998 to 2001, Vichare et al. [33] concluded that the main phase duration shows clear dependence on the duration of southward IMF but no

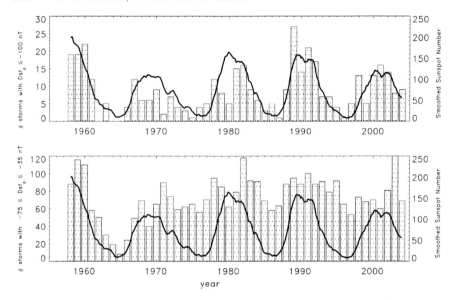

Fig. 5. Shows occurrence of magnetic storms per year between 1958 and 2004. The number of magnetic storms/year with Dst < -100 nT are given at the top and those with -35 > Dst > -75 nT on the bottom. The smoothed sunspot number is shown as a solid dark line. The number of major (Dst < -100 nT) magnetic storms follow the solar cycle sunspot number. There are ∼ 15 to 20 major magnetic storms/year during solar maximum and only ∼ 1 to 2 during solar minimum. The ratio is ∼ 15 to 20. For weak to moderate intensity magnetic storms, there is much smaller solar cycle dependence. CIRs/high-speed streams are presumably responsible for most of the weaker storms [taken from Tsurutani et al. [32]].

clear dependence on intensity of the storm. The underlined conclusion apparently disagrees with statistical study of Yokoyama and Kamide [31]. Vichare et al. [33] also showed that during the main phase of the storm, almost 5 percent of the total solar wind kinetic energy is available for redistribution in the magnetosphere, whereas during total storm period it reduces to 3.5 percent.

The intensity of the storm is controlled mainly by the magnitude of the peak of the southward component of the IMF B_z and its duration rather than the speed of the CME ejecta. For example, large southward IMF associated with slow CME of 18 November 2003 gave rise to more intense magnetic storm (Dst= - 491 nT) on 20 November 2003 than the very fast CMEs on 28 and 29 October 2003 which produced less intense magnetic storms (Dst= - 358 nT and - 406 nT) on 29 and 30-31 October 2003, respectively [34].

It is found that solar energetic particle (SEP) events with high flux levels or a 'plateau' after the shock passage produce much more intense storms than the events where the SEP flux levels decrease after the shock passage. An example of the SEP event is shown in Figure 6 [35]. SEP events having longer pre-shock southward IMF B_z duration produced stronger main phase storms

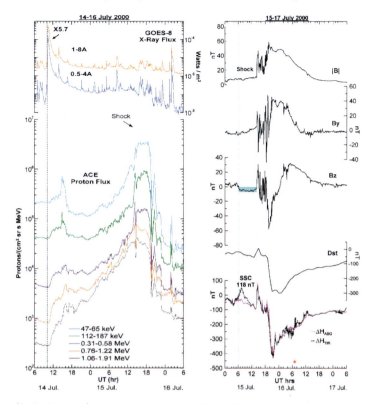

Fig. 6. (Left diagram), upper panel shows X-ray flux at two wavelengths 0.5-4 $A°$ and 1-8 $A°$. Proton flux characteristics in various energy levels of 47-65 keV, 112-187 keV, 0.31-0.58 MeV, 0.76-1.22 MeV and 1.06-1.91 MeV for 14-16 July 2000 are shown in lower panel. (Right diagram) shows from top to bottom: IMF $|B|$, IMF B_y, IMF B_z, hourly averaged Dst index, 1-minute digital magnetic data $\triangle H_{ABG}$ and $\triangle H_{TIR}$ for 15-17 July 2000. The red star indicates the local noon [after Rawat et al. [35]]

(see Figure 7). This result can be used as precursory signature for intense magnetic storms for space weather studies [35].

From the study of 8 intense storms (Dst < -200 nT), Rawat et al. [36] have found that the zonal component, B_y of IMF plays a substantial role for the development of intense main phase in the presence of significant southward B_z Component (see Figure 8).

5 Storm-Substorm Relationships

The magnetic storm-substorm relationship is an active topic of debate [37, 38].
In the earlier view proposed by Akasofu and Chapman [39], magnetic storms were presumably caused by frequent occurrence of intense sub-storms.

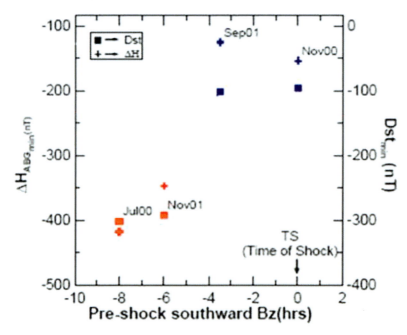

Fig. 7. Illustrates correspondence of $\triangle H_{ABG}$ minimum (filled '+' signs) with duration of southward B_z before shock and Dst_{min} (filled square sy-bols) for four SEP events. 'TS' indicates the shock time. The pre-shock hours with southward IMF B_z are marked with respect to the shock time. Red colour symbols depict two intense main phase events of 14-16 July 2000 and 4-6 November 2001. For intensity of these storms see the degree of depression in Alibag magnetic field and high level of auroral activity. Weak main-phase events of 8-10 November 2000 and 24-26 September 2001 have been shown by blue symbols. These two cases can be identified by lower values of $\triangle H_{ABG}$ minimum and AE values [after Rawat et al. [35]].

Observations of energetic particles in the outer region of the ring current region, which were presumably impulsively injected from the plasma sheet, during substorms lent support to this idea [40]. Sun et al. [41] and Sun and Akasofu [42] analyzed the magnetic data from the world wide observatory network and found two prominent current pattern consisting of one-cell and two-cell systems. The impulsive one-cell system, well-known to be associated with the substorms, was highly correlated with the Dst and poorly with the solar wind. On the other hand, the two-cell system, associated with the magnetospheric convection, correlated well with solar wind parameters but poorly with Dst. Such a separation of magnetic observatory network data into convection and substorm components provides a strong support to the idea of substorms driving the magnetic storms. Further support to this view comes from the numerical simulations by Fok et al. [43] of the ring current with impulsive electric fields mimicking the substorm effects. The simulation results

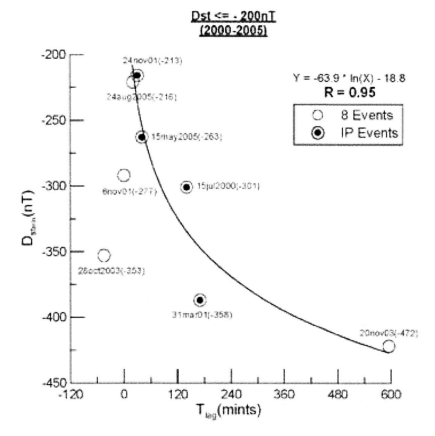

Fig. 8. Correspondence between T_{lag} and minimum Dst attained for 8 intense storm events. T_{lag} is the difference between onset times of significant duskward By and southward turning of Bz [after Rawat et al. [36]].

indicate a stronger ring current generation than the one that can be produced from only a convection electric field.

The modern view is that enhanced magnetospheric convection from sustained southward IMF can drive the magnetic storm. The particles residing in the plasma sheet can be transported closer to the Earth by a large magnetospheric electric field arising from the interaction of strong southward IMF with the geomagnetic field [44, 24]. This view is supported by the facts that the Dst index can be predicted well using interplanetary conditions alone [45, 46, 28] and the reported decrease in the rate of development of Dst index with substorm occurrence [47]. The observations by Gonzalez and Tsurutani [48] of intense magnetic storms during long duration (> 3-5 hrs) of southward IMF, a condition favoring strong dawn-to-dusk magnetospheric electric field, also support this idea. Additional support comes from the studies that found no substorms in a limited subset of magnetic storms caused by interplanetary magnetic clouds [49] but found intervals where there were very intense substorms without magnetic storms [50]. These latter events are now called as high intensity long duration continuous AE activity (HILDCAAs). Numerical simulations showing build-up of ring current during enhanced magnetospheric convection without including the impulsive injection from substorm [51] lend further support to this view.

It is now more or less established that interplanetary electric fields E_y (dawn-dusk component corresponding to southwards IMF) play important roles in both magnetic storms and substorms activity. Whereas the fluctuating E_y gives rise to substorms, the quasi-steady E_y can give rise to magnetic storms [52]. Further, it is believed that various plasma instabilities play important roles in substorm onset [53, 54, 55, 22, 57, 56, 58, 59] but essentially no role in driving the magnetic storms. Plasma instabilities, however, could be important during the main as well as the recovery phase of magnetic storm [60, 61, 62, 63, 64].

6 Conclusions

The interaction of solar wind with the geomagnetic field results in the formation of the Earth's magnetosphere which contains various plasma re-gions and the current systems. Magnetic reconnection occurring on the dayside magnetopause is believed to be the main plasma process which is responsible for the transfer of solar wind energy into the magnetosphere. This process is most efficient during southward IMF intervals corresponding to the condition when the dayside geomagnetic field and the inter-planetary field are oppositely directed. Magnetospheric substorms and storms are manifestation of the geomagnetic activity which is ultimately controlled by the solar wind conditions although the observations indicate that some internal plasma instability of the magnetotail may be triggering substorms. However, the excitation of the plasma instability depend on the magnetotail equilibrium which is controlled by the interplanetary parameters.

Magnetic storms form the main component of space weather studies [24]. Intense and super-intense geomagnetic storms create hostile space weather conditions that can generate many hazards to the spacecraft as well as technological systems at ground. Some adverse effects of intense magnetic storms are life-threatening power outages, failure and malfunctioning of satellite instruments due to deep dielectric charging by relativistic or "killer" electrons, satellite communication failure, data loss, and navigational errors, loss of low earth orbiting satellites, threat to astronauts and jetliner passengers due to high radiation dosage, damage to power transmission lines and corrosion of the long pipelines and cables due to geomagnetically induced currents (GICs). Some of these effects were noticed during the October-November 2003 intense magnetic storms.

An Overview of the Magnetosphere, Substorms and Geomagnetic Storms 307

Acknowledgements

Thanks are due to Council of Scientific and Industrial Research, Government India, for providing support to GSL under the Emeritus Scientist Scheme. The authors acknowledge CELIAS/MTOF experiment on SOHO, a joint ESA and NASA mission. We thank the ACE SWEPAM instrument team and the ACE Science Center for providing the ACE data. We acknowledge the information from the Space Environment Center, Boulder, CO, National Oceanic and Atmospheric Administration (NOAA), US Dept. of Commerce. We also thank WDC, Kyoto for providing the preliminary Quick look AE, AL and AU, Dst and SYM-H indices.

References

1. J. K. Hargreaves: *The Solar-Terrestrial Environment*, (Cambridge University Press, 1992)
2. M. G. Kivelson, and C. T. Russell (eds.): *Introduction to Space Physics* (Cambridge University Press, 1995)
3. Dungey, J. W.: Physical Review Letters **6**, 47 (1961)
4. D. G. Sibeck, R. E. Lopez, and E. C. Roelof: J. Geophys. Res. **96**, 5489 (1991)
5. G. Haerendel and G. Paschmann: Interaction of the solar wind with the dayside magnetosphere. In: *Magnetospheric Plasma Physics*, ed by A. Nishida (Cent. for Acad. Publ. Japan, Tokyo, 1982)
6. W. D. Gonzalez, J. A. Joselyn, Y. Kamide, H. W. Kroehl, G. Rostoker, B. T. Tsurutani, V. W. Vasyliunas: Journal of Geophysical Research **99**, 5771 (1994)
7. B. T. Tsurutani and R. M. Thorne: Geophys. Res. Lett. **9**, 1247 (1982)
8. R. Gendrin: Geophys. Res. Lett. **10**, 769 (1983)
9. B. T. Tsurutani, and W. D. Gonzalez: The interplanetary causes of magnetic storms; a review. In: *Magnetic Storms*, Ed by B. T. Tsurutani, W. D. Gonzalez, Y. Kamide, and J. K. Arballo, Geophysical Monograph Series, 98 (AGU, Washington D.C., 1997) pp 77-90.
10. W. I. Axford and C. O. Hines: Can. J. Phys., **39**, 1433 (1961)
11. A. Eviatar and R. A. Wolf: J. Geophys. Res. **73**, 5561 (1968)
12. Tsurutani, B. T., and W. D. Gonzalez: Geophys. Res. Lett. **22**, 663 (1995)
13. S. I. Akasofu, C. I. Meng, and D. S. Kimball: J. Atmos. Terr. Phys., **28**, 505 (1966)
14. J. R. Kan, T. A. Potembra, S. Kokubun and T. Iijima: Magnetospheric Substorms. In: *Geophysical Monograph 64*, (American Geophysical Union, Washington, D. C., 1991).
15. G. S. Lakhina: Some theoretical concepts of magnetospheric substorms. In: *Geomagnetic Studies at Low Latitudes*, Memoirs of Geological Society of India, No. 24, ed by G. K. Rangarajan and B. R. Arora (Geological Society of India, Bangalore, India, 1992) pp 307-355
16. L. F. Bargatze, D. N. Baker, R. L. McPherron, and E. W. Hones, Jr.: J. Geophys. Res., **90**, 6387 (1985).
17. D. N. Baker: Driven and Unloading aspects of magnetospheric substorms. In: *Proceedings of the International Conference on Substorms (ICS-1)*, Kiruna, Sweden, 23-27 March, ESA SP-335 (1992) pp 185-191.

308   G. S. Lakhina, S. Alex and R. Rawat

18. Clauer, C. R.: The technique of linear prediction filters applied to studies of solar wind-magnetosphere coupling. In: *Solar Wind- Magnetosphere Coupling*, ed by Y. Kamide and J. A. Slavin (Terra Sci. Pub. Co., Tokyo, 1986) p. 39
19. E.W. Hones Jr.: Space Sci. Revs. **23**, 393 (1979)
20. S. I. Akasofu: Magnetospheric substorms: a model. In *Solar Terrestrial Physics*, Part III, ed by E. R. Dryer (D. Reidel, Norwood, Mass., 1972) p. 131
21. R. L. McPherron, C. T. Russell and M. D. Aubry: J. Geophys. Res. **78**, 3131 (1973)
22. A. T. Y. Lui et al.: J. Geophys. Res. **96**, 11389 (1991)
23. A. T. Y. Lui, R. E. Lopez, B. J. Anderson, K. Takahashi, L .J. Zanetti, R. W. McEntire, T. A. Potemra, D. M. Klumpar, E. M. Greene, and R. Strangeway: J. Geophys. Res. **97**, 1461 (1992)
24. A. T. Y. Lui: Terres. Atmos. Oceanic Sci. (TAO) **14**, 221 (2003)
25. G. Rostoker Physics of magnetic storms. In: *Magnetic Storms*, ed by B. T. Tsurutani, W. D. Gonzalez, Y. Kamide, and J. K. Arballo, (Geophysical Monograph 98, AGU, Washington D C, 1997) p. 149
26. L. W. Klein and L. F. Burlaga: J. Geophys. Res. **87**, 613 (1982)
27. W. D. Gonzalez, B. T. Tsurutani, A. L. C. Gonzalez, E. J. Smith, F. Tang, and S.-I. Akasofu: J. Geophys. Res. **94**, 8835 (1989)
28. Y. Kamide, N. Yokoyama, W. D. Gonzalez, B. T. Tsurutani, I. A. Daglis, A. Brakke, S. Masuda: Journal of Geophysical Research **103**, 6917 (1998)
29. B. T. Tsurutani, W. D. Gonzalez,, G. S. Lakhina, and S. Alex: J. Geophys. Res. **108**, A7, 1268, doi:10.1029/2002JA009504 (2003a)
30. J. R. Taylor, M. Lester and T. K. Yeoman: Ann. Geophys. **12**, 612 (1994)
31. N. Yokoyama and Y. Kamide: J. Geophys. Res. **102**, A7, 14215 (1997)
32. B. T. Tsurutani, W. D. Gonzalez, A. L. C. Gonzalez, F. L. Guarnieri, N. Gopalswamy, M. Grande, Y. Kamide, Y. Kasahara, G. Lu, I.M., R. McPherron, F. Soraas, and V. Vasyliunas: J. Geophys. Res. **111**, A07S01, doi:10.1029/2005JA011273 (2006)
33. G. Vichare, S. Alex and G. S. Lakhina: J. Geophys. Res. **110**, A03204, doi:10.1029/2004JA010418 (2005)
34. S. Alex, S. Mukherjee and G. S. Lakhina: J. Atmos. Solar-Terr. Phys., **68**, 769 (2006)
35. R. Rawat, S. Alex, G. S. Lakhina: Ann. Geophys. **24**, 3569 (2006)
36. R. Rawat, S. Alex, G.S. Lakhina: Bull. Astron. Soc. India, in press (2007)
37. A.S. Sharma, Y. Kamide and G.S. Lakhina (eds): *Disturbances in Geospace: The Storm-Substorm Relationship*, Geophysical Monograph Series, 142, (AGU, Washington D.C., 2003)
38. G. S. Lakhina, S. Alex, S. Mukherjee, and G. Vichare: On Magnetospheric Storms and Substorms. In: *The Solar Influence on the Heliosphere and Earth's Environment: Recent Progress and Prospects*, Ed by N. Gopalswamy and A. Bhattacharyya (Quest Publications, Mumbai, 2006) pp. 320-327
39. Akasofu, S. I., and S. Chapman: J. Geophys. Res., **66**, 1321 (1961)
40. C. E. McIlwain: Substorm injection boundaries. In: *Magnetospheric Physics*, ed. by B. M. McCormac (D. Reidel, Hingham, Mass, 1974) p. 143
41. W. Sun, W. Y. Xu, and S. I. Akasofu: J. Geophys. Res., 103, 11695 (1998)
42. W. Sun and S.I. Akasofu: J. Geophys. Res. **105**, 5411 (2000)
43. M. C. Fok, T. E. Moore, and D. C. Delcourt: J. Geophys. Res. **104**, 14557 (1999)

An Overview of the Magnetosphere, Substorms and Geomagnetic Storms   309

44. Y. Kamide: J. Geomagn. Geoelectr. **44**, 109 (1992)
45. R. K. Burton, R. L. McPherron, and C. T. Russell: J. Geophys. Res., **80**, 4204 (1975)
46. R. L. McPherron, D. N. Baker, L. F. Bargatze, C. R. Clauer, and R. E. Holzer: Adv. Space Res. **8**, 71 (1988)
47. T. Iyemori and D. R. K. Rao: Ann. Geophysicae **14**, 608 (1996)
48. W. D. Gonzalez and B. T. Tsurutani: Planet. Space Sci. **35**, 1101 (1987)
49. B. T. Tsurutani, X.-Y. Zhou, and W. D. Gonzalez: A lack of substorm ex-pansion phases during magnetic storms induced by magnetic cloud. In: *Disturbances in Geospace: The Storm-Substorm Relationship*, ed by A.S. Sharma, Y. Kamide and G.S. Lakhina, Geophys. Monogr. Ser. 142 (AGU, Washington D.C., 2003b)
50. B. T. Tsurutani, W. D. Gonzalez, F. Guarnieri, Y. Kamide, X. Zhoua, J. K. Arballo: J. Atmos. Sol. Ter. Phys. **66**, 167 (2004)
51. M. W. Chen, M. Schulz, and L. R. Lyons: J. Geophys. Res., **99**, 5745 (1994)
52. Y. Kamide: J. Atmos. Sol. Ter. Phys. **63**, 4113 (2001)
53. K. Schindler: J. Geophys. Res. **70**, 2803 (1974)
54. G. S. Lakhina and K. Schindler: J. Geophys. Res. **93**, 8591 (1988)
55. A. P. Kakad, G. S. Lakhina, S. V. Singh: Planet. Space Sci. **51**, 177 (2003)
56. W. W. Liu: J. Geophys. Res. **102**, 4927 (1997)
57. S. Ohtani and T. Tamao: J. Geophys. Res. **98**, 19369 (1993)
58. Huba, J. D., N. T. Gladd, and K. Papadopoulos: The lower- hybrid-drift instability as a source of anamolous resistivity for magnetic field line re-connection, Geophys. Res. Lett., 4, 125, 1977.
59. G.S. Lakhina and B.T. Tsurutani: Geophys. Res. Letts. **24**, 1463 (1997)
60. R. B. Horne and R. M. Thorne: J. Geophys. Res. **99**, 17259 (1994).
61. M.C. Fok, T.E. Moore, and M.E. Greenspan: J. Geophys. Res. **101**, 15311, 1996.
62. G.S. Lakhina and S.V. Singh Role of plasma instabilities driven by Oxygen ions during magnetic storms and substorms. In: *Disturbances in Geospace: The Storm-Substorm Relationship*, Geophys. Monogr. Ser. 142, ed by A. S. Sharma, Y. Kamide and G. S. Lakhina (AGU, Washington, D.C., 2003) pp. 131-141
63. S. V. Singh, A. P. Kakad, R. V. Reddy, and G. S. Lakhina: J. Plasma Physics **70**, 613 (2004)
64. S.V. Singh, A. P. Kakad and G. S. Lakhina: Phys. Plasmas **12**, 012903 (2005)

Monte Carlo Simulation of Scattering of Solar Radio Emissions

G. Thejappa[1] and R. J. MacDowall[2]

[1] Department of Astronomy, University of Maryland, College Park, MD 20742
thejappa@astro.umd.edu
[2] NASA, Goddard Space Flight Center, Greenbelt, MD 20771
Robert.MacDowall@nasa.gov, gopals@gsfc.nasa.gov

Summary. We present Monte Carlo simulation techniques which are used to study the scattering of the solar and interplanetary radio emissions, both thermal and nonthermal, by random density fluctuations. We examine to what extent the scattering is responsible for (1) anomalous time delays, large source sizes, increased source heights, and widespread visibilities of type III and type II radio bursts, and (2) low brightness temperatures of the quiet sun radio emission and large apparent source sizes. Using the empirically derived models for the electron densities, and density fluctuations, we show that (1) scattering is mostly responsible for the unusual behavior of the solar and interplanetary thermal and nonthermal emissions, and (2) Monte Carlo simulation techniques can be used to extract the characteristics of the electron beams and CME driven shocks from the type III and type II radio burst observations and the electron temperatures from the quiet sun observations.

Key words: Sun: Monte Carlo Simulations, Scattering, Radio Bursts, and Quiet Sun

1 Introduction

Propagation effects, especially, the refraction by the spatially varying background the solar corona and the interplanetary medium, and the scattering by random density fluctuations distort low frequency solar and interplanetary radio emission characteristics. For kilometric type III radio bursts, the propagation effects are believed to be responsible for [1, 2, 3, 4] (1) the higher apparent source heights in comparison with emission levels derived from averaged in situ density measurements, (2) larger source sizes at a given frequency with increasing distances of the centroids from the Sun, (3) anomalous propagation time delays between signal arrival at widely separated spacecraft, (4) visibility to sensitive, kilometric wave radio instruments irrespective of the location of the radiating sources, be they in front of or behind the Sun, and (5) larger angular radius of the bursts that originate behind the Sun than that of those occurring in front of the Sun. The scattering is also believed to be the cause

312   G. Thejappa and R. J. MacDowall

of low brightness temperatures of the quiet sun radio emission, and large equatorial diameters [5, 6, 7, 8].

The regular refraction of radio waves in a spherically symmetric solar atmosphere has been thoroughly investigated using ray tracing methods [9, 10, 11]. The geometric optics method is usually used to study the scattering of the solar and interplanetary radio emissions, where the plasma can have a significant influence on the propagation [4, 12, 13, 14, 15, 16, 17]. In these calculations, the ray path is usually divided into linear steps of ΔS chosen in such a way that the conditions $\psi < 0.1$ radians, and $\Delta\mu/\mu < 0.1$ are satisfied over each step, where ψ is the deflection angle and μ is the refractive index. In other situations, such as interplanetary scintillations, where the influence of plasma on wave propagation can be neglected, the scattering and diffraction are studied using the parabolic equation method [17, 18, 19, 20, 21].

Fokker [13] was the first to introduce the Monte Carlo technique to solar radio astronomy, and [14, 16] added the regular refraction to this algorithm. The Monte Carlo methods were used extensively in solar radio astronomy [4, 5, 6, 7, 8, 15, 22, 23]. For example, [5, 6] used these methods to study the unusual behavior of the quiet sun radio emission at meter and decameter wavelengths. The main criticism of all these studies is that they assume idealized spherically-symmetric models for the electron density and a Gaussian spectrum for the electron density fluctuations, instead of using more realistic non-spherically symmetric models for the electron density, and power-law type spectral distribution for the density fluctuations.

We have developed an efficient Monte-Carlo simulation program. We have used this program to study not only the directivity, visibility, time profiles, source sizes, and East-West asymmetries of type II and type III radio bursts at kilometric and hectometric wavelengths [4], but also the effects of refraction and scattering on the quiet sun radio emission [8]. In this paper, we describe these methods and show how the scattering is mostly responsible for the unusual behavior of the solar and interplanetary thermal and nonthermal emissions, and how Monte Carlo simulation techniques can be effectively used to extract the characteristics of the electron beams and CME driven shocks from the type III and type II radio burst observations, as well as the electron temperatures from the quiet sun observations.

Electron Density

For the solar wind electron density, we use the empirical formula derived by [24]

$$N_e(r) = 6.14r^{-2.10}\text{cm}^{-3}, \tag{1}$$

where r is the heliocentric distance in units of AU. For the quiet sun, we use the empirical formula derived by [25], based on Skylab data obtained during the declining phase of solar cycle 20 (1973-1976)

$$N_e(r, \theta_{mg}) = N_p(r) + [N_{cs}(r) - N_p(r)]e^{-\theta_{mg}^2/w^2(r)} \text{ cm}^{-3}, \tag{2}$$

where, the radial distance r is in units of $R_\odot$. The electron densities at the current sheet $N_{cs}(r)$ and the poles $N_p(r)$ are defined as

$$N_e(r) = \Sigma_{i=1}^3 c_i r^{-d_i}. \tag{3}$$

Here c_1, c_2 and c_3 are 1.07, 19.94, and 22.10 for the current sheet, and 0.14, 8.02, and 8.12 for the pole, respectively. These coefficients are in units of 10^7. The coefficients

d_1, d_2 and d_3 are 2.8, 8.45, and 16.87, respectively. The angular distance of a point from the current sheet in a heliomagnetic coordinate system (heliomagnetic latitude) θ_{mg} is given by

$$\theta_{mg} = \sin^{-1}[-\cos\theta\sin\alpha\sin(\phi - \phi_0) + \sin\theta\cos\alpha], \qquad (4)$$

where θ and ϕ are the heliographic latitude and longitudes, respectively, $\alpha \simeq 15\,\mathrm{deg}$ is the tilt angle between the dipole axis and the rotation axis, and $\phi_0 \simeq 0$ is the angle between the heliomagnetic and heliographic equators. The functional form of the half-angular width of the current sheet, $w(r)$ is

$$w(r) = \Sigma\gamma_i r^{-\delta_i}, \qquad (5)$$

where γ_1, γ_2, and γ_3 are 16.3, 10.0, and 43.20 degrees, and, δ_1, δ_2, and δ_3 are 0.5, 7.31, and 7.52, respectively. We neglect the ambient magnetic field, since the electron cyclotron frequency f_{ce} is much smaller than the electron plasma frequency f_{pe}. We assume that the electron temperature T_e is 1×10^6 K in the solar corona, and 1.5×10^5 K in the solar wind. The f_{pe}, μ (refractive index), and the electron collision frequency ν are defined as

$$f_{pe}^2 = 80.6 \times 10^6 N_e, \qquad (6)$$

$$\mu^2 = 1 - \frac{f_{pe}^2}{f^2}, \qquad (7)$$

$$\nu = 4.36 N_e T_e^{-3/2}[17.72 + \ln(T_e^{3/2}/f)], \qquad (8)$$

where f is the frequency in Hz.

Electron Density Fluctuations

The power spectrum of density fluctuations in the solar corona and the solar wind is [26, 27]

$$P_n(q) = C_N^2 q^{-\alpha}; \quad q_o < q < q_i, \qquad (9)$$

where q is the magnitude of the spatial wavenumber, α is the power-law exponent, and $l_o = 2\pi q_o$ and $l_i = 2\pi q_i$ are the outer and the inner scales, respectively. It is [26, 27] shown that for scales larger than a few times 100 km, the α is 11/3 (Kolmogorov spectrum), for intermediate scales (a few km $\leq l \leq$ few times 100 km) α changes from 11/3 to ~ 3 (flat spectrum), and for the smallest scales of the order of 2 km (inner or dissipative scales) the spectrum becomes quite steep with $\alpha \simeq 4$. If this spectrum is normalized to the variance of the density fluctuations $< \Delta N_e^2 >$, the structural constant C_N^2 is [28]

$$C_N^2 = A(\alpha, q_o, q_i) < \Delta N_e^2 >, \qquad (10)$$

where

$$A(\alpha, q_o, q_i) = \begin{cases} \frac{(\alpha-3)\Gamma(\alpha/2)q_o^{\alpha-3}(2\pi)^{-3/2}}{\Gamma[(\alpha-1)/2]} & \text{for } 3 < \alpha < 4, \\ \frac{1}{4\pi\ln(\frac{2q_o}{q_i})} & \text{for } \alpha = 3. \end{cases}$$

Thus, for Kolmogorov spectrum with $\alpha = 11/3$, we obtain $C_N^2 = \frac{\epsilon^2 l_o^{-2/3} N_e^2}{6.6}$, where $\epsilon = \frac{\Delta N_e}{N_e}$. For example, [29] has derived $C_N^2 = \frac{\epsilon^2 l_o^{-2/3} N_e^2}{5.53}$. For the solar wind, α

is $\sim 11/3$ [29, 30, 31, 32, 33, 34]. Since, the inner scale l_i increases linearly with distance from the Sun as $l_i = (\frac{R}{R_\odot})^{\pm 0.1}$ km at $R \leq 100R_\odot$ and from 100 to 200 kms, $l_i \simeq 90 - 100$ kms [26, 35], we assume l_i as ~ 100 kms. For the outer scale l_0, we use the empirical formula derived by [34]

$$l_0 = 19r^{0.82}, \tag{11}$$

where r is in units of AU. Based on *Helios* observations, [36] deduced that ϵ is 0.07 during most of the time and 14% of the time it is 0.1 (see, also [29]). Therefore, we assume that ϵ is 0.07 throughout the solar wind. For quiet sun studies, we assume that most of the power is concentrated in the flat portion of the spectrum, with $\alpha = 3$ and obtain $C_N^2 = \frac{\epsilon^2 N_e^2}{4\pi \ln(\frac{2l_i}{l_o})}$. By substituting $L_i = 50$ km, and $L_o = 75$ km for l_i and l_o, we obtain $C_N^2 = 0.28\epsilon^2 N_e^2$. At radial distances, corresponding to meter and decameter wavelengths, we also assume that the longitudinal spatial scales are almost 10 times larger than the transverse scales. We assume that $\epsilon = 0.1$ throughout the corona.

Computational Scheme

We use the Cartesian coordinate system with origin at the center of the Sun, and the x-axis coinciding with the line of sight. For ray tracing, we use a set of 6 first-order differential equations [37]

$$\frac{d\mathbf{R}}{d\tau} = \mathbf{T} \tag{12}$$

$$\frac{d\mathbf{T}}{d\tau} = D(\mathbf{R}) = \frac{1}{2}\frac{\partial \mu^2}{\partial \mathbf{R}}, \tag{13}$$

with

$$T_x^2 + T_y^2 + T_z^2 = \mu. \tag{14}$$

Here

$$\mathbf{R} \equiv \begin{pmatrix} x \\ y \\ z \end{pmatrix} \text{ and } \mathbf{T} \equiv \begin{pmatrix} T_x \\ T_y \\ T_z \end{pmatrix}$$

are the position and direction vectors of the ray, respectively. The independent variable τ is related to actual path length s as $d\tau = \frac{ds}{\mu}$. We can write

$$D(\mathbf{R}) \equiv \frac{1}{2} \begin{pmatrix} \frac{\partial \mu^2}{\partial x} \\ \frac{\partial \mu^2}{\partial y} \\ \frac{\partial \mu^2}{\partial z} \end{pmatrix} = \frac{8.90 \times 10^{12}}{f^2} \frac{1}{r^4} N_e \mathbf{R}. \tag{15}$$

We use the 3rd order Runge-Kutta algorithm to integrate the ray tracing equations (12) and (13), which can be written in vectorial form as (see for example [38])

$$R_{n+1} = R_n + \Delta\tau[T_n + \frac{1}{6}(A + 2B)], \tag{16}$$

$$T_{n+1} = T_n + \frac{1}{6}(A + 4B + C), \tag{17}$$

$$A = \Delta\tau D(R_n), \tag{18}$$

$$B = \Delta\tau D(R_n + \frac{\Delta\tau}{2}T_n + \frac{1}{8}\Delta\tau A), \tag{19}$$

$$C = \Delta\tau D(R_n + \Delta\tau T_n + \frac{1}{2}\Delta\tau B). \tag{20}$$

We compute the optical depth τ and the transit time $\Delta t(s)$ at each step ΔS as

$$\tau_{i+1} = \tau_i + K\frac{f_{pe}^2}{f^2}\frac{\nu\Delta S}{\mu_i} \tag{21}$$

$$\Delta t_{i+1} = \Delta t_i + K\frac{\Delta S}{\mu_i}, \tag{22}$$

where K is 500 for the solar wind, and 2.32 for the solar corona.

Scattering

The mean-square angular deviation $< \Psi^2 >$ suffered by a ray due to scattering by Gaussian fluctuations is [12, 39]

$$< \Psi^2 >= b(f)\Delta S. \tag{23}$$

Here, the mean square deviation per unit length, $b(f)$ is defined as

$$b(f) = \frac{\sqrt{\pi}}{\mu^4}\frac{f_{pe}^4}{f^4}\frac{\epsilon^2}{h}, \tag{24}$$

where h is the scale height of the density fluctuations. For a power-law spectrum $P_n(q) = C_N^2 q^{-\alpha}$, the $< \Psi^2 >$ is derived as [17, 4]

$$< \Psi^2 >= \frac{r_e^2\lambda^4}{\pi\mu^2}\Delta S\frac{C_N^2}{4-\alpha}(q_i^{4-\alpha} - q_o^{4-\alpha}). \tag{25}$$

For $q_o << q_i$, $\alpha = 11/3$, $r_e = \frac{e^2}{mc^2}$ (the classical radius of the electron), $\lambda = \frac{c}{f}$ (the wavelength), $C_N^2 = \frac{\epsilon^2 l_o^{-2/3} N_e^2}{5.53}$, and $f_{pe}^2 = \frac{e^2 N_e}{\pi m_e}$ we obtain

$$b(f) = \pi\frac{f_{pe}^4}{f^4}\frac{\epsilon^2}{\mu^4 l_i^{1/3}l_o^{2/3}}. \tag{26}$$

For $\alpha = 3$, and for $Q_i = \frac{2\pi}{L_i}$ and $Q_o = \frac{2\pi}{L_o}$ corresponding to q_i and q_o we obtain

$$b(f) = \pi\frac{(1 - L_i/L_o)}{2\ln(2L_i/L_o)}\frac{f_{pe}^4}{f^4}\frac{\epsilon^2}{\mu^4 L_i}, \tag{27}$$

for $C_N^2 = \frac{\epsilon^2 N_e^2}{4\pi\ln(2L_i/L_o)}$. For $L_i = 50$ km, and $L_o = 75$ km, we obtain

$$b(f) \sim 0.6\pi\frac{f_{pe}^4}{f^4}\frac{\epsilon^2}{\mu^4 L_i}. \tag{28}$$

It is interesting to note that if we substitute an effective scale height $h = l_i^{1/3}l_o^{2/3}$ in the expression (26) corresponding to Kolmogorov spectrum, it coincides with expression (24) corresponding to the Gaussian spectrum. Similarly, the expression

316   G. Thejappa and R. J. MacDowall

(28) derived for $\alpha = 3$ almost coincides with the expression (24). The components of the random perturbation vector $< \mathbf{p} >$ are chosen from a Gaussian distribution of random numbers with a zero mean and a standard deviation of

$$\sigma = \mu \sqrt{b \Delta S}. \qquad (29)$$

For isotropic fluctuations, three independent Gaussian distributed random deviations of the direction cosines with the same standard deviation (29) are calculated. However, for anisotropic fluctuations, the standard deviation σ changes accordingly, with $\sigma_\parallel < \sigma_\perp$. This vector $< \mathbf{p} >$ is added to $\mathbf{T}$ at each step, and normalized to have a length of μ.

2 Type III Radio Bursts

We assume that both fundamental and harmonic emission sources are isotropic point sources. The initial direction vectors of the rays emitted by such sources end on a sphere of radius $\sim$ local μ_0. Since, the probability $p(\theta_0, \phi_0) = p_1(\theta_0)p_2(\phi_0)$ that a point belongs to an element of a spherical surface $\sin\theta_0 d\theta_0 d\phi_0$ is $\frac{\sin\theta_0 d\theta_0 d\phi_0}{4\pi}$, where $0 \le \theta_0 \le \pi$ and $0 \le \phi_0 \le 2\pi$, we can write $p_1(\theta_0) = \frac{\sin\theta_0}{2}$ and $p_2(\phi_0) = \frac{1}{2\pi}$. From $\int_0^{\phi_0} p_2(\phi_0)d\phi_0 = \frac{\phi_0}{2\pi} = \xi_1$, and $\int_0^{\theta_0} p_1(\theta_0)d\theta_0 = \frac{1}{2}\int_0^{\theta_0} \sin\theta_0 d\theta_0 = \frac{\cos\theta_0}{2} + \frac{1}{2} = \xi_2$, we obtain the azimuthal and elevation angles of the initial ray as

$$\phi_0 = 2\pi\xi_1 \qquad (30)$$

$$cos\theta_0 = 2\xi_2 - 1, \qquad (31)$$

where ξ_1 and ξ_2 are the random variables uniformly distributed between 0 and 1. Thus, the optical direction cosines become

$$T_{x0} = \mu_0 \sin\theta_0 \sin\phi_0 \qquad (32)$$

$$T_{y0} = \mu_0 \sin\theta_0 \cos\phi_0 \qquad (33)$$

$$T_{z0} = \mu_0 \cos\theta_0. \qquad (34)$$

which imply that $-\mu_0 \le (T_{x0}, T_{y0}, T_{z0}) \le \mu_0$. We assign a value of 120 kHz for both fundamental and harmonic emissions, and accordingly assign altitudes of 0.2097 AU (corresponding to the 115 kHz plasma level; this level is chosen to avoid severe deflection of rays at critical level) and 0.3895 AU (corresponding to the 60 kHz plasma level). These plasma levels are calculated using equation (1). We consider two cases for simulations (1) $\epsilon = 0$ and (2) $\epsilon = 0.07$. For each case, we launch 1000 randomly directed rays from the source, and trace them until they cross the sphere of 1 AU radius. In the case of $\epsilon = 0$, we take a constant step size of 0.002 AU, and in the case of $\epsilon = 0.07$, we take a variable step size $\Delta S = 10l$, where $l = l_i^{1/3} l_o^{2/3}$. At the exit point, we record three components of the position vector($\mathbf{R}$), three components of the direction vector ($\mathbf{T}$), total optical depth τ and the time delay Δt (calculated using equations 21 and 22). In Fig. 1, we present the typical trajectories of the traced rays, embedded in transparent spheres of 1 AU radius. The distributions ($\epsilon = 0$) clearly show that the regular refraction focuses the fundamental into a narrower cone and allows the harmonic to escape into a much wider cone, whereas, the scattering ($\epsilon = 0.07$) severely distorts the ray trajectories all the way from the source to the exit point and causes the destruction of the refractive focusing.

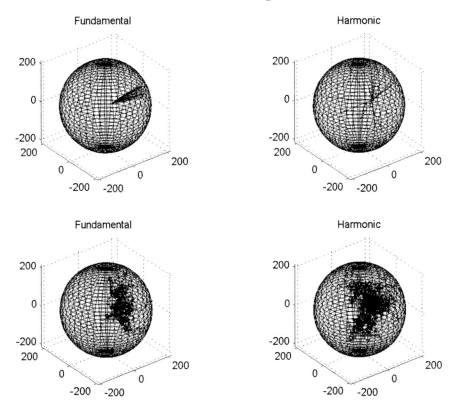

Fig. 1. *Typical distributions of the traced rays of the fundamental and harmonic emissions.*

Directivity

In a spherically symmetric case, the axis through the source and the center of the Sun is also the axis of the cylindrical symmetry. The angle β, subtended at the center by the traced portion of the ray from the source to 1 AU determines the distance of the apparent source from the center of the solar disk. In heliographic degrees, β, which is the longitude for an equatorial source is defined as

$$\beta = \frac{\cos^{-1}(\mathbf{R}.\mathbf{x})}{|\mathbf{R}|}, \tag{35}$$

where, the x-axis is the axis of symmetry, and $\mathbf{R}$ is the position vector at the exit point. The directivity, which is the ratio of the power received (i.e., number of rays) in a range of angles from β to $\beta + d\beta$ from the source embedded in a scattering medium, to the power received from the same source at the same position emitting the same total power isotropically in a vacuum is defined as

$$D(\beta) = \frac{4\pi \Sigma_{n_\beta} e^{-\tau_{i_\beta}}}{\Delta \Omega N_T}, \tag{36}$$

318   G. Thejappa and R. J. MacDowall

where n_β is the number of rays escaping in the angles from β to $\beta+d\beta$, and N_T is the total number of rays, the τ_β is the optical depth (equation 21), and $e^{-\tau}$ represents the losses suffered by the ray. The solid angle $\Delta\Omega$ spanned by grid separation in the β direction around the annular ring is defined as

$$\Delta\Omega = 4\pi \sin[(i_\beta + 0.5)\Delta\beta] \sin(\Delta\beta/2), \tag{37}$$

where $\Delta\beta$ and i_β are the angular width and index of the group, respectively.

We have computed the directivities for $\epsilon = 0$ as well as for $\epsilon = 0.07$ by counting the number of harmonic and fundamental rays in groups of 5 and 1 degree intervals, respectively, and normalized them by dividing by the largest number of rays in each case (see, Fig. 2). The refraction focuses the fundamental and harmonic emissions into narrow and broad cones of ~ 18 and ~ 80 degree angular widths, respectively (first row of Fig. 2). The intense "shoulders" at the edges of the limiting cones probably are due to ingoing rays from the source [14, 15]. For $\epsilon = 0$, the ratio of intensities at 5 and 15 degrees is 0.5 for the fundamental, and the ratio of intensities at 5 and 80 degree longitudes is 0.75 for the harmonic. For $\epsilon = 0$, the fundamental is very intense and directive for an observer located within its limiting cone, indicating that during low level of density fluctuations, the positive identification of the mode of emission as the fundamental, or a mixture of a strong fundamental and a weak harmonic can serve as a good warning signal for an imminent arrival of the flare accelerated electrons or the CME driven shock accelerated electrons at the spacecraft. The directivity diagrams ($\epsilon = 0.07$) of the scattered fundamental and harmonic emissions (second row of Fig. 2) do not show the "shoulders" at the edges of the limiting cones, i.e., the scattering has completely destroyed the shoulders. More importantly, the scattering has widened the fundamental and harmonic limiting cones from ~ 18 to ~ 100 degrees, and from ~ 80 to ~ 150 degrees, respectively, and extended their visibilities beyond 90 degree longitude which would have been completely occulted at the same longitude in the absence of scattering. The symmetry of the directivity diagrams with respect to the radial direction as seen in Fig. 2 is consistent with the observed directivity of type III radio bursts at 3 MHz as reported by [40] with a peak at the central meridian followed by an exponential fall at higher longitudes. The comparison of these directivity diagrams indicates that the fundamental is generally weaker at high longitudes than the harmonic, implying that the type II and type III bursts visible at higher longitudes probably correspond to the harmonic emission. Thus the computed directivities suggest that (1) the second harmonic is easily visible to the widely separated spacecrafts, (2) the type III and type II bursts visible at higher longitudes beyond 40 degrees probably correspond to the second harmonic, and (3) the emissions visible only at low longitudes from 0 to 30 degrees can be a mixture of the fundamental and second harmonic; fundamental being more intense than the harmonic.

Time Profiles

The time taken by the ray to travel from the source to the point of exist is the sum of all the time steps Δt_i as given in equation (22). The histogram of these arrival times at 1 AU gives the time profile. In Fig. 3, we present the time profiles of the fundamental and harmonic emissions constructed using the arrival times of the rays in the longitude range from 0 to 30, and 0 to 90 degrees, respectively. The

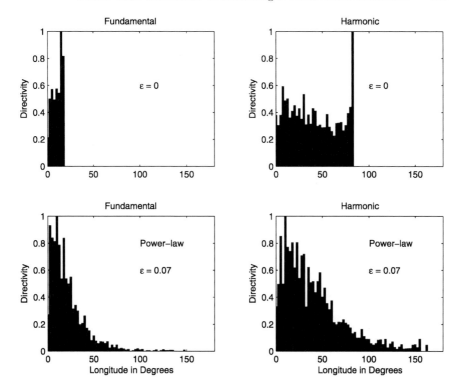

Fig. 2. *The left column shows the directivities of the fundamental and the right column shows the directivities of the harmonic emission. Here ϵ is equal to the level of relative density fluctuations $\frac{\Delta N_e}{N_e}$*

total durations of the unscattered and scattered emissions are 50 (top panel), and ~ 3000 seconds, and ~ 350 and ~ 3000 seconds, for the fundamental and harmonics, respectively. The time profile of both scattered fundamental as well as harmonic emissions look like idealized type III burst time profiles, with total durations of ~ 3000 s comparable to the observed durations of type III bursts which can last for an hour or longer [1]. This indicates that the propagation effects should be removed from the observed time profiles before using them to extract any parameters, such as the duration of the source or the decay constant due to collisional damping [16]. The computed time profiles also show that the fundamental (F) and harmonic (H) emissions arrive at the spacecraft at different times, by indicating that they may contain two peaks. However, the excitation of the harmonic emission is delayed with respect to the fundamental by

$$\Delta t \simeq \frac{r(f_{pe}) - r(f_{pe}/2)}{v_b}, \qquad (38)$$

where v_b is the beam speed, and $r(f_{pe}) = 19.2 f_{pe}^{-0.952}$ AU. For example, a beam with $v_b = \alpha c$ takes $\frac{89.9}{\alpha}$ seconds to travel from $f \sim f_{pe}$ layer located at 0.2097

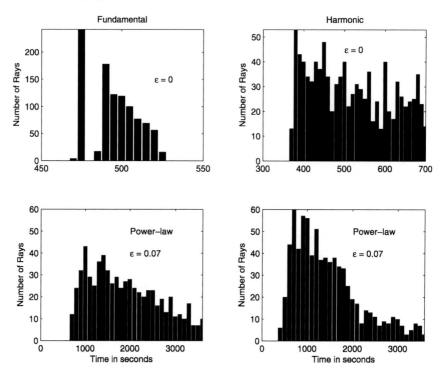

Fig. 3. The computed time profiles of the fundamental (left column) and harmonic emissions (right column) at 120 kHz for various cases (see text), where ϵ is equal to the level of relative density fluctuations $\frac{\Delta N_e}{N_e}$

AU (corresponding to $f_{pe} = 115$ kHz) to the harmonic layer located at 0.3895 AU (corresponding to $f_{pe} = 60$ kHz). Thus, when α takes values of 0.1 and 0.5, the time delay is ~ 15 and ~ 3 minutes, respectively. These values are comparable with the observed time delays of the fundamental-harmonic pairs at high frequencies [41], as well as with those of interplanetary type III radio bursts. For typical beam velocities, the harmonic peak always occurs later than the fundamental. However, due to superposition only a single peak may appear in the time profile, which may correspond to the fundamental at the low longitudes, and to the harmonic at high longitudes. The occurrence of the fundamental emission followed by the harmonic is consistent with observations [42, 43].

Source Size and Displacement

The sizes and displacements of the apparent sources can be determined using the distribution of projected points of the exit points (**R**) of the rays in the angular range β and $\beta + d\beta$ onto a plane passing through the source S and perpendicular to the direction, **T**. The equation of the plane through the source with radius vector R_s and normal to **T** can be written as

$$T_1 x + T_2 y + T_3 z = D, \tag{39}$$

where $D = T_1x_1+T_2y_1+T_3z_1$, $\mathbf{R_s} = (x_1, y_1, z_1)$ and $\mathbf{T} = (T_1, T_2, T_3)$. The projection of the point (x_2, y_2, z_2) on this plane can be obtained from

$$\frac{x-x_2}{T_1} = \frac{y-y_2}{T_2} = \frac{z-z_2}{T_3} = p, \quad (40)$$

as $(pT_1 + x_2, pT_2 + y_2, pT_3 + z_2)$, where p is a parameter. By substituting these coordinates in the equation of the plane (39), we obtain

$$\mathbf{T}.(p\mathbf{T} + \mathbf{R}) = D = \mathbf{T}.\mathbf{R_s}, \quad (41)$$

yielding

$$p = \frac{\mathbf{T}.(\mathbf{R_s} - \mathbf{R})}{\mathbf{T}.\mathbf{T}}. \quad (42)$$

The distributions of projected points (Fig. 4) scattered into the longitude range of 0

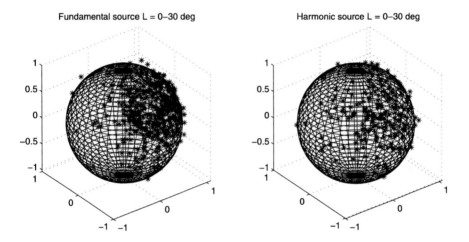

Fig. 4. *The projected images of the scattered fundamental and harmonic sources.*

to 30 degrees represent the sizes of the apparent sources, with altitudes equal to the heliocentric distances of the centroids of these distributions. For example, the 120 kHz fundamental source located at 0.2097 AU (corresponding to $f_{pe} = 115$ kHz) is displaced inward to the radial distance of 0.2033 AU (the critical layer corresponding to ~ 120 kHz) when $\epsilon = 0$, and displaced radially outward to a distance of 0.5950 AU (critical layer corresponding to 38.8 kHz) when $\epsilon = 0.07$. Thus, the radial distance of the apparent source corresponds to $\sim \frac{f}{3}$ layer, agreeing very well with the observed heights of $f/2$ and $f/5$ layers for type III radio bursts [2]. The computed centroid of the apparent harmonic source is displaced inwards from 0.3895 AU corresponding to 60 kHz plasma level to 0.3152- 0.3329 AU, and to 0.3993 AU corresponding to $\sim \frac{f}{2}$ critical level when $\epsilon = 0$, and $\epsilon = 0.07$, respectively.

When $\epsilon = 0$, the size of the apparent fundamental source, computed as the half-power widths of these distributions is 0.2429, and 1.1233 degrees parallel and

322   G. Thejappa and R. J. MacDowall

perpendicular to the radial direction. When $\epsilon = 0.07$, this size is increased respectively to 24 and 27 degrees. When $\epsilon = 0$, the size of the apparent harmonic source is ~ 6.5, and ~ 2 degrees along and across the radial directions in both longitude ranges, and increased to 36 and 38 degrees when $\epsilon = 0.07$. These computed sizes and heights agree very well with the observations of [1].

Comparison with Observations

In Fig. 5, we present an example of a multi spacecraft detection of a type II and a couple of type III radio bursts by the Unified Radio and Plasma Wave (URAP) experiment on Ulysses [44] and the Waves investigation on Wind [45]. Ulysses is in a highly elliptical orbit out of the ecliptic plane with aphelion (perihelion) at ~ 5.4 AU (~ 1.3 AU), the trajectory of the Wind takes it from near Earth orbits to the Lagrange point (L1), about 230 R_E upstream of Earth. The Ulysses data presented in the top panel show an intense type III burst after 12:00 on 1997/11/6 and several other weaker type III bursts. The data early on 1997/11/6 are corrupted by a poor telemetry link. The type II emission is the weaker activity (see color bar scale) starting at 18:00 and continuing to 12:00 on the next day while drifting from 200 to 100 kHz. The flare site related to these events was at S18 and W63. The bottom panel shows similar data from the Wind spacecraft, where the same type II and type III bursts are detected slightly earlier because Wind is closer to the Sun (at 1 AU near the Earth) than Ulysses (at 5.3 AU). The additional emission features in the Wind data are Auroral kilometric Emission and the electron thermal noise (the horizontal feature seen in the bottom half of the panel). In the middle panel, single frequency data from the two spacecraft at ~ 120 kHz is plotted. The signal levels are not the same at the two spacecraft because the radio bursts are directive, and the distances of the sources are different for different spacecrafts. However, the time profiles are remarkably similar at both spacecraft. During these events the two spacecraft were separated by more than 100 degrees in heliographic longitude. Note that the Ulysses data plotted in the middle panel have been shifted in time by about 35 minutes to correct for the longer propagation distance to Ulysses. (ULYSSES: Heliographic latitude = 2.0 degrees, Heliographic longitude = 53.9 degrees, Range to Sun = 5.3 AU; EARTH/WIND: Heliographic latitude = 3.8 degrees, Heliographic longitude = 301.2 degrees, Range to Sun = 1.0 AU). The distributions of rays emitted by the fundamental and harmonic sources located at (S18, W63) at altitudes of 0.2050 and 0.3895 AU for $\epsilon = 0$ as well as for $\epsilon = 0.07$ are shown in Fig. 6, where the locations of the Ulysses and Wind spacecraft are also shown. It is clear from these distributions that (1) when $\epsilon = 0$ the fundamental is highly beamed and visible only to Ulysses spacecraft, (2) when $\epsilon = 0.07$ the scattering causes the fundamental to be visible to both spacecraft, and (3) the harmonic emission is visible to both spacecraft for $\epsilon = 0$ as well as for $\epsilon = 0.07$. Thus, except for the refracted fundamental, the rest of the emissions, namely scattered fundamental, unscattered harmonic as well as scattered harmonic can be visible to both Ulysses and Wind spacecraft. This indicates that the visibility depends on the coordinates of their sources.

3 Quiet Sun Component

The brightness temperature of the quiet Sun radio emission is computed as

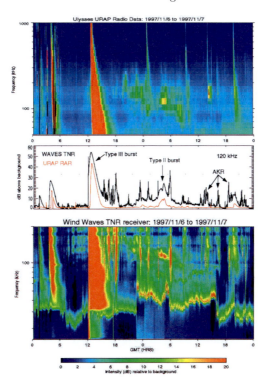

Fig. 5. *Top: In the top panel, Ulysses URAP data show an intense type III burst after 12:00 on 1997/11/6, as well as several other weaker type IIIs. The type II emission is the weaker emission band (see color bar scale) from 18:00 to 12:00 on the next day and drifting from 200 to 100 kHz. The bottom panel shows similar data from the Wind Waves instrument. In the middle panel, single frequency data from the two spacecraft at approximately 120 kHz is plotted. Because of differing distances from the sources to the spacecraft as well as the effects of directivity, the signal levels seen for a given emission is different at the two spacecraft. The two spacecraft are separated by more than 100 deg in heliographic longitude, providing an ideal angular separation for studying these events.*

$$T_b = T_e(1 - e^{-\tau}). \tag{43}$$

Here $\tau = \int_{s_1}^{s_2} \xi ds$ is the optical depth, where s_1 and s_2 are the heliocentric distances of the source and the observer, respectively. The $\xi = \frac{f_{pe}^2 \nu}{f^2 \mu c}$ is the absorption coefficient per centimeter of path length. The optical depth τ is calculated by tracing the rays (initially launched toward that point). The rays are traced from a distance of $2.5 R_\odot$ toward the Sun until τ reaches a large value of ~ 10, or the ray is traveled at least $5 R_\odot$. The rays are launched only in the equatorial plane, where the x-axis

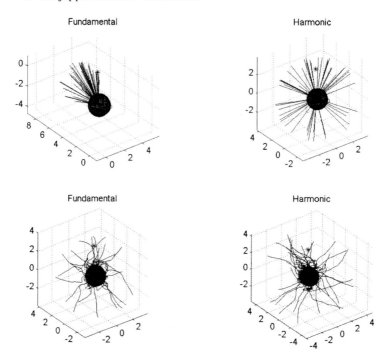

Fig. 6. *The typical distributions of the refracted and scattered rays from the actual source location until they reach distances of Ulysses spacecraft. The locations of Ulysses and Wind are also shown as * and o, respectively, in these distribution diagrams.*

is directed toward the observer. Here, the positive y direction represents the west longitude. In Fig. 7, we present the typical trajectories of the traced rays, where red and blue trajectories correspond respectively to the cases, where only the refraction is considered, and when both refraction as well as scattering are considered. It is interesting to note that the scattered rays in the top panel turn back before reaching the critical layer, i.e., much earlier than the refracted rays, by indicating that the scattering raises the apparent East-West diameter of the Sun at 34.5 MHz. On the other hand, the turning points of the scattered rays in the bottom panel almost coincide with the critical layer, similar to the refracted rays, by indicating that the scattering may not affect the apparent size of the radio sun at 73.8 MHz.

To calculate the brightness temperatures for different longitudes, corresponding to different values of y, we trace fifty rays at intervals of 0.25 solar radii for each y, and calculate the average brightness temperature using the computed total optical depth of the ray. By the principle of reciprocity, this value represents the brightness temperature of thermal emission from that point on the disk. The error bars are estimated by using the variance of the measures contributing to the mean. In Fig. 8, we present the distributions of the brightness temperature at 50 MHz, where the solid curve corresponds to the case, where only the refraction is considered, and the points with error bars correspond to the case, where both the refraction and scat-

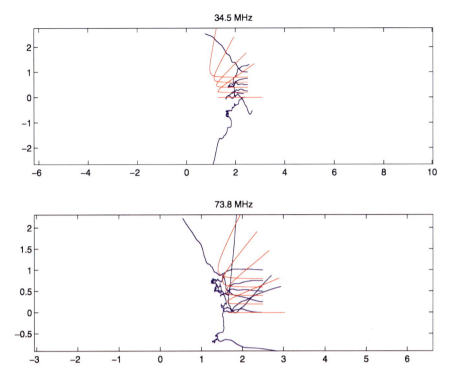

Fig. 7. *Typical ray trajectories traced in the refracting (red) and refracting and scattering (blue) non-spherical symmetric corona. The rays are initially directed toward points on the solar disk in the intervals of* $0.25 R_\odot$.

tering are included in the calculations. These brightness temperature distributions resemble very much to those computed by previous authors [22, 5, 6]. The half-power angular diameter (angular width at half-maximum) in units of arc minutes calculated using a cubic polynomial interpolation technique is $\sim 48'$ at 50 MHz when only the refraction is included, and it is increased to $\sim 56'$ when the scattering is added in the calculations. This value compares reasonably well with observed value of $50'$ within the limits of error bars. The brightness temperature is reduced from 9×10^5 K ($\epsilon = 0$) to $\sim 2.3 \times 10^5$ K when the scattering was included, i.e., T_B is reduced almost by 50%. The central brightness temperatures (zero longitude, or $y = 0$) for four different frequencies (30.9, 50, 60, 73.8 MHz) are presented as the computed spectrum in Fig. 9. The shape of the spectrum of thermal emission is preserved in the considered frequency range, even in the presence of density fluctuations by remaining almost steady at these frequencies in the limits of error bars. The central brightness temperatures are 3.5×10^5 K, 1.4×10^5 K, 1.7×10^5 K, and 1.4×10^5 K at 30.9, 50, 60 and 73.8 MHz, respectively.

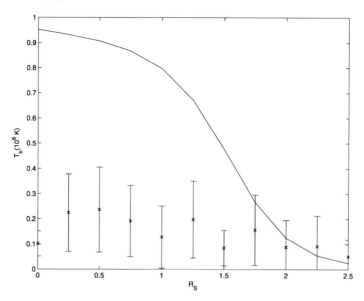

Fig. 8. Brightness temperature (T_B) distributions for 50.0 MHz radiation. The error bars correspond to the r.m.s deviation from the mean of the computed T_B for individual rays.

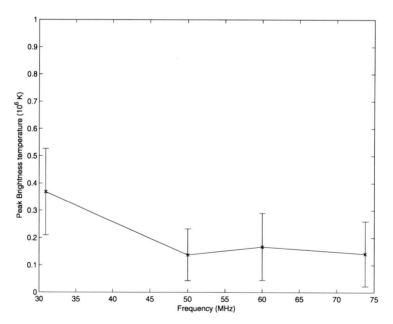

Fig. 9. The brightness temperature (T_B) spectrum of the quiet Sun

4 Conclusions

The Monte Carlo simulations show that: (1) the widespread visibility of radio bursts is due to scattering of radio waves by density fluctuations, (2) the scattered fundamental and harmonic emissions produce time profiles which look very much like the idealized type III radio bursts indicating that the duration of the beam and collisional decay constants can be extracted from the observations only if they are corrected for the propagation effects, (3) the identification of the emission modes in the type III burst time profile, namely the fundamental in the rise part followed by the harmonic in the decay part of the time profile can be accounted for by the scattering, (4) the sizes and heights of the apparent sources derived using the distributions of scattered rays from the isotropic point sources agree very well with observed values, (5) the scattering at meter-decameter wavelengths leads to a considerable reduction in the central brightness temperatures, (6) although scattering causes the reduction in the central brightness temperatures, the resultant spectrum, i.e., the peak brightness temperature as a function of the frequency remains very similar to the thermal spectrum of electromagnetic radiation, and (7) by knowing the density distribution, and the parameters of density fluctuations during the radio observations, we can determine the electron temperatures of the solar corona using the Monte Carlo simulations.

The Monte Carlo simulation method developed in these studies can be used to study the propagation of waves in any refracting and scattering medium. These techniques can be very useful to calculate the propagation related corrections in the direction finding algorithms, used by the STEREO spacecraft to track the radio sources, as well as to extract the physical parameters, such as the speed and dispersion of the electron beams, electron temperatures, arrival times of the CME driven shocks and flare electrons etc.

References

1. Steinberg, J.-L., Dulk, G. A., Hoang, S., Lecacheux, A., & Aubier, M. G. 1984, A&A, 140, 39.
2. Steinberg, J.-L., Hoang, S., & Dulk, G. A. 1985, A&A, 150, 205.
3. Lecacheux, A., Steinberg, J.-L., Hoang, S., & Dulk, G. A. 1989, A&A, 217, 237.
4. Thejappa, G., MacDowall, R. J., & Kaiser, M. L. 2007, ApJ, 671, 894.
5. M. Aubier, Y. Leblanc, & A. Boischot, A. 1971, A&A, 12, 435.
6. Thejappa, G., & Kundu, M. R. 1992, Sol. Phys., 140, 19.
7. Thejappa, G., & Kundu, M. R. 1994, Sol. Phys., 149, 31.
8. Thejappa, G. & MacDowall, R.J., April 2008, ApJ, v.
9. Jaeger, J. C., & Westfold, K. C. 1950, Austr. J. Res., (A), 2, 322.
10. Smerd, S.F. 1950, Australian J. Sc., Res., A3, 34.
11. Bracewell, R.N., & Preston, G.W. 1956, ApJ, 123, 14.
12. Hollweg, J. 1968, AJ., 73, 972.
13. Fokker, A.D. 1965, Bull. Astr. Inst. Netherlands, 18, 111.
14. Steinberg, J.-L., Aubier-Giraud, M., Leblanc, Y., & Boischot, A. 1971, A&A, 10, 362.
15. Steinberg, J. L. 1972, A&A, 18, 382.
16. Riddle, A. C. 1974, Sol. Phys., 35, 153, 1974.

328   G. Thejappa and R. J. MacDowall

17. Cairns, I. H. 1998, ApJ, 506, 456.
18. Lee, L. C., & Jokipi, J. R. 1975, ApJ, 196, 695.
19. Rickett, B. J. 1977, Ann. Rev. Astr. Astrophys., 15, 479.
20. Rytov, S.M., Kravtsov, Yu. A., & Tatarskii, V. I. 1989, Principles of Statistical Radiophysics. vol. 4. Wave Propagation Through Random Media. Springer-Verlag, 1989.
21. Bastian, T. S. 1994, ApJ, 426, 774.
22. Riddle, A. C. 1974, Sol. Phys., 36, 375.
23. Hoang, S., & Steinberg, J. L. 1977, A&A, 58, 287.
24. Bougeret, J.-L., King, J. H., & Schwenn, R. 1984, Sol. Phys., 90, 401.
25. Guhathakurta, M., Holzer, T. E., & MacQueen, R. M. 1996, ApJ, 458, 817.
26. Coles, W. A., & Harmon, J. K. 1989, ApJ, 337, 1023.
27. Coles, W. A., Liu, W., Harmon, J. K. & Martin, C. L. 1991, J. Geophys. Res., 96, 1745.
28. Efimov, A. I., Chashei, I. V., Bird, M. K., Samoznaev, L. N., & Plettemeier, D. 2005, Astro. Rep., 49, 485.
29. Spangler, S. R. 2002, ApJ, 576, 997.
30. Tu, C. Y., & Marsch, E. 1994, JGR, 99, 21, 481.
31. Spangler, S. R., & Sakurai, T. 1995, ApJ, 445, 999.
32. Spangler, S. R., Kavars, D. W., Kortenkamp, P. S., Bondi, M., Mantovani, F., & Alef, W., 2002, A&A, 384, 654.
33. Woo, R., Armstrong, J. W., Bird, M. K., & Patzold, M. 1995, Geophys. Res. Lett., 22, 329.
34. Wohlmuth, R., Plettemeier, D., Edenhofer, P., Bird, M. K., Efimov, A. I., Andreev, V. E., Samoznaev, L. N., & Chashei, I. V., 2001, Space Sci. Rev., 97, 9.
35. Manoharan, P. K., Ananthakrishnan, S., & Rao, A. P. 1988, Proc. Sixth International Solar Wind Conf. Vol. 1 (Boulder, NCAR), 55.
36. Bavassano, B., & Bruno, R. 1995, JGR, 100, 9475.
37. Haselgrove, J. 1963, J. Atmos. Terr. Phys., 25, 397.
38. Sharma, A., Vizia Kumar, D., Ghatak, A.K. 1982, Appl. Opt., 21.
39. Chandrasekhar, S. 1952, MNRAS, 112, 475.
40. Fainberg, J., & Stone, R. G. 1970, Sol. Phys., 15, 222.
41. Caroubalos, C., & Steinberg, J. L. 1974, A&A, 32, 245.
42. Kellogg, P. J. 1980, ApJ, 236, 696.
43. Dulk, G. A., Steinberg, J. L., & Hoang, S. 1984, A&A, 141, 30.
44. Stone, R. G., et al., 1992, A&AS, 92, 291.
45. Bougeret, J.-L., et al., 1995, Space Sci. Rev., 71, 231.

Evolution of Magnetic Helicity in NOAA 10923 Over Three Consecutive Solar Rotations

Sanjiv Kumar Tiwari, Jayant Joshi, Sanjay Gosain and P. Venkatakrishnan

Udaipur Solar Observatory, Physical Research Laboratory, P. Box - 198, Dewali, Bari Road, Udaipur-313 001, Rajasthan, India
stiwari@prl.res.in

Summary. We have studied the evolution of magnetic helicity and chirality in an active region over three consecutive solar rotations. The region where it first appeared was named NOAA10923 and in subsequent rotations it was numbered NOAA 10930, 10935 and 10941. We compare the chirality of these regions at photo-spheric, chromospheric and coronal heights. The observations used for photospheric and chromospheric heights are taken from Solar Vector Magnetograph (SVM) and H-α imaging telescope of Udaipur Solar Observatory (USO), respectively. We discuss the chirality of the sunspots and associated H-α filaments in these regions. We find that the twistedness of superpenumbral filaments is maintained in the photo-spheric transverse field vectors also. We also compare the chirality at photospheric and chromospheric heights with the chirality of the associated coronal loops, as observed from the HINODE X-Ray Telescope.

Key words: Sun : Helicity – chirality – superpenumbral whirls – sigmoids

1 Introduction

Magnetic fields exhibit chirality and is observed in most of the solar features like filament channels, filaments, sunspots, coronal loops, coronal X-Ray arcades and interplanetary magnetic clouds (IMCs) ([1], [2], [3] and the references therein). First of all G.E. Hale in 1925 ([4]) observed vortices in H-α around sunspots and he called these features as 'sunspot whirls'. He investigated the data extending over three solar cycles and found no relationship between the direction of these vortices and the polarity of the sunspots. Also, he found no reversal of the whirl direction together with the general reversal of the sunspot polarities with cycle. He found that about 80 % of sunspot whirls have counterclockwise orientation in the northern hemisphere and clockwise in the southern hemisphere, now known as the helicity hemispheric rule. Richardson ([5]) confirmed Hale's results after doing the same type of investigation with the data for four solar cycles. Seehafer ([1]) also found that the negative helicity is dominant in the northern hemisphere and positive in the southern hemisphere. This is known as the hemispheric helicity rule and is

330   S. Tiwari et al.

independent of the solar cycle. Since 90s the subject is highly taken into account by most of the researchers in the field. Pevtsov, Canfield and Metcalf ([2]) demonstrated the existence of chirality in the active regions, after analyzing vector magnetograms from Mees Solar Observatory. Although, the entire active region may not show the same type of chirality everywhere yet a dominant sense of chirality can be found for most of the active regions.

The chirality of the H-α filament can be directly recognized by looking at the filament barbs. If the orientation of the barbs is counterclockwise the chirality of the filament is known to be dextral and if it is clockwise the chirality is sinistral. Martin ([6]) mentioned that the chirality of the solar features can be used for resolving 180 degree azimuthal ambiguity in the solar vector magnetic field. It is believed that there is one-to-one correspondence between the filament chirality and the magnetic helicity sign. A right-handed twist and a clockwise rotation of the loops, when viewed from the above implies positive helicity or chirality and vice-versa. The magnetic helicity is a quantitative measure of the chiral properties of the solar magnetic structures ([7], [8]). It is given by a volume integral over the scalar product of the magnetic field B and its vector potential A.

$$H = \int \mathbf{A} \cdot \mathbf{B} dV. \tag{1}$$

It is well known that the vector potential A is not unique. Thus the helicity can't immediately be calculated from the above given equation. Seehafer ([1]) pointed out that the helicity of magnetic field can best be characterized by force-free parameter alpha also known as the helicity parameter. The force free condition is given as

$$\nabla \times \mathbf{B} = \alpha \mathbf{B}. \tag{2}$$

Taking the z-component of the magnetic field

$$\alpha = (\nabla \times \mathbf{B})_z / \mathbf{B}_z \tag{3}$$

The magnetic helicity density can be given as

$$H_m = B^2 / \alpha \tag{4}$$

but except for potential fields.

And the current helicity density will be given in terms of alpha as

$$H_c = B^2 \alpha \tag{5}$$

In this paper, we identify the chirality of a sunspot in an active region named NOAA 10923 when it first appeared and NOAA 10930, 10935, 10941 in successive rotations. The associated H-α filaments in the three consecutive solar rotations were obtained from the Udaipur Solar Observatory (USO) high resolution H-α data. We have calculated the helicity parameter of the sunspots NOAA 10935 and NOAA 10941 and we found that the value of helicity parameter has increased after one solar rotation. These active regions are not following the helicity hemispheric rule. There are theories ([9], [10],[11]) which discuss about the 'wrong' sign of helicity in the beginning of the solar cycle. Sokoloff et al ([12]) observes a significant excess of active regions with the 'wrong' sign of helicity just at the beginning of the cycle. But there is no discussion found about the 'wrong' sign of helicity during the end of the solar cycle. The sign of helicity is supposed to follow the helicity hemispheric rule which our result doesn't show. We compare the helicity of the active regions with the sign of associated sigmoids obtained from the Hinode.

2 Data and Instruments used

We use high resolution H-α images taken from Udaipur Solar Observatory (USO) from Spar Telescope and the vector magnetograms from the USO Solar Vector Magnetograph (SVM) ([13], [14]). The Spar Telescope uses 1392×1024 ccd with the pixel resolution of 0.395 arcsec and the SVM has 1024×1024 ccd with pixel resolution of 0.98 arcsec. Our H-α observations, after the monsoon break, started on 23rd November 2006. So we don't have USO H-α image of the NOAA 10923 in its first appearance. After one rotation we observe the same sunspot in the name of NOAA 10930 on Dec 11 2006. And in other two consecutive rotations we have observed the same sunspot in the name of NOAA 10935 on 09 Jan 2007 and NOAA 10941 on 06 Feb 2007. We use the available vector magnetograms of NOAA 10935 (09 Jan 07) and NOAA 10941 (06 Feb 07) and compare the vectors with the whirls of the sunspot images taken in H-α wavelength.

The Spar Telescope has f/15 doublet lens with focal length of 2.25 meters and objective 0.15 meters. It uses a H-α Halle lyot type filter with FWHM of 500 mÅ operating at the wavelength of 6563 Å. The telescope utilizes a 1392×1024 CCD with the pixel size of 6.45 μm. The pixel resolution of the CCD is 0.395 arc-sec and the field of view it covers is 9 arc-min × 7 arc-min. The H-alpha images of the active regions NOAA 10930, NOAA 10935 and NOAA 10941 are shown in the Figure 1.

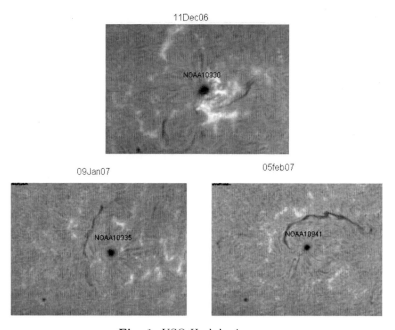

Fig. 1. USO H-alpha images

The magnetograms are taken from Solar Vector Magnetograph (SVM), which has recently become operational at USO. SVM is basically an instrument which makes two-dimensional spatial maps of solar active regions in the Zeeman induced polarized

332   S. Tiwari et al.

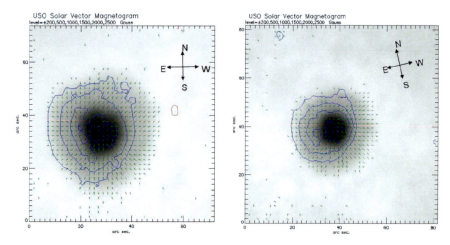

Fig. 2. USO solar vector magnetograms (09Jan07 and 06Feb07 respectively)

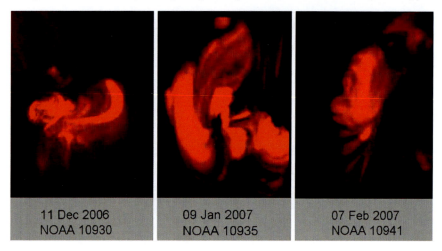

Fig. 3. XRT (Hinode) data reverse-S sigmoid showing the dextral chirality in comparison with that of photospheric and chromospheric data.

light of the solar spectral lines. SVM has the following main components : a Schmidt-Cassegrain telescope tube, rotating wave-plate polarimeter, tunable narrow-band Fabry-Perot filter, calcite analyzer (Savart plate) and a cooled CCD camera. The primary imaging is done by using a Celestron C-8 (TM) Schmidt-Cassegrain telescope of 8 inch aperture. The focal length of the telescope is 2032 mm and the resulting output beam is a f/10 beam. The telescope has a pre-filter in front with a 15nm pass-band centered at 630nm wavelength. A circular aperture of 2 arc-min diameter selects the field of view (FOV) at the prime focus. This FOV is then modulated by the rotating waveplates of the polarimeter. The modulated beam is now collimated by a 180mm focal length lens. This modulated and collimated beam now enters the Fabry-Perot etalon and order sorting pre-filter. Now the re-imaging lens makes the image on the CCD camera. Just before the CCD camera a combination of two crossed calcite beam-displacing crystals is placed for the analysis of polarization. So we get two orthogonal polarized images of the selected FOV onto the CCD camera. The vector magnetograms of the two active regions NOAA 10935 and NOAA 10941 are shown in the Figure - 2. We have taken the Hinode (XRT) data for looking at the sigmoidal structure of corresponding active regions.

3 Analysis

We observed the sunspot NOAA 10923 (Nov 06) which sustains in the three consecutive solar rotation in the name of NOAA 10930 (Dec 06), NOAA 10935 (Jan 07) and NOAA 10941 (Feb 07). There was no major activity associated with the active region NOAA 10923. Looking at the NOAA 10930 in H-Alpha (Fig(1)) we ensure that there is no particular orientation of the whirls and we can't recognize the helicity of the sunspot. There was no filament seen associated with this active region. There were X3.4 (at 02:00 UT) and X1.5 (at 21:07 UT) class flares as well as strong CMEs associated with the active region NOAA 10930 on 13 December 2006. After next rotation the same sunspot is found in the name of NOAA 10935 (Jan 07) (Fig(1)). We now observe the whirls with counter-clockwise orientation are dominating. A filament associated with the active region is observed. The end of this active region filament is curving towards the sunspot with counter - clockwise whirls and according to Rust and Martin ([15]) it should (not necessarily) be dextral. Still in the image here the orientation of the filament barbs are not clearly recognized to be dextral. There was not any major event associated with this active region.

In the next rotation we are able to recognize clearly the orientation of the filament barbs. The active region NOAA 10941 (Feb 07) has come in its third consecutive rotation of the sun. It has the barbs with counter - clockwise orientation close to filament and this type of chirality is dominating in the whole active region. The associated filament is dextral is clear now. No event was observed associated with this active region. We calculated the helicity parameter for the sunspots NOAA 10935 and NOAA 10941 which comes out to be $-1.1\pm0.12 \times 10^{-9}\ m^{-1}$ and $-3.77\pm2.14\times10^{-8}m^{-1}$ respectively. The value of helicity has increased after next rotation which can also verified by directly looking at the H-alpha images. The helicity sign doesn't support the helicity hemispheric rule. Both the active regions are found in the southern hemisphere and should bear positive chirality. For the calculation of helicity we have calculated the helicity parameter alpha best which will give one value of the alpha for the whole sunspot instead of the value at each pixel. It reduces the noise in the data. We use our SVM data to plot the transverse field upon the associated H-α image. First of all, the H-α data is re-binned according to the size of SVM data and then the related vector field is plotted over the H-α image. The 180 degree azimuthal ambiguity has been resolved by using acute angle method. We can see in Figure - 2 that the direction of H-α super-penumbral whirls show the same chirality as the photospheric transverse field vectors. So, by combining the photospheric and chromospheric data one can use the method of chirality to resolve 180 degree azimuthal ambiguity. Also by looking at the HINODE (XRT) data we find the reverse-S structure in the sigmoids associated with the active regions.

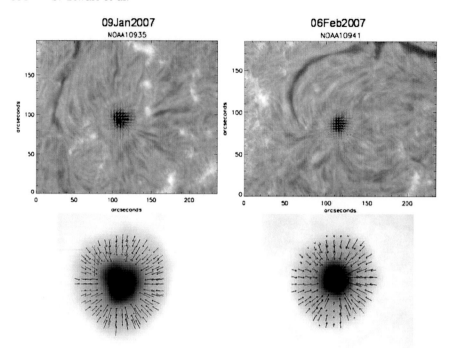

Fig. 4. Plot of vector magnetic fields of the sunspots upon the respective H-alpha images. In the lower part of the image blue arrows show the radial direction and the red arrows show the actual vector field direction. We can see the shear of the field.

4 Conclusion and Discussion

We find that the active region NOAA 10923 in its different appearances doesn't follow the hemispheric helicity rule, neither its associated filaments. But their association in the terms of chirality follow the well known result of Rust and Martin ([15]). We calculated the helicity parameter which was found to be negative as expected from the sunspot with dextral whirls. The helicity increases in the last appearance but there was no major activity observed associated with the active region in its last appearance. Hinode (XRT) images also show the same type of chirality in the associated sigmoids. Thus the sign of helicity derived from photosphere, chromosphere and corona are strongly correlated. Figure - 3 shows the vector fields plotted over the associated active regions in the chromospheric H-α observations. We find the good matching of the vectors with the super-penumbral whirls of the active regions. By combining the photospheric and chromospheric data the axial field direction at neutral line can be derived and this method can be used to resolve 180 degree azimuthal ambiguity. Sara F. Martin et al ([6]) has already mentioned that the 180 degree azimuthal ambiguity can be resolved by using this method of chirality.

Acknowledgements

Hinode is a Japanese mission developed and launched by ISAS/JAXA, collaborating with NAOJ as a domestic partner, NASA and STFC (UK) as international partners. Scientific operation of the Hinode mission is conducted by the Hinode science team organized at ISAS/JAXA.

This team mainly consists of scientists from institutes in the partner countries. Support for the post-launch operation is provided by JAXA and NAOJ (Japan), STFC (UK), NASA (USA), ESA, and NSC (Norway). One of us (Jayant Joshi) acknowledge financial support under ISRO/CAWSES – India programme.

References

1. Seehafer, N., 1990, Solar Phys., **125**, 219.
2. Pevtsov, A., Canfield, R. C. and Metcalf, T. R., 1995, ApJ, **440**, L109
3. Martin, S. F., 1998, ASP Conference Series, **150**, 419.
4. Hale, G. E., 1925, Publ. Astron. Soc. Pacific, **37**, 268.
5. Richardson, R. S., 1941, Astrophys. J., **93**, 24
6. Martin, S. F., Lin, Y., Engvold, O., 2006, American Astronomocal Society, SPD Meeting, **37**, 129
7. Berger, M. A. and Field, G. B., 1984, J. Fluid Mech., **147**, 133.
8. Berger, M.A. 1999, Magnetic Helicity in Space and Laboratory Plasmas, Geophys. Monograph, Vol. **111**, American Geophysical Union, p. 1.
9. Choudhuri, A. R., et al., 2004, Astrophys. J., **615**, L57.
10. Zhang, H., et al., 2006 MNRAS, **365**, 276 (Paper I)
11. Chaterjee, P. 2006, J. Astron. Astrophys., **27**, 89.
12. Sokoloff, D., 2006, Astronomische Nachrichten, **327, Issue 9**, 876
13. Gosain, S., Venkatakrishnan, P. and Venugopalan, K. 2005, Exp. Astron., **18**, 31.
14. Gosain, S., Venkatakrishnan, P. and Venugopalan, K. 2006, J. Astrophys. Astron., **27**, 285.
15. Rust, M.D. and Martin, S.F. 1994, ASP Conference Series, **68**, 337.

Stability of Double Layer in Multi-Ion Plasmas

A.M. Ahadi[1], S. Sobhanian[2]

[1] Physics Department, Shahid Chamran University, Ahvaz, Iran
  `E.Mail:ahadi.am@gmail.com`
[2] Faculty of Physics, Tabriz University, Tabriz, Iran
  `E-Mail:sobhanian@tabrizu.ac.ir`

Summary. Langmuir condition shows the existence of a net current through the double layer (DL). Using Langmuir criterion, one may determine the double layer's position, but it can not represent alone if this position is stable or not. We obtain the generalized Langmuir condition in multi-ion plasmas by using a simple model. Then by analyzing the results, the stability of the determined position is discussed and the DL stability criterion is obtained. This criterion assures that, when some perturbations are imposed on the DL, it returns to the stable position.

Key words: Double Layer, Stability, Multi-Ion Plasmas

1 Introduction

The study of Double layers (DLs) is similar to that of plasma sheath. Considering their special properties and their role as an accelerating system, they have found to be attractive research interests for several years. Even today, their formation mechanisms and their effects on different plasma properties constitute one of the advanced research branches in plasma physics ([1]-[7]). Double layer (DL) structure is determined in a self-consistent behavior by the particle dynamics in the electric field set up by the net charge distribution produced by the particles [2]. The power supplied to accelerated particles is simply the product of the current crossing the double layer and the potential across it [2].

Langmuir showed in 1929 for the first time that for establishing the pressure equilibrium in electron-ion plasma, the condition $\psi_i/\psi_e = \sqrt{m_i/m_e}$ must be satisfied [8]. Later, Carlqvist in 1979 and 1984 extended Langmuir results to the case of relativistic DLs [9, 10].

Computer simulations [12, 13] and experimental investigations [14, 15] show that when a current inequilibrium exists in DLs, Langmuir condition would not hold there in the rest frame, so they have to set themselves to motion. To compensate the inequilibrium and in this way to settle down the stability in current and to restore the Langmuir condition is satisfied. Although DL exhibits fluctuations produced by oscillations of the flowing current through plasma, but for strong DL, these fluctuations are relatively small [16].

In this paper attempt has been done to study a strong moving DL in a multispecies plasma and to obtain the generalized Langmuir condition. This kind of plasma with strong DLs is observed in the astrophysical plasmas. By analyzing the obtained results, some criteria are found for the stability of these DLs. We employ here the method used by Song [16].

2 Model and Basic Equation

We consider a DL with a large transverse potential fall and assume that its thickness is small compared with other dimensions. We showed this in Fig. 1. In general

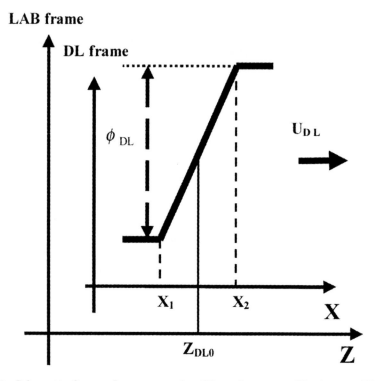

Fig. 1. Schematic figure of strong moving DL and two possible frames. The DL moves toward high potential side. Its boundaries are at x_1 and x_2 in moving frame and initial equilibrium position is z_{DL0} in lab frame.

condition the DL can move, so to study it we need two systems: the laboratory system of reference and the moving frame, which is assumed to be attached to the DL. If the relative velocity in the laboratory system is given by U_{DL}, the following relation will be held between the position coordinates:

$$x = z - \int U_{DL} dt \tag{1}$$

In order to find the Langmuir condition and the location of the DL formation, we assume that the plasma is under the action of an inhomogeneous magnetic field, that it is confined by a magnetic flux tube with cross section S. We start with the equation of continuity and momentum equations:

$$n_\alpha m_\alpha \left(\frac{\partial u_\alpha}{\partial t} + u_\alpha \frac{\partial u_\alpha}{\partial z} \right) + \rho_\alpha m_\alpha u_\alpha + \frac{\partial (n_\alpha T_\alpha)}{\partial z} = n_\alpha q_\alpha E \tag{2}$$

$$\frac{\partial (S n_\alpha u_\alpha)}{\partial z} = S \rho_\alpha - S \frac{\partial n_\alpha}{\partial t} \tag{3}$$

where the index α represent the particles type and ρ_α and u_α are the rate of production of the particles and particles velocity of type α, respectively. Also m_α indicate mass of particles and and T_α is the particle temperature.

Expressing the particles velocity, density and electric field in moving reference frame as functions of the corresponding quantities in the rest reference frame, we arrive at the following equation:

$$P_i = P_i(z - \int_0^t U(t')dt') = P_i(z,t) \tag{4}$$

where P_i could be velocity, particle density or the electrical field.

3 Generalized Langmuir Condition

Using the mentioned transformation in equation (4), equations (2) and (3) are rewritten in the moving frame as:

$$m_\alpha n_\alpha v_\alpha \left(\frac{\partial v_\alpha}{\partial x} \right) + \rho_\alpha m_\alpha v_\alpha + \rho_\alpha m_\alpha U_{DL} + \frac{\partial (n_\alpha T_\alpha)}{\partial x} = n_\alpha q_\alpha E \tag{5}$$

$$\frac{\partial (n_\alpha v_\alpha)}{\partial x} + n_\alpha u_\alpha \frac{\partial (\ln S)}{\partial x} = \rho_\alpha , \tag{6}$$

where we have made use of $v_\alpha = u_\alpha - U_{DL}$.

We can combine, in the moving reference frame (DL frame) the equations of continuity and momentum. Thus, we will have for each species:

$$m_\alpha n_\alpha v_\alpha u_\alpha \left(\frac{\partial \ln S}{\partial x} \right) + \rho_\alpha m_\alpha U_{DL} + \frac{\partial (m_\alpha n_\alpha v_\alpha^2 + n_\alpha T_\alpha)}{\partial x} = n_\alpha q_\alpha E \tag{7}$$

On the other hand, Poisson equation makes possible to express the role of particles in one equation, we have $\frac{\partial E}{\partial z} = \frac{\partial E}{\partial x}$, so that:

$$\epsilon \left(\frac{\partial E}{\partial x} \right) = \sum_\alpha q_\alpha n_\alpha \tag{8}$$

Multiplying both sides of this equation by E and using equation (7), we can reach to:

$$\epsilon E(\frac{\partial E}{\partial x}) = \sum_{\alpha}(m_\alpha n_\alpha v_\alpha u_\alpha(\frac{\partial \ln S}{\partial x}) + \rho_\alpha m_\alpha U_{DL} + \frac{\partial(m_\alpha n_\alpha v_\alpha^2 + n_\alpha T_\alpha)}{\partial x}) \quad (9)$$

We can now integrate equation (9), and write:

$$\sum_{\alpha} \int_{x_1}^{x_2} ((m_\alpha n_\alpha v_\alpha u_\alpha(\frac{\partial \ln S}{\partial x}) + \rho_\alpha m_\alpha U_{DL})dx + \sum_{\alpha}(m_\alpha n_\alpha v_\alpha^2 + n_\alpha T_\alpha)]_{x_1}^{x_2} = 0 \quad (10)$$

Since we have supposed that the DL is strong, the acquired energy by particles is much bigger than their thermal fluctuation. Neglecting small terms of second order in equation (10), and reminding that electrons are accelerated in opposite directions relative to positive ions, we have:

$$\sum_i m_i n_i (v_{if} + U_{DL})^2|_{x_2} = m_e n_e (v_{ef} - U_{DL})^2|_{x_1} \quad (11)$$

where v_{if}, v_{ef} in the last equation represent the final velocities of the free ions and free electrons inside the DL respectively, and n_i and n_e are their densities. It is also assumed that the DL moves in the direction of electron motion, toward the higher potential. Experimental and simulation results confirm the direction of this motion [2].

On the other hand, electrons and ions must acquire similar kinetic energies, since they fall in the same potential of DL, Then:

$$\frac{1}{2}\sum_i m_i(v_{if} + U_{DL})^2|_{x_2} = \frac{1}{2}m_e(v_{ef} - U_{DL})^2|_{x_1} \quad (12)$$

From equations (11) and (12), the neutrality condition becomes $\sum_i n_i|_{x_2} = n_e|_{x_1}$. Now by using equation (11) and defining the electron and ions fluxes as $\psi_e = n_e v_{ef}^2|_{x_1}$ and $\psi_i = n_i v_{if}^2|_{x_2}$ respectively, We arrive at:

$$\sum_i \frac{m_i}{n_i}(\psi_i + n_i U_{DL})^2|_{x_2} = \frac{m_e}{n_e}(\psi_e - n_e U_{DL})^2|_{x_1} \quad (13)$$

Since $m_e \ll m_i$, we can neglect in the last equation, the small terms and obtain the generalized Langmuir condition as:

$$\sum_i \frac{m_i}{n_i|_{x_2}}\psi_i^2 + 2\psi_i n_i U_{DL}|_{x_2} = \frac{m_e}{n_e|_{x_1}}\psi_e^2 \quad (14)$$

This equation shows explicitly that the net flux inside a strong DL moving through plasma of electron and multi-ion species depends not only upon the mass ratio but also to the number density of each ion species.

4 Stability of Double Layer

As mentioned earlier, Langmuir condition indicates simply the location of the DL. To see if the obtained location is stable, the equation (14) for the initial equilibrium point (z_{DL0}) must be written as:

Stability of Double Layer in Multi-Ion Plasmas   341

$$\left(\frac{m_e}{n_e|_{x_1}}\psi_e^2 - \sum_i \frac{m_i}{n_i|_{x_2}}\psi_i^2\right)\Big|_{z_{DL0}} = 0 \tag{15}$$

Now if we impose some fluctuation to the DL, it dislocates from its initial position and we can write:

$$\frac{d}{dz}\left(\frac{m_e}{n_e|_{x_1}}\psi_e^2 - \sum_i \frac{m_i}{n_i|_{x_2}}\psi_i^2\right)\Big|_{z_{DL0}}\partial z = \sum_i 2n_i|_{x_2}\psi_i\left(\frac{d\partial z}{dt}\right) \tag{16}$$

where ∂z has been taken as a small deviation from the equilibrium position. Integration of this gives:

$$\partial z = A\exp\left(\frac{d}{dz}\left[\frac{\left(\frac{m_e}{n_e|_{x_1}}\psi_e^2 - \sum_i \frac{m_i}{n_i|_{x_2}}\psi_i^2\right)}{\sum_i 2n_i|_{x_2}\psi_i}\right]t\right) \tag{17}$$

In order to have stable DL, the term inside the parentheses must be negative, so:

$$\frac{d}{dz}\left(\frac{m_e}{n_e|_{x_1}}\psi_e^2 - \sum_i \frac{m_i}{n_i|_{x_2}}\psi_i^2\right) < 0 \tag{18}$$

This expresses the stability condition of the DL. When this condition holds, the motion of the DL will be backward, towards the initial equilibrium position.

5 Conclusion

In this work, using the fluid and Poisson equations, Langmuir condition is obtained for a strong DL moving through a multi-species plasma. The obtained results confirm and reproduces those obtained in other works. Also, it is shown that the presence of various types of ions have direct influence on the pressure equilibrium and on location of the DL formation. From the stability study of DL formation place, a criterion has been obtained for the establishment of the stability and for the production of a restoring force.

References

1. M.A. Raadu, Phys Reports, **178, No.2**, 25 (1989).
2. M.A. Raadu, J.J. Rasmussen, Astrophys and Space Sci, **144**, 43 (1988)
3. L.P. Block, Astrophys and Space Sci, **55**, 59 (1978)
4. W.M. Moslem, S. Ali, P.K. Shukla, R. Schlickeiser, Phys Plasma, **14**, 042107/1-8 (2007)
5. F. Verheest, M.A. Hellberg, J. Plasma Phys, **57, Part.2**, 465 (1997)
6. B. Sahu, R. Roychoudhury, Phys Plasma, **11, No.5**, 1947 (2004)
7. F.F. Chen, Phys Plasma, **33**, 34502 (2006)
8. I. Langmuir, Phys Rev, **33**, 954 (1929)
9. P. Carlqvist, Solar Phys, **63**,353 (1979).
10. P. Carlqvist, Proceedings 2st Syms on Plasma DLs, Innsbruck, (1984), p 340
11. A.C. Williams, IEEE Trans on Plasma Sci, **Ps.14, No.14**, 800 (1986)
12. T. Yamamoto, J. Plasma Phys, **34**, 271 (1985)

342   A.M. Ahadi, S. Sobhanian

13. N. Sing, R. W. Schunk, J. Geophysics Res, **87**, 3561 (1982)
14. N. Hershkowitz, Space Sci Rev, **41**, 351 (1985)
15. N.Sato, Proc on Conference on Plasma Physics, Lausanne, Switzerland, June 27- July 3, Vol.II, p 555 (1984)
16. B. Song, R.L. Merlino, N. D'Angelo, Phys Scripta, **45**, 391 (1992)